Biochemistry, Nutrition, and Therapeutics of Black Cumin Seed

BIOCHEMISTRY, NUTRITION, AND THERAPEUTICS OF BLACK CUMIN SEED

Edited by

ABDALBASIT ADAM MARIOD

Indigenous Knowledge and Heritage Centre, Ghibaish College of Science & Technology, Ghibaish, Sudan

ELSEVIER

ACADEMIC PRESS

An imprint of Elsevier

Academic Press is an imprint of Elsevier
125 London Wall, London EC2Y 5AS, United Kingdom
525 B Street, Suite 1650, San Diego, CA 92101, United States
50 Hampshire Street, 5th Floor, Cambridge, MA 02139, United States
The Boulevard, Langford Lane, Kidlington, Oxford OX5 1GB, United Kingdom

ISBN: 978-0-323-90788-0

For information on all Academic Press publications visit our website at https://www.elsevier.com/books-and-journals

Publisher: Nikki Levy
Acquisitions Editor: Nina Bandeira
Editorial Project Manager: Catherine Costello
Production Project Manager: Kiruthika Govindaraju
Cover Designer: Matthew Limbert

Typeset by TNQ Technologies

Contents

List of contributors *xiii*

1. Introduction, agronomy, varieties, and production of black cumin seed **1**

Samia Osman Yagoub

1.1 Introduction 1
1.2 Importance of black cumin 1
1.3 Origin of black cumin 3
1.4 Adaptation of black cumin 3
1.5 Varieties of black cumin 4
1.6 Production of black cumin 4
1.7 Conclusion 5
References 5
Further reading 6

2. The history of medicinal use of black cumin seed and its position in various cultures **7**

Abdalbasit Adam Mariod and Haroon Elrasheid Tahir

2.1 Introduction 7
2.2 History of medicinal use of black cumin seed 7
2.3 Black cumin seed and its position in various cultures 8
2.4 Conclusion 10
References 10

3. Effects of climate, soil type, irrigation and selected agricultural practices on yield and seeds quality bioactive compounds of black cumin **11**

Abbas E. Rahma

3.1 Introduction 11
3.2 Bioactive compounds of black cumin 12
3.3 Black cumin ecology and growing conditions 14
3.4 Irrigation and water management on bioactive compounds of black cumin 15
3.5 Effect of fertilizers and soil type on bioactive ingredients of black cumin 15

3.6 Conclusion 16
References 17

4. Black cumin: morphology, physiology, growth, and agricultural yield **19**

Samia Osman Yagoub

4.1 Botanical aspect of black cumin 19
4.2 Black cumin physiology 22
4.3 Conclusion 23
References 24

5. Nutritional, biochemical, and functional characteristics of black cumin seeds **27**

Newlove Akowuah Afoakwah and Gustav Komla Mahunu

5.1 Introduction 27
5.2 Black cumin biochemical components 29
5.3 Nutritional value of black cumin seeds 30
5.4 Carbohydrates in black cumin seeds 31
5.5 Proteins in black cumin seeds 32
5.6 Vitamins and minerals in black cumin seeds 33
5.7 Black cumin seeds contain lipids 33
5.8 Black cumin seed polysaccharide 35
5.9 Structural attributes of BCSP 36
5.10 Black cumin seed polysaccharide functional properties 36
5.11 The polysaccharide of black cumin extraction 37
5.12 Conclusion 39
References 39

6. Black cumin seed oil, protein, protein concentrates, and hydrolysate **43**

Ahmed A.M. Elnour, Mohamed E.S. Mirghani, Abdelgadir A. Elfadil, El Bashier Sallam and Ibrahim Bani

6.1 Introduction 43
6.2 Black cumin and its importance 45
6.3 Black cumin seed oil 46
6.4 Protein of black cumin seeds and its extraction 50
6.5 Black cumin seed protein hydrolysate 55

6.6 Conclusion 56
Acknowledgments 57
References 57

7. Functional properties and applications of black cumin (*Nigella sativa* L) seeds 61

Zanele Pamela Msibi

7.1 Introduction 61
7.2 Botanical overview of *Nigella sativa* L. 62
7.3 Medicinal application of black cumin seeds 62
7.4 Food application of black cumin seeds 64
7.5 Functional application of black cumin seed press waste material 66
7.6 Conclusion 67
References 67

8. Black cumin (*Nigella sativa* L.) essential oil and aroma quality 71

Islam Md Rizwanul Fattah, Rahman S.M. Ashrafur and Arslan Ahmed

8.1 Introduction 71
8.2 Essential oil and its extraction 74
8.3 Aroma composition of *Nigella sativa* L. 80
8.4 Benefits and use of *Nigella sativa* L. 81
8.5 Conclusion 82
Acknowledgment 83
References 83

9. Effect of different food processing techniques on the composition of black cumin seed and seed oil 89

Richard Osae, Maurice Tibiru Apaliya, Emmanuel Kwaw, Newlove Akowuah Afoakwah, Patrick Owusu-Ansah and Raphael N. Alolga

9.1 Introduction 89
9.2 Traditional or conventional processing techniques of black cumin seed and seed oil 92
9.3 Current/advanced food processing techniques on the composition of black cumin seed oil, their advantages and disadvantages 101
9.4 Purification techniques for black cumin seed and seed oil 106
9.5 Conclusion 108
References 109

10. Biological activities: antiinflammatory, antihyperlipidemic, and antimicrobial of black cumin seeds **113**

Mosab Nouraldein Mohammed Hamad

10.1 Introduction 113
10.2 Chemical composition 113
10.3 Biological activities 114
10.4 Antiinflammatory activity 114
10.5 Antihyperlipidemic properties 115
10.6 Antimicrobial properties 115
10.7 Conclusion 116
References 116

11. Black cumin seeds as animal feed **119**

Suzy Munir Salama

11.1 Introduction 119
11.2 Discussion 119
11.3 Conclusion 124
References 124

12. Therapeutic potential of *Nigella sativa* **127**

Nor Zaihana Abdul-Rahman and Nurul Szawani Mohd-Zubri

12.1 Introduction 127
12.2 Clinical therapeutic potential of *Nigella sativa* 127
12.3 Nonclinical study 130
12.4 Conclusion 138
References 139
Further reading 142

13. Antihypercholesterolemic and antihyperglycemic activity of *Nigella sativa* **143**

Siti Aisyah Abd Ghafar, Muhammad Fahmi Yakop, Rohazila Mohamad Hanafiah and Norsharina Ismail

Abbreviations 143

13.1 Introduction 144
13.2 Antihypercholesterolemic activity of *Nigella sativa* in hypercholesterolemic-induced animal model 144
13.3 Hypocholesterolemic effects of *Nigella sativa* in clinical trial 147
13.4 Antiatherogenic effects of *Nigella sativa* 149

13.5 Antihyperglycemic effects of *Nigella sativa* 150
13.6 Mechanism of action of *Nigella sativa* as antihypercholesterolemic and antihyperglycemic agent 153
References 156

14. Anticancer activity of black cumin seed bioactive compounds **161**
Suzy Munir Salama
Abbreviations 161
14.1 Introduction 161
14.2 Discussion 162
14.3 Conclusion 165
References 165

15. Antioxidant activity and phenolic content of black cumin seeds **169**
Norsharina Ismail, Siti Aisyah Abd Ghafar and Md Zuki Abu Bakar
Abbreviations 169
15.1 Introduction 169
15.2 Antioxidant and oxidative stress 171
15.3 Phenolic compounds 172
15.4 Antioxidant properties and phenolic constituents of black cumin seeds 172
15.5 Effect of processing methods on the antioxidant activity and phenolic content of black cumin seeds 181
15.6 Phenolic compounds in black cumin seeds 182
15.7 Conclusion 184
References 184

16. Antibacterial, antimicrobial, and antimalarial effect of black seed oil **189**
Rohazila Mohamad Hanafiah, Nur Fatimah Zaharah Salehuddin, Siti Aisyah Abd Ghafar and Mohd Asyraf Kassim
Abbreviations 189
16.1 Introduction 189
16.2 Antibacterial activities of black seed oil 190
16.3 Antimicroorganism activities of black seed oil 192
16.4 Antimalarial activities of black seed oil 197

References 198
Further reading 200

17. Neuroprotective effects of black cumin seed and seed oil 201

Hakan Ak, İsmail GÜLŞEN and İhsan CANBEK

17.1 Introduction 201
17.2 Effects of thymoquinone on neuroinflammation 201
17.3 Effects of thymoquinone on migraine pain 205
17.4 Effects of thymoquinone on depression 205
17.5 Effects of thymoquinone on epilepsy 206
17.6 Effects of thymoquinone on Parkinson's disease 208
17.7 Effects of thymoquinone on Alzheimer's disease 209
17.8 Effects of thymoquinone on cerebral ischemia 211
17.9 Impacts of thymoquinone traumatic brain injury 212
17.10 Effects of thymoquinone on multiple sclerosis 212
17.11 Impacts of thymoquinone on the spinal cord injury 213
17.12 Effects of thymoquinone on peripheral nerve injury 214
17.13 Neuroprotective effects of thymol 215
17.14 Effects of thymol on Alzheimer's disease 215
17.15 Effects of thymol on anxiety 215
17.16 Effects of thymol on dementia 216
17.17 Effects of thymol on depression 216
17.18 Impacts of thymol on epilepsy 217
17.19 Effects of thymol on Parkinson's disease 217
17.20 Other thymol's impacts on nervous system 218
17.21 Neuroprotective effects of p-cymene 219
17.22 Neuroprotective effects of carvacrol 220
17.23 Conclusion 221
References 221
Further reading 230

18. Black cumin (*Nigella sativa*) to treat bronchial asthma, ischemia, and cardiovascular diseases 231

Ahmed Abdul-Sabour Ahmed Bader

18.1 Introduction 231
18.2 NS to treat ischemia and cardiovascular diseases 232
18.3 Conclusion 234
References 234

19. Potential medicinal properties of black cumin (*Nigella sativa*) in traditional medicine **237**

Sarya F. Amin

Abbreviations 237

19.1 Introduction 237

19.2 The miracle herb *Nigella sativa* 239

19.3 Medicinal properties of *Nigella sativa* in traditional medicine 240

19.4 Conclusion 242

References 242

Index 247

List of contributors

Siti Aisyah Abd Ghafar
Department of Basic Sciences, Faculty of Dentistry, Universiti Sains Islam Malaysia, Kuala Lumpur, Selangor, Malaysia

Nor Zaihana Abdul-Rahman
Department of Basic Sciences, Faculty of Dentistry, Universiti Sains Islam Malaysia, Kuala Lumpur, Selangor, Malaysia

Ahmed Abdul-Sabour Ahmed Bader
Department of Clinical Pharmacology, Faculty of Medicine, Menoufiya University, Shibin el Kom, Egypt

Md Zuki Abu Bakar
Natural Medicines and Products Research Laboratory, Institute of Bioscience, Universiti Putra Malaysia, Serdang, Selangor, Malaysia; Department of Veterinary Pre-Clinical Sciences, Faculty of Veterinary Medicine, Universiti Putra Malaysia, Serdang, Selangor, Malaysia

Newlove Akowuah Afoakwah
Department of Food Science and Technology, Faculty of Agriculture, Food and Consumer Sciences, University for Development Studies, Tamale, Ghana

Arslan Ahmed
Department of Mechanical Engineering, Comsats University Islamabad, Wah Campus, Rawalpindi, Punjab, Pakistan

Hakan Ak
Kõrşehir Ahi Evran University, School of Medicine, Department of Neurosurgery, Kõrşehir, Turkey

Raphael N. Alolga
State Key Laboratory of Natural Medicines, Department of Pharmacognosy, China Pharmaceutical University, Nanjing, China

Sarya F. Amin
Indigenous Knowledge and Heritage Center, Ghibaish College of Science and Technology, Ghibaish, Sudan

Maurice Tibiru Apaliya
Cape Coast Technical University, School of Applied Science and Technology, Department of Food and Postharvest Technology, Cape Coast, Ghana

Rahman S.M. Ashrafur
Biofuel Engine Research Facility, Queensland University of Technology, Brisbane, QLD, Australia

Ibrahim Bani
College of Medicine, Ajman University, Ajman, United Arab Emirates; Centre of Medical and Bio-Allied Health Sciences Research, Ajman University, Ajman, United Arab Emirates

İhsan CANBEK
Afyonkarahisar Health Sciences University, Department of Neurosurgery, Afyonkarahisar, Turkey

Abdelgadir A. Elfadil
Department of Environmental Science, Faculty of Science and Technology, Al-Neelain University, Sudan

Ahmed A.M. Elnour
Department of Biology, Faculty of Science and Mathematics, Sultan Idris Education University (UPSI), Tanjung Malim, Perak, Malaysia; International Institute for Halal Research and Training (INHART), International Islamic University Malaysia (IIUM), Gombak, Kuala Lumpur, Malaysia

İsmail GÜLŞEN
Lokman Hekim Hospital, Clinic of Neurosurgery, Van, Turkey

Mosab Nouraldein Mohammed Hamad
Faculty of Health Sciences, Elsheikh Abdallah Elbadri University, Berber, Sudan

Norsharina Ismail
Natural Medicines and Products Research Laboratory, Institute of Bioscience, Universiti Putra Malaysia, Serdang, Selangor, Malaysia

Mohd Asyraf Kassim
School of Industrial Technology, Universiti Sains Malaysia, Gelugor, Pulau Pinang, Malaysia

Emmanuel Kwaw
Cape Coast Technical University, School of Applied Science and Technology, Department of Food and Postharvest Technology, Cape Coast, Ghana

Gustav Komla Mahunu
Department of Food Science and Technology, Faculty of Agriculture, Food and Consumer Sciences, University for Development Studies, Tamale, Ghana

Abdalbasit Adam Mariod
Indigenous Knowledge and Heritage Centre, Ghibaish College of Science & Technology, Ghibaish, Sudan

Mohamed E.S. Mirghani
International Institute for Halal Research and Training (INHART), International Islamic University Malaysia (IIUM), Gombak, Kuala Lumpur, Malaysia

Rohazila Mohamad Hanafiah
Department of Basic Sciences, Faculty of Dentistry Universiti Sains Islam Malaysia, Kuala Lumpur, Selangor, Malaysia

Nurul Szawani Mohd-Zubri
Department of Basic Sciences, Faculty of Dentistry, Universiti Sains Islam Malaysia, Kuala Lumpur, Selangor, Malaysia

Zanele Pamela Msibi
Food Science Nutrition and Technology, University of Eswatini, Luyengo, Eswatini; AgroFood Chain Innovation and Sustainability, Ecole Nationale Supérieure de Formation de l'Enseignement Agricole, Auzeville-Tolosane, France

Richard Osae
Cape Coast Technical University, School of Applied Science and Technology, Department of Food and Postharvest Technology, Cape Coast, Ghana

Patrick Owusu-Ansah
University of Development Studies, Department of Food Processing Technology, Tamale, Ghana

Abbas E. Rahma
Sudan University of Science and Technology, College of Agricultural Studies, Department of Agricultural Engineering, Khartoum, Sudan

Islam Md Rizwanul Fattah
Centre for Green Technology (CGT), School of Civil and Environmental Engineering, Faculty of Engineering and IT, University of Technology Sydney, Ultimo, NSW, Australia

Suzy Munir Salama
Indigenous Knowledge and Heritage Center, Ghibaish College of Science and Technology, Ghibaish, Sudan

Nur Fatimah Zaharah Salehuddin
Department of Basic Science, Faculty of Dentistry Universiti Sains Islam Malaysia, Kuala Lumpur, Selangor, Malaysia

El Bashier Sallam
Freelance Global and Public Health Expert (FGPHE), Sudan

Haroon Elrasheid Tahir
School of Food and Biological Engineering, Jiangsu University, Zhenjiang, Jiangsu, China

Samia Osman Yagoub
Department of Agronomy, College of Agricultural Studies, Weed Science Centre, Sudan University of Science and Technology, Khartoum, Sudan

Muhammad Fahmi Yakop
Department of Basic Sciences, Faculty of Dentistry, Universiti Sains Islam Malaysia, Kuala Lumpur, Selangor, Malaysia

CHAPTER ONE

Introduction, agronomy, varieties, and production of black cumin seed

Samia Osman Yagoub
Department of Agronomy, College of Agricultural Studies, Weed Science Centre, Sudan University of Science and Technology, Khartoum, Sudan

1.1 Introduction

Black cumin (*Nigella sativa* L.) is an herbal plant belongs to the Ranunculaceae, the buttercup family, with about 70 genera and at least 3000 species. The *Nigella* genus is diploid (2n = 12) and contains about 14 species of annual plant herbs (Weiss, 2002). There are about 13 species in Turkey; it includes some important species (e.g., *N. sativa* L, *Nigella damascena L*, and *Nigella arvensis L*), with aromatic and medicinal properties, the most popular of which is *N. sativa* L (Kökdil et al., 2016). The name *Nigella* comes from the Latin *niger, or nigellus,* meaning black. It is known by many different names around the world, including black cumin, black caraway, kalonji, Roman coriander, and fennel flower. The fact that black cumin is found in different countries is proved by its different names in various countries. Its English name is *black cumin*. Its Italian name is *nigella*; *neguilla* or *pasinara* in Spanish; *cheveux de venus, nigelle, cumin noir,* or *poivrette* in French; *schwarzkummel* in Germany; *kolongi* in Turkish; *jinten hitan* in Indonesia and Malaya; *kala zira, kalongi, krishanjirka, mangrail,* and many other vernacular names in India. Black seed plants and seeds are known to all Arabian and Islamic countries and carry various colloquial names. It is known generally by the names Habbat Albarakah (blessed seeds), Alhabahat Alsawda, and Alkamoun Alaswad (black seeds).

1.2 Importance of black cumin

Black cumin is an herb that is said to have magical qualities. Its seeds are bitter in taste, and consumption of seeds, even in small quantities, gives a feeling of constriction in the throat (Sultana et al., 2018). The fatty oil from

Biochemistry, Nutrition, and Therapeutics of Black Cumin Seed
ISBN: 978-0-323-90788-0
https://doi.org/10.1016/B978-0-323-90788-0.00011-1

black cumin is not used for cooking owing to its nigellon content, which causes a bitter taste in oil. This plant is a life-saving herb and has highly protein-containing properties. It has various properties that have extremely good effects on an individual's health and well-being. The popularity of the plant came from the ideological belief in the herb as a cure for multiple diseases as well as the medicinal value of the immense, varied, and powerful pharmacological traits. In fact, the Prophet Mohammed even said that this herb could cure all diseases except death. Black cumin seeds have been used as a natural remedy for over 4000 years in various parts of the world. Evidence of the importation of black cumin seeds were found in an archeological site of a second century Roman settlement near the Rhine River in Germany. Both Galen and Hippocrates, two ancient renowned physicians who lived in Rome in 200 A.D. and Greece in 400 B.C., wrote about the uses of black cumin (Vandorpe, 2010).

The seeds of black cumin were found in the Egyptian pharaoh, King Tut's tomb, which reflects the degree of significance the seeds had during the ancient Egyptians time. The Egyptians used it mainly to preserve bodies before mummifying them. Ibn Sina extensively praised the black cumin seed for its restorative and preventative qualities. Ibn Sina believed that the black cumin helped people recover from dispiritedness or fatigue, and also stimulated the body's energy. In his book, Al-Biruni (973—1048), the Muslim scholar revealed that the black seed is called alwanak, a kind of grain in the Sigzi dialect. This reference to black seed as "grains" aims at the seed's possible nutritional use during the 10th and 11th centuries. A Greek physician, Dioscorides, of the 1st century, pointed out that black seeds were used to treat headaches, nasal congestion, toothaches, and intestinal worms. Also, it was used as a diuretic to promote menstruation and increase milk production (Mollazadeh et al., 2017).

The ancient Romans used it as both a spice and a medicine. In India and the Arab world, it is used as medicine and food (Yarnell and Abascal, 2011). It was used to enhance digestion by boiling it in water and using it as medicine for edema, heart disease, stomach ailments, and to remove gas from the stomach and intestine. It had antimicrobial and antifungal properties. Black cumin is characterized by its preservative quality and is often used in meats like sausage. It is a great source of iron and other important minerals. It contains dietary fiber and vitamins. Black seeds are used in food as a flavoring additive in breads and pickles (Hassanien et al., 2015).

Black cumin pharmacological functions in seed fixed volatile oils are due to the presence of secondary metabolites (stearic, palmitic, and linoleic acid) as well as phenolic compounds in roots and shoots (Mathaus and Ozcan, 2011). It was used to remedy illnesses such as tumors, diabetes, irritation, asthma, high blood pressure, digestive disturbances, and women's disorders (Ramadan, 2007).

The seeds contain alkaloids and protein, saponin (-herderin), (nigellicines and nigelledine), eight important fatty acids including linoleic, oleic, and palmitic acids (Tuncturk et al., 2012), essential and fixed oil (Ozel et al., 2009). Essential oils contain varying bioactive molecules such as moqunone, thymol, tocopherol, trans retinol, and selenium (Yarnell and Abascal, 2011).

1.3 Origin of black cumin

Black cumin is wild species, native to the Mediterranean region, spread through West Asia to Northern India, Pakistan, Syria, Turkey, Iran, Saudi Arabia, some parts of Europe, Eastern Asia, and some region of Iran (Onofrei et al., 2017; Kiralan, 2014). Also, it is cultivated from Morocco to sub-Saharan Africa, in Eastern Ethiopia and Niger (Gashaw, 2020). Black cumin is widely cultivated in Southern Europe, the Middle Eastern Mediterranean region, and Northern India. Shewaye (2011) stated that black cumin originated in East Mediterranean and Egypt and commonly cultivated in Iran, Japan, China, and Turkey.

1.4 Adaptation of black cumin

Black cumin is generally short-lived, annual, and typical of disturbed soils or natural communities in semiarid areas with a dominance of therophytes. It is sensitive to soil conditions and climate. Its production extends primarily throughout the Middle East and the Mediterranean Basin, which includes India, Bangladesh, Turkey, Iran, Pakistan, Egypt, Sudan, and Iraq. It is cultivated in some parts of Europe, Eastern Asia, and some regions of Iran (Kirlan, 2014; Majeed et al., 2020). In Iran, black cumin is grown in arid and semiarid regions with a short growth period of about 100–120 days with low input requirements (Merijipoor, 2020). Rostaei et al. (2018) reviewed that black cumin seeds were mature about 133 days after sowing and harvested after the maturity stages. Black cumin is known as one of the most important species growing widely in various regions of Iran. It is cultivated in some areas (Sajedi et al., 2017). Black cumin in Ethiopia is found in indifferent parts at various latitude ranges in a wide agroecology (Gashaw, 2020).

1.5 Varieties of black cumin

There are 12 black cumin species naturally grown in Turkey's flora and only the species of black cumin (*N. sativa L.*) is traded, and the production is limited compared to many crops and used for domestic consump tion (Kar et al., 2007). Ethiopia is one of the centers of black cumin diversity. Eighty-four accessions of black cumin local cultivars have been collected from potential production areas, agroecologies, and under maintenance from entries of black cumin local cultivars grow. Improved varieties such as Darbera, Dershaye, and Eden have been released to users. In Turkey, local populations are generally used as seed production and breeding materials by breeders and farmers due to a lack of cultivars (Kara et al., 2015).

1.6 Production of black cumin

The major producing and exporting countries of black cumin are India, Sri Lanka, Bangladesh, Afghanistan, Pakistan, Egypt, Iran, Iraq, Syria, and Turkey (Sultana et al., 2018). The global average productivity in major producing countries such as India is 2.2 tons per hectare, but in Ethiopia (Kifelew et al., 2017; Gashaw, 2020), it is reported to be 0.79 and 0.64 tons per hectare, which is very low production compared with other producing countries. Hammo (2008) refers to the low productivity of the varieties and the poor agronomic practices applied by the farmers. Toncer and Kizil (2004) mentioned that the seed yield of black cumin was 828 kg/ha, and the content of essential oil was 0.27% −0.35%. Ozel et al. (2009) pointed out that the seed yield of black cumin is 2482.3−−1406.6 kg/ha, and the content of essential oil is between 0.24% and 0.43%. In Ethiopia, it is the second cash crop exported next to ginger and has a share of about 12% of the world market and 99% of the product consumed locally. Ethiopia produced 5.887 tons of black cumin in 2003 (Moard, 2003). Also, the annual yield of black cumin increased rapidly from the previous low production level to over 18,000 metric tons of seed produced in the 2014/15 production season (Gashaw, 2020). About 98% of local exports go to Arab and Muslim countries. The value of cumin is very low because the export is made in the form of whole grain. In Iran, Golkar and Nourbakhsh stated that there was no significant market share.

1.7 Conclusion

Production methods and increasing production and productivity are of great importance from an economic point of view for any crop, as well as the varieties and their knowledge and determining the best and highest productivity in addition to their suitability for the conditions of this production. Despite the importance of black cumin, there are research gaps to increase the yield potential of this crop, especially the management of nutrients, weeds, and diseases and insect pests. Therefore, it is important to improve the yield of the black cumin crop to meet the need of the consumer.

References

Gashaw, Z., 2020. Status of black cumin (Nigella sativa L.) research and production in Ethiopia; A review. International Journal of Forestry and Horticulture 6 (1), 20–29.

Hammo, Y.H., 2008. Effect of high levels of nitrogen and phosphorus fertilizer on growth, yield and yield components of Nigella sativa L. Mesopotamia Journal of Agriculture 36, 2–11.

Hassanien, M.F., Assiri, A.M., Alzohairy, A.M., Oraby, H.F., 2015. Health-promoting value and food applications of black cumin essential oil: an overview. Journal of Food Science & Technology 52 (10), 6136–6142. https://doi.org/10.1007/s13197-015-1785-4.

Kar, Y., Sen, N., Tekeli, Y., 2007. Investigation of black cumin (*Nigella sativa L.*) seeds cultivated in region Samsun and country of Egypt in terms of antioxidant activity. Suleyman Demirel University Journal of Science. 2 (2), 197–203.

Kara, N., Katar, D., Baydar, H., 2015. Yield and quality of black cumin (*Nigella sativa L.*) population: the effect of Ecological condition. Turkish Journal of Field Crops 20 (1), 9–14.

Kifelew, H., Fikre, D., Luleseged, T., Bekele, D., Mitiku, H., Getachew, W., 2017. Seed Spices Production Guideline. Technical Research Report. Ethiopian Institute of Agricultural Research (EIAR), Addis Ababa.

Kirlan, M., 2014. Change in volatile compounds of black cumin (*Nigella sativa L.*) seed oil during thermal oxidization. International Journal of Food Properties. 17, 1482–1489.

Kökdil, A.İ., Özbilgin, B., Uygun, C., 2016. Morphology and stem anatomy of some species of genus nigella L. in Turkey. Ankara Eczacılık Fakültesi Dergisi 35 (1), 19–41.

Majeed, A., et al., 2020. *Nigela sativa* uses in traditional and contemporary medicine –an overview. Acta Ecologica Sinica 20, 20.

Matthaus, B., Ozcan, M.M., 2011. Fatty acids, tocopherol, and sterol contents of some nigella species seed oil. Czech. Journal of Food Sciences 29 (2), 145–150.

Merajipoor, M., Movahned, M., Salehi, D.A., Yadavi, A., 2020. Physiological responses of black cumin to chemical and biological nitrogen fertilizers under different irrigation regimes. Journal of Agriculture, Science and Technology 22 (4), 1023–1037.

Moard, Ministry of Agriculture and Rural Development, 2003. Development of Black Cumin and White Cumin Production, p. 25.

Mollazadeh, H., Afshari, A.R., Hosseinzadeh, H., 2017. Review on the potential therapeutic roles of *Nigella sativa* in the treatment of patients with cancer: involvement of apoptosis: - black cumin and cancer. Journal of Pharmacopuncture 20 (3), 158–172. https://doi.org/10.3831/KPI.2017.20.019.

Onofrei, V., et al., 2017. Organic foliar fertilization increases polyphenol content of Calendula officinalis L. Industrial Crops and Products 109 (7), 509–513.

Ozel, A., Demirel, U., Guler, I., Erden, K., 2009. Effect of different row spacing and seeding rate on black cumin (*Nigella sativa L.*) yields and some agricultural characters. Harran University Journal of Agriculture Faculty. 13 (1), 17–25 (in Turkish).

Ramadan, M.F., 2007. Nutritional value, functional properties and nutraceutical applications of black cumin (*Nigella sativa L.*): an overview. International Journal of Food Science and Technology 42 (10), 1208–1218.

Rostaei, N., Fallah, S., Lorigooini, Z., Surki, A.A., 2018. Crop productivity and chemical compositions of black cumin essential oil in sole crop and intercropped with soybean under contrasting fertilizers. Industrial Crop and Products 125, 622–629.

Sajedi Shaker, S., Mohammadi, A., Shahli, M.K., 2017. Cytological studies on some ecotypes of (*Nigella sativa L*). in Iran. The Japan Mendel Society. Cytologia 82 (2), 123–126.

Sultana, S., Das, B., Rudra, B.C., Das, G., Alam, M.B., 2018. Effect of date of sowing on productivity of black cumin. International Journal of Current Microbiology and Applied Sciences 7 (1), 1796–1800. https://doi.org/10.20546/ijcmas.2018.701.217.

Shewaye, L., 2011. Antifungal Substances from Essential Oils. M.Sc. Thesis. Addis Ababa University 3, 8.

Toncer, O., Kizil, S., 2004. Effect of seed rate on agronomic and technologic characters of (*Nigella sativa L).* International Journal of Agriculture and Biology 3, 529–532.

Tuncturk, R., Tuncturk, M., Ciftci, V., 2012. The effects of varying nitrogen doses on yield and some yield components of black cumin (*Nigella sativa L.*). Advances in Environmental Biology 6 (2), 855–858.

Vandorpe, P., 2010. Plant Macro Remains from the 1st and 2nd Cent. AD in Roman Oedenburg/Biesheim-Kunheim (F): Methodological Aspects and Insights into Local Nutrition, Agricultural Practices, Import and the Natural Environment. Universität Basel, Basel. Unpublished Thesis.

Weiss, E.A., 2002. Spice Crops. CABI Publishing, CABI International, Wallingford, Oxon, UK.

Yarnell, E., Kathy Abascal, K., 2011. Undervalued Herbs: Use in Clinical Practice and Need for Validating Research. Alternative and Complementary Therapies 17 (4), 220–224. https://doi.org/10.1089/act.2011.17406.

Further reading

Golkar, P., Nourbakhsh, V., 2019. Analysis of genetic diversity and population structure in *Nigella sativa L*, using agronomic traits and molecular markers (SRAP and SCoT). Industerial Crops and Products 130, 170–178.

Maleki, J., shourabadi, E.S.,A., Mirza, M., Heydari, H., Lebaschy, M.H., 2021. Improving the productivity and quality of black cumin (*Nigella sativa L*) by using soil fertility management practice. Journal of Plant Nutrition 44 (12), 1741–1757.

CHAPTER TWO

The history of medicinal use of black cumin seed and its position in various cultures

Abdalbasit Adam Mariod[1], Haroon Elrasheid Tahir[2]
[1]Indigenous Knowledge and Heritage Centre, Ghibaish College of Science & Technology, Ghibaish, Sudan
[2]School of Food and Biological Engineering, Jiangsu University, Zhenjiang, Jiangsu, China

2.1 Introduction

The black seed (*Nigella sativa*) has become famous in recent years in the Arab world, especially the Gulf countries, and its name has spread everywhere with the people's demand for natural and healthy materials, herbs, and what were beneficial to health in particular (Ahmad et al., 2013). The black seed topped the throne of spices and these substances because it treats every disease, depending on what came in one of the hadiths of the Messenger, may God bless him and grant him peace, which was transmitted by Al-Bukhari and others. On the authority of Aisha, may God be pleased with her, she said: The prophet Mohamed, may God bless him and grant him peace, said: "This black seed is healing from every disease except death." For this very reason, the information available on the benefits of the black seed in the Arabic language on the Internet and books is hundreds of times greater than what is available from the rest of the materials and spices (Ali et al., 2018). This chapter is an attempt to shed light on recent literature in a clear and concise text for researchers to understand the history of pharmacological properties of *N. sativa* and its position in different cultures.

2.2 History of medicinal use of black cumin seed

It is considered a Mediterranean spice that grows in India, Pakistan, and Afghanistan. Its historical references are found in ancient religious and medical texts. Where both Hippocrates and Dioscorides referred to it as "melanthion," while the Bible describes it as "the healing black cumin." That is why the Pharaohs describe it as from life here forever, as it is known to Muslims as the blessing seed (Padhye et al., 2008).

Biochemistry, Nutrition, and Therapeutics of Black Cumin Seed
ISBN: 978-0-323-90788-0
https://doi.org/10.1016/B978-0-323-90788-0.00004-4

In any case, the interest in the black bean does not depend on the Arabs alone. Many peoples and civilizations have taken care of it, and it has been respected by many who are familiar with the science of spices and food for its special features. Black seed was one of the main spices in the ancient world, and people used it extensively to improve the taste of food and for many types of remedies medical (Padhye et al., 2008).

Available information indicates that the use of black seed was common in the days of the Assyrians and Babylonians, and it was mentioned in some herbal books as a treatment for diseases of the stomach, eye, ear, mouth, and skin, such as itching and other difficult diseases.

Muslims have been using and instructing the use of *N. sativa* for many years, and several articles have been formed on the subject. Black seed has also been used around the world for more than 3000 years. The black seed represents a prophetic herb and also holds a prominent place in the Prophet's medicines. Interestingly, it was not used in abundance before the Prophet Muhammad made its use known. Although there are more than 400 herbs used by the Prophet Muhammad and recorded in the Herbs of Galen and Hippocrates, black seed was not an obscure remedy among most of the prevalent remedies at the time. Because of the way Islam spread, the use and spread of the black seed are widely known as the "Prophet's remedy." (Ali et al., 2018).

2.3 Black cumin seed and its position in various cultures

Although, we do not know exactly why the pharaohs loved the black seed and cared for it and the way they used it. The black-colored oil was found in the holdings of the young king the Pharaonic King Tutankhamun, which was later known as Nigella or black seed oil. Some experts say that the black seeds found in the tomb were and are still usable despite the passage of that long period of time in the tomb. The black seed was used to help the pharaonic king cross into the afterlife after death (Padhye et al., 2008).

Cleopatra used the black seed for cosmetic and health purposes, and Queen Nefertiti was also addicted to black seed oil, and doctors listed the days of the pharaohs in general by using it to treat colds, headaches, digestive problems, dental diseases, infections, and allergies. The black bean received great interest in Greece, the Arab world, and the Islamic world, especially in the medical world. It was of great importance in the treatment of diseases of the liver and digestive system (Hussain, 2016).

Historians indicate that the father of medicine and the greatest physician, the Greek Hippocrates (fifth century BC), considered the black seed very useful in treating disorders of the liver and digestive system (Yimer et al., 2019). In the first century AD, it was mentioned and its many uses in the Natural History Encyclopedia of Pliny or Pliny the Elder, which was concerned with knowledge in the ancient world. Pliny referred to it as "Git," and according to the encyclopedia, it was used to treat snake and scorpion bites and various types of tumors, boils, rashes, colds, and infections. Then came the mention of the black seed in the book of Dioscorides (a famous Greek doctor who lived in the first century AD). In the Middle Ages, black seed was used in the treatment of headaches, nasal congestion, and toothache; in addition to its use to expel worms, it was also used as a diuretic and milk (Al-Attass et al., 2016).

According to Ibn Sina in "*The Canon of Medicine*," *N. sativa* can stimulate energy and help overcome fatigue and stress. And it came in the book: "The shonies (*N. sativa*) are sour, cut for phlegm, clears, analyzes gases and blows, and purifies it thoroughly. It is placed with vinegar on milky pimples, and it dissolves phlegm and solid tumors, and with vinegar on phlegmatic sores and ulcerated scabies, and it is useful for colds, especially powdered ones" (Yarnell and Abascal, 2011).

Black cumin seeds are wrapped in a linen bundle and then rubbed on the forehead of someone who has a headache, and if it is soaked in vinegar for a night and then crushed and given to the patient to inhale, it is beneficial for the chronic aches in the head, and for killing worms, even if it is coated on the navel, and it induces menstruation if used for days, and it is irrigated with honey and hot water for pebbles in the bladder and kidneys.

As it came in Daud Antaki book *Tazkira Ulil Albab* about the nigella seed: "Using nigella every morning cooked with raisins turns the skin red and clears it, and its ashes cut the hemorrhoids as a coating. Or anointing the front of the head with it prevents the regression of the common cold, and with bitter melon and wormwood, the parasites of the abdomen are removed as a coating on the navel, and it is an antidote to poisons, so that its smoke repels vermin" (Daud Antaki 2010).

The black seed was mentioned in the German medical encyclopedia in the sixth and eighteenth centuries, and based on the important encyclopedia in Europe, the benefits of this seed are removing sweat, opening appetite, strengthening the digestive system, expelling worms, combating constipation, and expelling fever (Ahmad et al., 2013). Black seed is traditionally used in the East to treat digestive system, flatulence, diarrhea, constipation, jaundice, get rid of gallstones, mental atrophy, stimulate the kidneys, increase urine flow, infection, congestion, bronchial tubes, menstrual complaints, promote lactation, skin parasites, exterminate worms, skin care, treat dandruff, and hair loss (Hussain, 2016).

2.4 Conclusion

The black seed has miraculous therapeutic benefits, as it was ranked among the best medicines based on evidence and evidence by pharmacists and doctors. The different parts of it give different therapeutic and medicinal properties to treat many diseases, thus maintaining public health. The main active ingredient of *N. sativa* extract is thymoquinone, which has a wide range of medicinal uses.

References

Ahmad, A., Husain, A., Mujeeb, M., Khan, S.A., Najmi, A.K., Siddique, N.A., Damanhouri, Z.A., Anwar, F., 2013. A review on therapeutic potential of Nigella sativa: a miracle herb. Asian Pacific Journal of Tropical Biomedicine 3, 337−352.

Al-Attass, S.A., Zahran, F.H.M., Turkistany, S.A., 2016. Nigella sativa and its active constituent thymoquinone in oral health. Saudi Medical Journal 37, 235−244.

Ali, S.A., Parveen, N., ALI, A.S., 2018. Links between the Prophet Muhammad (PBUH) recommended foods and disease management: a review in the light of modern superfoods. International Journal of Health Sciences 12, 61−69.

Daud, A., 2010. Tazkira Ulil Albab, India, Published by CCRUM, Ministry of Health and Family Walfare, Government of India.

Hussain, D.A.S., Hussain, M.M., 2016. Nigella sativa (black seed) is an effective herbal remedy for every disease except death − a Prophetic statement which modern scientists confirm unanimously: a review. Advancement in Medicinal Plant Research 4, 27−57.

Padhye, S., Banerjee, S., Ahmad, A., Mohammad, R., Sarkar, F.H., 2008. From here to eternity - the secret of Pharaohs: therapeutic potential of black cumin seeds and beyond. Cancer Therapy 6, 495−510.

Yarnell, E., Abascal, K., 2011. Nigella sativa: holy herb of the Middle East. Alternative and Complementary Therapies 17, 99−105.

Yimer, E.M., Tuem, K.B., Karim, A., Ur-Rehman, N., Anwar, F., 2019. Nigella Sativa L. (Black Cumin): a promising natural remedy for wide range of illnesses. Evidence-Based Complementary and Alternative Medicine: eCAM 2019, 1528635.

CHAPTER THREE

Effects of climate, soil type, irrigation and selected agricultural practices on yield and seeds quality bioactive compounds of black cumin

Abbas E. Rahma
Sudan University of Science and Technology, College of Agricultural Studies, Department of Agricultural Engineering, Khartoum, Sudan

3.1 Introduction

Nigella sativa (L.), a Ranunculaceae is an annual herbaceous flowering plant native to the northern regions of Africa, the middle East, Southwest Asia, and southern parts of Europe (Hussain and Hussain, 2016). However, the plant has spread and becomes naturalized in many countries across the globe (Van Kleunen et al., 2015). *N. sativa* has gained importance as a multipurpose plant where it has been used in culinary since time immemorial across cultures and civilization. The first record of the plant in culinary dated back to ancient Mesopotamia as recorded in the oldest world recipe collection known Yale culinary tables which dates back to 1750 BC. The plant is known by different vernacular names across countries and regions, which reflects interwoven popularity into different cultures. The plant is renowned for its high nutritive value (Mathur et al., 2011); however, its importance stems from its medicinal uses, which are well entrenched in traditional systems of medicine including Ayurveda, Unani, Chinese, and Arabic (Khan et al., 2011). The plant has shown potential to suppress several human ailments spanning headache and cancer (Fatima Shad et al., 2021). Recently, intensive research is ongoing to verify scientific viability of traditional uses and on plausibility of using the plant and/or product as a treatment for COVID-19 (Fatima Shad et al., 2021).

Biochemistry, Nutrition, and Therapeutics of Black Cumin Seed
ISBN: 978-0-323-90788-0
https://doi.org/10.1016/B978-0-323-90788-0.00015-9

3.2 Bioactive compounds of black cumin

Phytochemicals in oils and seeds of black cumin have shown a substantial number of bioactive ingredients with potentials to cure various ailments (Fatima Shad et al., 2021). Among these bioactive ingredients, thymoquinone (TQ) (Fig. 3.1) is the most prominent. The bioactive compounds, through suppression of the inflammatory mediators and leukotrienes, exhibit potent antiinflammatory effects on several inflammation-based models including experimental encephalomyelitis, colitis, peritonitis, edema, and arthritis (Chakrabarty et al., 2003). Recent research indicated the plausibility of developing black cumin and TQ as therapeutic agents for combating contemporary inflammatory and infectious diseases including COVID-19 (Fatima Shad et al., 2021).

N. sativa seed fixed oil contents (28−36%w/w) are chiefly composed of a plethora of unsaturated fatty acids including arachidonic, eicosadienoic, and

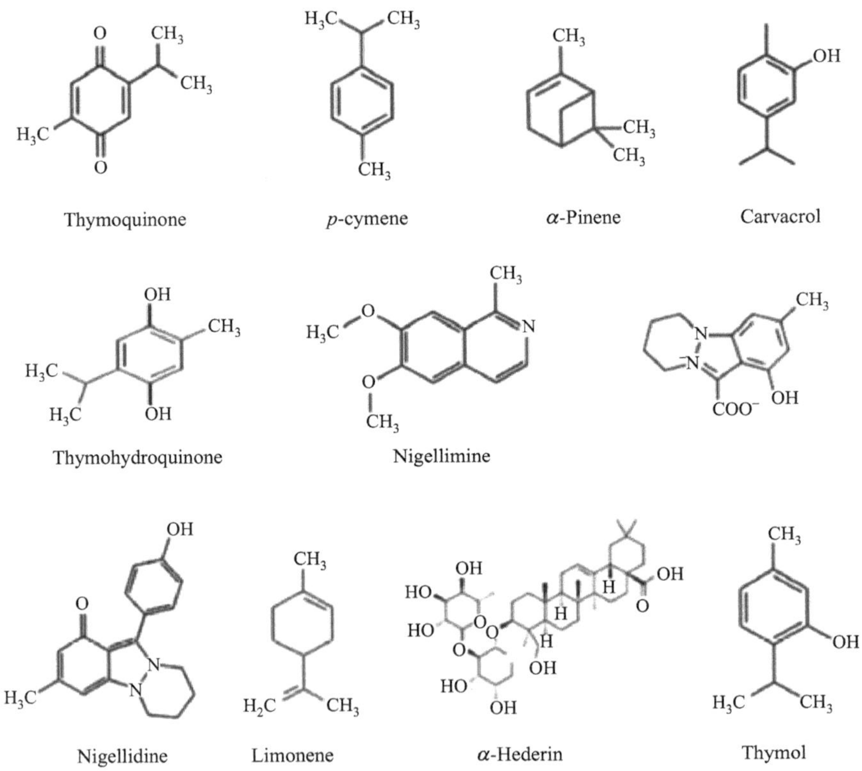

Figure 3.1 Chemical structures of some main constituents of *Nigella sativa* seeds (Kooti et al., 2016).

Table 3.1 Minerals, proteins, and sugars content, expressed as a mass percentage of dry matter for cumin seeds from four different geographic origins.

Nutritional characteristics	Lebanon	France	Algeria	Syria
Minerals %	10.5	5.4	8.0	8.7
Proteins %	24.7	22.6	20.9	24.0
Soluble sugars %	10.8	8.9	11.2	13.2

Adopted from Merah, O., Sayed-Ahmad, B., Talou, T., Saad, Z., Cerny, M., Grivot, S., Evon, P. & Hijazi, A., 2020. Biochemical composition of cumin seeds, and biorefining study. Biomolecules 10, 1054.

linoleic together with saturated congeners palmitic, stearic, and myristic acids (Hajhashemi et al., 2004). The seed oil contains cholesterol, campesterol, stigmasterol, β-sitosterol, α-spinasterol, (+)-citronellol, (+) limonene, p-cymene, citronellyl acetate, carvone, nigellone, arachidic, linolenic, linoleic, myristic, oleic, palmitic, palmitoleic, and stearic acids (Dinagaran et al., 2016). Seed oil contains fixed oils like linoleic acid (55.6%), oleic acid (23.4%), and palmitic acid (12.5%) and volatile oils like trans-anethole (38.3%), p-cymene (14.8%), limonene (4.3%), and carvone (4.0%) (Nickavar et al., 2003). *N. sativa* seeds are renowned for their high contents of vitamins A, B, and C and the minerals Cu, P, Zn, and Fe. However, the composition and proportions of phytochemicals, proteins, and mineral constituents in the seeds and oil may differ within provinces and regions (Tembhurne et al., 2014) (Table 3.1).

3.2.1 Plausible impact of climate change on productivity of black cumin

In general, crop production can be influenced by several biotic and abiotic stress factors, including availability of nutrients or water (Ma et al., 2020). With a changing climate, heat and drought stress are predicted to be among the main challenges for crop production, particularly in semiarid areas, which are already exposed to high temperatures and water scarcity (Zubair et al., 2013).

For consistent performance of a medicinal plant, maintenance of quantity and quality of the bioactive ingredients is a necessity (Can et al., 2021). The bulk black cumin is produced mainly under dry farming conditions where drought and high temperature are common production constraints. Growing black cumin under these conditions often results in low seed yield and quality losses due to limited rainfall, prolonged dry spells, and inability to compete with weeds (Güllü and Gülcan, 2013; Can et al., 2021). However,

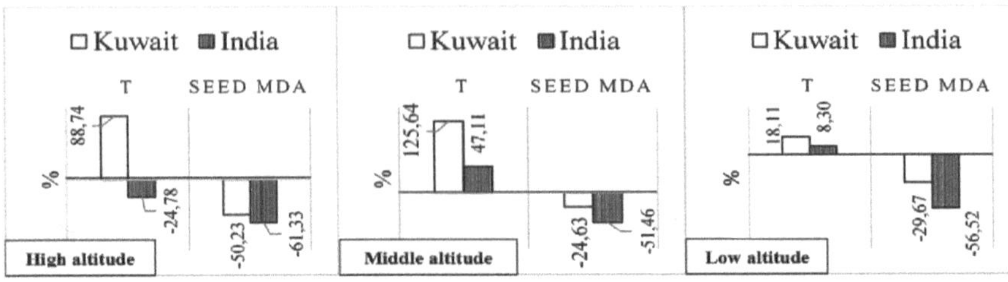

Figure 3.2 Percentage of THQ, THY, and MDA S compared with seed from the country of origin in three altitudes (Aziz et al., 2017b). *MDA S*, seed malondialdehyde; *THQ*, thymoquinone; *THY*, thymol.

available evidence indicates that variability in yield and quality are influenced by variety, altitude, and geographical location, thus suggesting crucial roles for geographical and genetic diversity in determining yield and quality characteristics of black cumin. Further, reductions in yield and seeds quality of black cumin were strongly associated with drought conditions (Ghamarnia and Jalili, 2013; Can et al., 2021).

Air temperature is crucial in determining growth, seeds yield, and phytochemical composition of black cumin (Aziz et al., 2017a). Temperature under field conditions cannot be controlled. However, it could be modified to simulate environmental conditions prevalent at black cumin plant center of origin (Ahrens and Henson, 2021). A study undertaken in Indonesian tropical regions showed that production of the bioactive ingredients, TQ, and thymol and the oxidative stress biomarker malondialdehyde (MD) in black cumin was influenced by altitude and accession origin (Fig. 3.2). The difference due to altitude was attributed to disparity temperature, whereas differences between accessions were attributed to differential tolerance to temperature stress as designated by MD contents. It is worth mentioning that both supraoptimal temperature and drought induce reactive oxygen species accumulation in planta (Wormuth et al., 2007). However, differences due to niche differentiation cannot be ruled out (Carroll et al., 2011). Thus, it would appear that for sustainable production of good quality black cumin, the microclimate at production sites should not to be overlooked (Aziz et al., 2017a).

3.3 Black cumin ecology and growing conditions

For maintenance of better growth, yield, and quality, black cumin requires an abstemiously cool and dry climate. At harvest, the crop is prone to damage by rains. Rainfall during harvest reduces yield and crop quality

(Sastry and DM, 2013). Crop quality is badly affected by diseases which negatively impact prices once the seeds have turned black. The crop is most vulnerable to frost damage, especially at flowering and early seed formation stages. Control measures include: spraying with sulfuric acid (0.1%), irrigating the crop prior to frost incidence, setting up windbreaks against cool waves, or creating an early morning smoke cover. The best growth of black cumin is attained on well-drained loamy to sand loamy soils with pH around neutral (7–8) (Anandaraj and Sudharshan, 2011).

3.4 Irrigation and water management on bioactive compounds of black cumin

One third of the world lands are classified as arid and semiarid regions, and the remains are faced with water seasonal or local fluctuations (Lange, 2012).Water shortage rather than land is the major constraint facing agricultural production in the arid and semiarid regions (Bannayan et al., 2008).

Black cumin requires less water than most other spices. In all growing countries, a light irrigation is given immediately after sowing to moisten the surface soil. Coinciding with germination, a second irrigation is applied some 8–10 days after the first irrigation. Depending upon edaphic and climatic conditions, subsequent irrigations may be given at an interval of 15–25 days. The last heavy irrigation is to coincide with seed formation (Sastry and Anandaraj, 2013).

Available research revealed that black cumin is tolerant to moderate water deficit (Mozaffari et al., 2000). However, irrigation scheduling, frequency, and synchronization with developmental stages improved growth and productivity (Bannayan et al., 2008; Karim et al., 2017).

3.5 Effect of fertilizers and soil type on bioactive ingredients of black cumin

Black cumin is generally adapted to disturbed soils and semiarid regions (D'Antuono et al., 2002). In these regions, nitrogen (N) deficiency is the most important factor that limits productivity, and consequently N fertilization is a necessity. Several findings have shown the positive effects of synthetic fertilizers, farmyard manure, and biofertilizers on seed yield and seed oil quality of black cumin (Özgüven and Şekeroğlu, 2007; Aytac et al., 2017).

Soil incorporation of about 10–15 tons/ha of farmyard manure or compost prior to sowing increase seed yield and quality (REF). In India, phosphate at 20 kg/ha is commonly applied at sowing, followed by 30 kg N/ha as a top dressing, either in a single dose 30 days later or in two equal splits at, respectively, 30 and 60 days after sowing. In Syria, the common practice is to apply 50 kg/ha triple superphosphate and 50 kg/ha of urea at planting. In addition, when spring rains are abundant, 50 kg/ha of top-dressed ammonium nitrate is recommended (Sastry and Anandaraj, 2013).

N at 60 kg/ha with supplemental potassium (K) affected high seed yield and improved seed quality and oil content of black cumin (Aytac et al., 2017). A further increase in N rate increased oil content. However, employment of higher N rates is discouraging for nonsignificant effects on the main components of oil, linoleic (54.5%), oleic (23.5%), and palmitic acids (13.5%) and for plausible jeopardy of environmental safety. The combination of N and K should be encouraged. K improves N uptake and use efficiency, increases roots development and plant tolerance to water deficit and curtails wilting, and lodging (Stewart, 2008; Milford and Johnston, 2009).

An experiment, comprising organic manure and synthetic and biofertilizers, conducted at the Kingdom of Saudi Arabia on several medicinal and cosmetic plants including black cumin, showed the superiority of the combinations in promoting growth, seed yield, and oil content over the individual treatment when each was applied alone. Oil content and chemical constituents of all plants were significantly increased (Ali and Hassan, 2014).

Further investigations on black cumin revealed the positive impact of the combinations of organic and biofertilizers on N and phosphorous uptake, vegetative growth, seed yield, seed quality, and quality and contents of fixed and volatile oils (Darakeh et al., 2021). Further, the combinations of organic manure and biofertilizers were found to reduce production cost, minimize reliance on synthetic fertilizers, improved soil health, and are environmentally benign. Among the treatments, the combination cattle manure (30 m^3/ feddan), N fixing, and phosphorus solubilizing bacteria were the most effective (Ali and Hassan, 2014).

3.6 Conclusion

Black cumin has gained importance as a nutritive and medicinal plant since time immemorial. Recent research indicates its potentials as a cure and/or suppressor of difficult to control contemporary inflammatory and infectious diseases spanning cancer and COVID-19. As a medicinal plant, consistent performance depends on the maintenance of quantity and quality of the active ingredients, which are influenced by both genetics and epigenetics factors. The plant is cultivated, mainly, in the semiarid regions, which are characterized by degraded soils and are most vulnerable to climate change. Climate change and soil degradation constitute serious constraints to black cumin production, quality, and values as a medicinal plant. Intensive research on production practices that preserve both yield and quality of the produce are therefore imperative.

References

Ahrens, C.D., Henson, R., 2021. Meteorology Today: An Introduction to Weather, Climate, and the Environment. Cengage Learning.

Ali, E., Hassan, F., 2014. Bio-production of Nigella sativa L. seeds and oil in Taif area. International Journal of Current Microbiology and Applied Sciences 3, 315–328.

Anandaraj, M., Sudharshan, M., 2011. Cardamom, Ginger and Turmeric. Encyclopedia of Life Support Systems (EOLSS)—Soils, Plant Growth and Crop Production. EOLSS Publishers, Oxford, UK.

Aytac, Z., Gulmezoglu, N., Saglam, T., Kulan, E.G., Selengil, U., Hosgun, H.L., 2017. Changes in N, K, and fatty acid composition of black cumin seeds affected by nitrogen doses under supplemental potassium application. Journal of Chemistry 7, 3162062.

Aziz, S.A., Kurniawati, A., Faridah, D.N., 2017a. Changes of black cumin (Nigella sativa L.) thymoquinone, thymol, and malondialdehyde content in response to Indonesia tropical altitude variation. Hayati Journal of Biosciences 24, 156–161.

Aziz, S.A., Kurniawati, A., Faridah, D.N., 2017b. Changes of thymoquinone, thymol, and malondialdehyde content of black cumin (Nigella sativa L.) in response to Indonesia tropical altitude variation. Hayati Journal of Biosciences 24, 156–161.

Bannayan, M., Nadjafi, F., Azizi, M., Tabrizi, L., Rastgoo, M., 2008. Yield and seed quality of Plantago ovata and Nigella sativa under different irrigation treatments. Industrial Crops and Products 27, 11–16.

Can, M., Katar, D., Katar, N., Bagci, M., Subasi, I., 2021. Yield and fatty acid composition of black cumin (Nigella sativa L.) populations collected from regions under different ecological conditions. Applied Ecology and Environmental Research 19, 1325–1336.

Carroll, I.T., Cardinale, B.J., Nisbet, R.M., 2011. Niche and fitness differences relate the maintenance of diversity to ecosystem function. Ecology 92, 1157–1165.

Chakrabarty, A., Emerson, M., Levine, S., 2003. Hemeoxygenase-1 in SJL mice with experimental allergic encephalomyelitis. Multiple Sclerosis Journal 9, 372–381.

D'antuono, L.F., Moretti, A., Lovato, A.F., 2002. Seed yield, yield components, oil content and essential oil content and composition of Nigella sativa L. and Nigella damascena L. Industrial Crops and Products 15, 59–69.

Darakeh, S.A.S.S., Weisany, W., Diyanat, M., Ebrahimi, R., 2021. Bio-organic fertilizers induce biochemical changes and affect seed oil fatty acids composition in black cumin (Nigella sativa Linn). Industrial Crops and Products 164, 113383.

Dinagaran, S., Sridhar, S., Eganathan, P., 2016. Chemical composition and antioxidant activities of black seed oil (Nigella sativa L.). International Journal of Pharmaceutical Sciences and Research 7, 4473.

Fatima Shad, K., Soubra, W., Cordato, D.J., 2021. The role of thymoquinone, a major constituent of Nigella sativa, in the treatment of inflammatory and infectious diseases. Clinical and Experimental Pharmacology and Physiology 48, 1445–1453.

Ghamarnia, H., Jalili, Z., 2013. Water stress effects on different Black cumin (Nigella sativa L.) components in a semi-arid region. International Journal of Agronomy and Plant Production 4, 545–554.

Güllü, E.B., Gülcan, A., 2013. Timokinon: Nigella Sativa'nın biyoaktif komponenti. Kocatepe Veterinary Journal 6, 51–61.

Hajhashemi, V., Ghannadi, A., Jafarabadi, H., 2004. Black cumin seed essential oil, as a potent analgesic and antiinflammatory drug. Phytotherapy Research: An International Journal Devoted to Pharmacological and Toxicological Evaluation of Natural Product Derivatives 18, 195–199.

Hussain, D.A., Hussain, M.M., 2016. Nigella sativa (black seed) is an effective herbal remedy for every disease except death-a Prophetic statement which modern scientists confirm unanimously: a review. Advancement in Medicinal Plant Research 4, 27–57.

Karim, M., Himel, R., Ferdush, J., Zakaria, M., 2017. Effect of irrigation levels on yield performance of black cumin. International Journal of Environment, Agriculture and Biotechnology 2, 238754.

Khan, A., Chen, H., Tania, M., Zhang, D.Z., 2011. Anticancer activities of Nigella sativa (black cumin). African Journal of Traditional, Complementary and Alternative Medicines 8, 226–232.

Kooti, W., Hasanzadeh-Noohi, Z., Sharafi-Ahvazi, N., Asadi-Samani, M., Ashtary-Larky, D., 2016. Phytochemistry, pharmacology, and therapeutic uses of black seed (Nigella sativa). Chinese Journal of Natural Medicines 14, 732–745.

Lange, O.L., 2012. Physiological Plant Ecology II: Water Relations and Carbon Assimilation. Springer Science & Business Media.

Ma, Y., Dias, M.C., Freitas, H., 2020. Drought and salinity stress responses and microbe-induced tolerance in plants. Frontiers of Plant Science 11, 1750.

Mathur, M.L., Gaur, J., Sharma, R., Haldiya, K.R., 2011. Antidiabetic properties of a spice plant Nigella sativa. Journal of Endocrinology and Metabolism 1, 1–8.

Merah, O., Sayed-Ahmad, B., Talou, T., Saad, Z., Cerny, M., Grivot, S., Evon, P., Hijazi, A., 2020. Biochemical composition of cumin seeds, and biorefining study. Biomolecules 10, 1054.

Milford, G., Johnston, A., 2009. Potassium and nitrogen interactions in crop production. Nawozy i Nawożenie (Fertilisers and Fertilization) 34, 143–162.

Mozaffari, F., Ghorbanli, M., Babai, A., Sepehr, M.F., 2000. The effect of water stress on the seed oil of Nigella sativa L. Journal of Essential Oil Research 12, 36–38.

Nickavar, B., Mojab, F., Javidnia, K., Amoli, M.A.R., 2003. Chemical composition of the fixed and volatile oils of Nigella sativa L. from Iran. Zeitschrift für Naturforschung C 58, 629–631.

Özgüven, M., Şekeroğlu, N., 2007. Agricultural practices for high yield and quality of black cumin (*Nigella sativa* L.) cultivated in Turkey. International Symposium on Medicinal and Nutraceutical Plants 756, 329–338.

Sastry, E., Anandaraj, M., 2013. Cumin, fennel and fenugreek. Soil, Plant Growth and Crop Production. Encyclopedia of Life Support Systems 1–10.

Stewart, M., 2008. Fertilizer Sources for Irrigated Corn. International Plant Nutrition Institute, pp. 1–6.

Tembhurne, S., Feroz, S., More, B., Sakarkar, D., 2014. A review on therapeutic potential of Nigella sativa (kalonji) seeds. Journal of Medicinal Plants Research 8, 167–177.

Van Kleunen, M., Dawson, W., Essl, F., Pergl, J., Winter, M., Weber, E., Kreft, H., Weigelt, P., Kartesz, J., Nishino, M., 2015. Global exchange and accumulation of non-native plants. Nature 525, 100–103.

Wormuth, D., Heiber, I., Shaikali, J., Kandlbinder, A., Baier, M., Dietz, K.-J., 2007. Redox regulation and antioxidative defence in Arabidopsis leaves viewed from a systems biology perspective. Journal of Biotechnology 129, 229–248.

Zubair, H., Khan, H., Sohail, A., Azim, S., Ullah, M., Ahmad, A., Sarkar, F., Hadi, S., 2013. Redox cycling of endogenous copper by thymoquinone leads to ROS-mediated DNA breakage and consequent cell death: putative anticancer mechanism of antioxidants. Cell Death & Disease 4 e660-e660.

CHAPTER FOUR

Black cumin: morphology, physiology, growth, and agricultural yield

Samia Osman Yagoub
Department of Agronomy, College of Agricultural Studies, Weed Science Centre, Sudan University of Science and Technology, Khartoum, Sudan

4.1 Botanical aspect of black cumin

4.1.1 Taxonomy of black cumin

Domain Eukarya, common name black cumin, fennel flower, Roman coriander, nutmeg flower, black seed, black caraway, damascena, devil in the bush, wild onion seed.

Kingdom	Plantae
Subkingdom	Viridiplantae (green plants)
Infrakingdom	Streptophyta (land plants)
Superdivision	Embryophytea (vascular plant)
Division	Tracheophytes
Subdivision	Spermatophytina
Class	Magnoliopsida
Superorder	Ranunculanae
Order	Ranunculales
Family	Ranunculaceae
Genius	Nigella
Specious	*Nigella sativa* L.

4.1.2 Distribution of black cumin

The genus Nigella is come from a Latin word Niger which means black. The term *nigellus* means "blackish" or "dark"; the seed coat is black in color. The species name *sativa* means cultivated, planted, sown, and not wild (Marzell, 1977). *Nigella sativa* L is the most known member of the genus. This species has pale blue to blue purple flowers that bloom quickly and produce seed capsules (fruit) with black seeds. The black cumin is cultivated in

Biochemistry, Nutrition, and Therapeutics of Black Cumin Seed
ISBN: 978-0-323-90788-0
https://doi.org/10.1016/B978-0-323-90788-0.00016-0

different parts of Asia, Europe, Africa, and the Americas. *N. sativa* is found growing wild in regions of northern Africa, Turkey, Syria, Iraq, and Iran (Saad and Said, 2011). The species is also cultivated on a commercial scale in Southern Asia in India (Ved, 2008), Iran (Ghouzhdi, 2010), Pakistan (Rabbani, 2011), Western Asia, Iraq (Toma et al., 2013), Israel (Botnick et al.,2012), Jordan (Abu Hammur, 2008), Lebanon (Paarach, 2010), Syria (Hoppe et al.,2013), Turkey , Yemen (Al Nageep et al., 2009), Northern Africa, Sudan (Teucher, 2006), Egypt, Tunisia (Tomaet al., 2013), eastern Africa, and Ethiopia (Gashaw, 2020).

The major producing countries are India, Sri Lanka, Bangladesh, Afghanistan, Pakistan, Iran, Iraq, Turkey, Syria, and Ethiopia. The market size of black cumin oil in 2018 was over 15 million USD, and by 2025, a significant increase is expected. The natural average productivity of black cumin was reported to be 0.79 tons/ha (Kifelew et al., 2017). The global average and major producing countries were India (2.2 tons/ha). In India, black cumin is widely used in medicine, so the plant is cultivated on a large scale and estimated to be traded in annual quantities in more than 100 metric tons. Gashaw (2020) reported 0.64 tons/ha in Ethiopia, which is very low compared with producing countries. Ermias et al. (2015) attributed the low production to the low productivity of the cultivars, while Hammo (2008) attributed it on poor agronomic practices applied by the farmers.

4.1.3 Morphology of *Nigella sativa* L

Black cumin is an annual hardy crop that is not parasitic, has no spine, is not toxic, anthropogenic, manmade, or disturbed, and grows to a height of 60–70 cm. The plant does not appear to have bulbs, no branch tendrils, and the stem has hairs between nodes. There is at least one full leaf at the base of the flowering stem. The leaves or young stems are green. The leaf blades do not have glandular dots or scale. Sap is clear and watery.

Flowers are solitary, with 5–10 petals; both the petals and sepals are separate and not fused. The fruit is dry and splits open when ripe, with length up to 1.2 mm. Flower (Fig. 4.1) anthers have narrow slits or furrows that run lengthwise along the anthers and have no spurs (a tapering, sometimes curving projection). The calyx is readily symmetrical. The flower does not have an epicalyx. Carpels are fused to the ovary; the filament is smooth with no hair or scales. The flowers have a superior ovary, no hypanthium, a narrow style, and are arranged in a spiral around the inflorescence axis or branch, occurring singly or in several ranks.

Figure 4.1 Black cumin (*Nigella sativa*) flowers. *https://commons.wikimedia.org*

Nigella seeds have been employed for thousands of years in medicine as a spice (Benkaci et al.,2013). The seeds remain viable in extreme temperatures and sun exposure, and this may be due to their antioxidant content. The seeds are small, matt-black grains with a rough surface and an oily white interior, and they are roughly triangulate, measuring 1.5—3.0 mm long. The seeds of *N. sativa* are small (1—5 mm) and have corrugated integuments (Benkaci et al., 2007). The seeds are rich in oil (Fig. 4.2). Because it possesses an off flavor, the *N. sativa* seed cake (meal) has yet to be efficiently utilized in

Figure 4.2 Seed of black cumin. *https://commons.wikimedia.org*

animal feeds. Like most herbs, the composition of black cumin varies with the geographic distribution, time of harvest, and agronomic practices (Sultana et al., 2009). Sener et al. (1985) reported that there was 0.26% lauric, 1.06% myristic, 20.4% palmitic, 1.56% stearic, 4.75% oleic, 64.6% linoleic, and 7.18% arachidonic acid in the stable oil. Nergiz and Otles (1993) stated that there was about 32% oil in black cumin seeds.

4.2 Black cumin physiology

The physiological growth of plants is affected mainly by ecological and genetic factors. As aromatic plants, black cumin physiological growth is an interaction of active substances that determines the amount and quality of essential oil. Yasar (2005) advised that black cumin as an aromatic plant should only be cultivated in the convenient regions for the ecology of this plant due to the importance of the quality of aromatic plants. Those of a certain quality cannot be cultivated even if they have a very high yield.

Kara et al. (2015) discovered that location and season had a significant effect on black cumin seed yield, some yield components, essential oil, and fixed oil content. The average seed yield varied between 201.0 and 407.1 kg/ha and 458.9 and 790.3 kg/ha in two different locations and consecutive seasons in Turkey. On the other hand, Ozol et al. (2009) detected a yield of 1406.3–2482.3 kg/ha.

4.2.1 Nigella seed agriculture, growth, and yield

Black cumin grows as a rain-fed crop within 1500–2400 m of altitude on heavy black soils and prefers loamy sand soils (Datta et al., 2001) on soil pH of 7.0–7.5 (Orgut 2007). It is a hardy crop growing in all types of soil (Jansen, 1981), sandy, and loam. It is rich in microbial activity, with a pH of 5–8, sloppy, leveled, and well-drained soils (Takrun and Dameh, 1998).

Black cumin is propagated by seed. The seedling is very fragile for transplanting. Germination is epigeal and takes 10–15 days. Black cumin flowers about 100 days after sowing; seeds mature 50 days later. Flowering is protandrous, and pollination takes place by insects. Self-pollinated. Flowering starts from the tip and progresses down to the bottom. Sowing time varies as a function of the local environment. Sultana et al. (2018) found that the different dates of sowing had significant differences in morphology, growth, yield, and quality attributes of black cumin. The lack of information on agromanagement practices is due to weather variation in West Bangladesh. Melnyk and colleagues (2015).

Black cumin seed has low volume and can be a good rotation crop for cereal and pulses. It requires well-prepared land. The soil should be fine, free of logging, loam, light, and with good nutrient composition for black cumin growth. Or it can be improved by fertilizer application. So good plowing, harrowing, and leveling must be done. The distance between rows was 20 cm, and within rows it was 5 cm. Seeds were sown at 1–2 cm depth in the soil. In Ethiopia, black cumin is cultivated as a rain-fed crop in high land of 1500–2500 m, with annual rainfall of 120–400 mm and an average temperature of 12–20°C (Gahaw, 2020). It is sown with a row spacing of 15–40 cm and frequently intercropped with barley and wheat (Ahmed and Haque, 1986), given at a 10-day interval depending on the moisture condition of the soil. A first thinning was done to ensure an optimum plant population and adequate space, 15 days after sowing, followed by a second thinning up to 10 days after the first thinning. Hand weddings are needed according to the presence of weeds in the area. The optimum sowing date is October 15th (Malhautra and Vashinshtha, 2008). The seeds were sown from March to August in Turkey (Kara et al., 2015). In India, crops mature in 135–150 days after sowing, with a seed rate of 5.7 kg/ha in different states, sown from mid-September to mid-October. MaleKi et al. (2021). In Iran, black cumin is sowed in late April. In Ethiopia, it is sown in the rain season from August to September, and by irrigation from April to May (Gashaw, 2020). Recommended seed rates for broadcasting are 15–20 kg/ha and 5–7.5 kg/ha for drilling with row spacing of 15–40 cm in Ethiopia. The crop harvesting was done by pulling the plants. The harvested plants were sundried for three days, and threshing was done by beating with sticks. The seeds of black cumin are stored well for one year in low light conditions to prevent the loss of aroma. It is advised to keep it separate from other species because the species has an overpowering flower and aroma that interferes with the flavor of other species (Animesh et al., 2012).

4.3 Conclusion

Black cumin is an annual medicinal plant belonging to the family Ranunculaceae. It is native to Mediterranean regions, India, and the Middle East. It is an annual herb up to 30 cm high, with erect branched stems, delicate, deep-lobed leaves, blue to gray flowers, pods, and toothed seeds. It is cultivated in different soils with different climates.

References

Al-Naqeep, G.N., Ismail, M.M., Al-Zubairi, A.S., Esa, N.M., 2009. Nutrients composition and minerals content of three different samples of Nigella sativa L. cultivated in Yemen. Asian Journal of Biological Sciences 2 (2), 43–48.

Abu-Hammour, K., 2008. Pollination of Medicinal Plants (*Nigella sativa* and *Coriandrum sativum*) and *Cucurbita pepo* in Jordan. Institut fur Nutzpflanzenwissenschaften und Ressoucenschutz, Rheinischen Friedrich-Wilhelms-Universitat, Bonn, Germany, 2008.

Ahmed, N.U., Haque, K.R., 1986. Effect of row spacing and time of sowing on the yield of black cumin (*Nigella sativa*). Bangladish Journal of Agriculture 11, 21–24.

Animesh, K.D., Saha, A., Bhattacharya, A., Mandal, A., Paul, R., Sengupta, S., 2012. A Review on Black Cumin *Nigella sativa* L.). Department of Botany, Cytogenetics and Plant Breeding Section, Kalyani University, Kalyani 741235, West Bengal, India.

Benkaci-Ali, F., Baaliouamer, A., Meklati, B.Y., Chemat, F., 2007. Chemical composition of seedessential oils from Algerian Nigella sativa extracted by microwave and hydrodistillation. Flavour and Fragrance Journal 22, 148–153.

Benkaci-Ali, F., Akloul, R., Boukenouche, A., De Pauw, E., 2013. Chemical composition of the essential oil of *Nigella sativa* seeds extracted by microwave steam distillation. Journal of Essential Oil Bearing Plants 16 (6), 781–794.

Botnick, I., Xue, W., Bar, E., Ibdah, M., Schwartz, A., Joel, D.M., Lev, E., Fait, A., Lewinsohn, E., 2012. Distribution of primary and specialized metabolites in Nigella sativa seeds, a spice with vast traditional and historical uses. Molecules 17 (9), 10159–10177.

Datta, S., Sharanghi, A.B., Pariari, A., Chatterjee, R.R., 2001. Prospect of minor spices cultivation in West Bengal. In: Paper Presented in VIII State Science Congress, 28th Feb–2 nd March, 2001. Kalyyani Publisher, West Bangal, India.

Ermias, A., 2015. Adaptability study of black cumin (*Nigella sativa* L) varieties in the mid and high land areas of Kaffa Zone, South West Ethiopia. Agriculture, Forestry and Fisheries 15 (6), 14–17.

Gashaw, Z., 2020. Status of black cumin (*Nigella sativa* L.) research and production in Ethiopia; A review. International Journal of Forestry and Horticulture 6 (1), 20–29.

Ghouzhdi, H.G., 2010. Indigenous knowledge in agriculture with particular reference to black cumin (*Nigella sativa*) production in Iran. Scientific Research and Essays 5 (25), 4107–4109, 2010.

Hammo, Y.H., 2008. Effect of high levels of nitrogen and phosphorus fertilizer on growth, yield and yield components of *Nigella sativa* L. Mesopotamia Journal of Agriculture 36, 2–11.

Hoppe, B., Biertümpfel, A., Echim, T., Graf, T., Schwarzkümmel, E., 2013. *Nigella sativa* L. In: Hoppe, B. (Ed.), Handbuch des Arznei- und Gewürzpflanzenbaus, Band 5. Bernburg, Germany: Verein für Arznei- und Gewürzpflanzen SALUPLANTA e.V. Bernburg; 2013, pp. 514–525.

Jansen, P.C.M., 1980. Spices, Condiments and Medicinal Plants in Ethiopia. Their Taxonomy and Agricultural Significance. Addis Ababa: Center for Agricultural Publishing and Documentation, pp. 111–120, 1981.

Kara, N., Katar, D., Baydar, H., 2015. Yield and quality of black cumin (*Nigella sativa* L.) population: the effect of Ecological condition. Turkish Journal of Field Crops 20 (1), 9–14.

Kifelew, H., Fikre, D., Luleseged, T., Bekele, D., Mitiku, H., Getachew, W., 2017. Seed Spices Production Guideline. Technical Research Report". Ethiopian Institute of Agricultural Research (EIAR), Addis Ababa.

Marzell, H., 1977. Wörterbuch der deutschen Pflanzennamen. 3. Band. S. Hirzel Verlag/ Franz Steiner Verlag, Stuttgart/Wiesbaden, Germany, 1977.

Malhotra, S.K., Vashishtha, B.B., 2008. Response of nigella (*Nigella sativa* L) variety NRCSS AN 1 to different agrotechniques. Journal of Spices and Aromatic Crops 17 (2), 190–193.

Maleki, J., Ashourabadi, E.S., Mirza, M., Heydari, H., Lebaschy, M.H., 2021. Improving the productivity and quality of black cumin (*Nigella sativa* L) by using soil fertility management practice. Journal of Plant Nutrition 44 (12), 1741–1757.

MelnyK, H.I., Stalyus, L.V., Kozak, I.I., 2015. The perspective of *Nigella sativa* L, growing in the climatic conditions of the Precarpathian Region. The Pharma Innovation Journal 4 (3), 24–26.

Nergiz, C., Otles, S., 1993. Chemical composition of *Nigella sativa* L. seeds. Food Chemistry 48, 259–261.

Ozel, A., Demirel, U., Guler, I., Erden, K., 2009. Effect of different row spacing and seeding rate on black cumin (*Nigella sativa* L.) yields and some agricultural characters. Harran University Journal of Agriculture Faculty. 13 (1), 17–25 (inTurkish).

Orgut, 2007. Market Assessment Study Ethiopian Nile Irrigation and Drainage Project, Main Report and Annexes. Ministry of WATER resources, Addis Ababa.

Paarakh, P.M., 2010. *Nigella sativa* Linn.—a comprehensive review. Indian Journal of Natural Products and Resources 1 (4), 409–429, 2010.

Rabbani, M.A., Ghafoor, A., Masood, M.S., 2011. NARC-Kalonji: an early maturing and high yielding variety of Nigella sativa released for cultivation in Pakistan. Pakistan Journal of Botany 43 (SI), 191–195.

Saad, B., Said, O., 2011. Greco-Arab and Islamic Herbal Medicine: Traditional System, Ethics, Safety, Efficacy, and Regulatory Issues. John Wiley & Sons, Inc., Hoboken, NJ, 2011.

Sener, B., Kusmenoglu, S., Mutlugil, A., Bingol, F., 1985. A study with the seed of *Nigella sativa*. Gazi University Journal of Pharmacy Faculty. 2 (1), 1–8.

Sultan, M.T., Butt, M.S., Anjum, F.M., Jamil, A., Akhtar, S., Nasir, M., 2009. Nutritional profile of indigenous cultivar of black cumin seeds and antioxidant potential of its fixed and essential oil. Pakistan Journal of Botany 41, 32–40.

Sultana, S., Das, B., Rudra, B.C., Das, G., Alam, M.B., 2018. Effect of date of sowing on productivity of black cumin. International Journal of Current Microbiology and Applied Sciences 7 (1), 1796–1800.

Takrun, H.R.H., Dameh, M.A.F., 1998. Study of the nutritional value of Black cumin seeds (Nigella sativa L.). Journal of Scientific Agriculture 76, 404–410.

Teuscher, E., 2006. Medicinal Spices: A Handbook of Culinary Herbs, Spices, Spice Mixtures and Their Essential Oils. Medpharm Scientific Publishers, Stuttgart, Germany, 2006.

Toma, C.C., Simu, G.M., Hanganu, D., et al., 2013. Chemical composition of the Tunisian *Nigella sativa*. note II. Profile on fatty oil. Farmacia 61 (3), 454–458, 2013.

Ved, D.K., Goraya, G.S., 2008. Demand and Supply of Medicinal Plants in India. Bishen Singh Mahendra Pal Singh, Dehra Dun, India, 2008.

Yasar, S., 2005. Determination of Fixed and Essential Oil Contents and Soil Characteristic of Some Perennial Medical Plants That Grow Naturally in the Campus of Cukurova University. M. Sc. Thesis, vol 43p. Department of Biology Institute of Natural and Applied Sciences University of Cukurova, Adana (in Turkish).

CHAPTER FIVE

Nutritional, biochemical, and functional characteristics of black cumin seeds

Newlove Akowuah Afoakwah and Gustav Komla Mahunu
Department of Food Science and Technology, Faculty of Agriculture, Food and Consumer Sciences, University for Development Studies, Tamale, Ghana

5.1 Introduction

Because of its nutritional and medicinal characteristics, black cumin (*Nigella sativa*) seeds are widely utilized in various cultures. These seeds have a fragrant flavor that is intense and bitter on the tongue, and they are small and black in appearance. It's an herbaceous plant that blooms every 12 months. It originated from Africa and South-West Asia, but it is now widely grown worldwide (Ahmad et al., 2021). Egypt is a good source of high-quality black cumin seeds (BCSs) because Egypt's climatic conditions are ideal for its propagation (Naz, 2011). The seeds are readily available for use as a condiment and is a traditional medicine, which can be found in markets (Cheikh-Rouhou et al., 2007).

BCS is known to have a religious background, as mentioned in Isaiah's book (Bible), it is a spice used for the preparation of pastries (Naz, 2011). It can be found in works from the past. It has a long history as a medication in China (Ahmad et al., 2021). It was used to treat various illnesses and was seen as a necessary seed to treat Indians. The BCS is a type of Islamic medicine known as "Habb-e-Sauda." BCS is also a prophetic medication, according to the prophet (SAW). Furthermore, it has been demonstrated to heal various ailments in the book "*Bukhari-5687*" (Ahmad et al., 2021).

The leaves of the *N. sativa* plant are alienated and green in color. Its flowers are pale blue in some cases and yellowish in others (Fig. 5.1A1—A4). There are around five to ten petals on it. A swollen capsule contains the fruits. Three to seven united follicles contain these capsules. Several oval black seeds, each measuring about one millimeter in diameter, are contained within these follicles (Gholamnezhad et al., 2016). Its name is black seed in

Biochemistry, Nutrition, and Therapeutics of Black Cumin Seed
ISBN: 978-0-323-90788-0
https://doi.org/10.1016/B978-0-323-90788-0.00009-3

Figure 5.1 A1 and A2: *Nigella sativa* flower; A3: *Nigella sativa* plant; A4: Black cumin seed.

English; Habba Al-Sauda or Habba Al-Bar-akah in Arabic; Kalonji in Urdu; Siyah Danch in Persian; Cork out in Turkish; Love in a mist, Habatul-Barakah, So-nez, Kri-shana, Jira-ka, and Sidad-anah in Turkish.

BCS is commonly used as a flavoring in curries, as a component of vinegar, and as a pepper substitute in culinary and pastry preparations because it has a strong spicy peppery flavor and scent (Babayan et al., 1978). Depending on geographical distribution, harvest period, and agronomic practices, the nutritional profile of BCS varies.

Moisture, oil, proteins, ash, and carbohydrates make up the BCS (Atta 2003; Takruri and Dameh, 1992). Linoleic acid (40.3%–70.8%), oleic (15.2%–28.1%), palmitic (9.47%–13.34%), and stearic (2.6%–3.1%) acids make up the fatty acid of BCS (Matthaus, 2011; Tulukcu, 2011; Cheikh-Rouhou, 2007; Kabir et al., 2019). Monounsaturated and polyunsaturated fatty acids such as oleic and linoleic acids are greater in BCS (Kabir et al., 2019). The seed-fixed oil is primarily composed of polyunsaturated fatty acids. This accounts for over half of the total fatty acid content (48%–70%). Monounsaturated (18%–29%) and saturated (12%–25%) fatty acids, on the other hand, make up a smaller fraction of the seed (Kabir et al., 2019; Anjum et al., 2009; Cheikh-Rouhou et al., 2007). Many biological actions of BCS extracts have been discovered in several studies (Trigui et al., 2018), while the antioxidant properties are known to promote human health by neutralizing free radicals, which have been linked to heart disease, abnormal bodily states such as tumors, inflammation, and aging (Trigui et al., 2018). The increased concentration of essential oils in BCS is linked to its health-promoting properties (Ramadan, 2007). Aside from having a superior fatty acid composition, it also contains many levels of tocopherol and other biochemicals responsible for humans' overall antioxidation properties (Valko et al., 2006). The presence of phytosterols in the seed also improves its hypoglycemic and hypercholesterolemic properties (Atta, 2003; Cheikh-Rouhou et al., 2007; Ramadan and Mörsel, 2002). Similarly, essential oils and their active biomarker thymoquinone were able to prevent oxidative stress, tumor growth, immunological irregularity, and diabetes complexity (Gali-Muhtasib et al., 2006; Hussein et al., 2005).

5.2 Black cumin biochemical components

Greenish stated the chemical components of BCS in 1880 (Ahmad et al., 2021). BCS was discovered to be made up of fatty acids, proteins, carbs, and fibers in that article (Ahmad et al., 2021). However, recent research has revealed that *N. sativa* has curative properties due to the presence of the quinone (thymoquinone) component (Sahak et al., 2016). Thymoquinone, a part of volatile oil, has various pharmacological effects (Ahmad et al., 2021). Thymoquinone's molecular mechanism of action has been demonstrated in the literature (Gholamnezhad et al., 2016; Ahmad et al., 2021). Additional chemicals identified in *N. sativa* include p-cymene carvacrol, thymohydroquinone, dihydro-thymoquinone, -thujene, thymol, t-anethole, -pinene, and -terpinene (Fig. 5.2A and B). These chemicals serve an essential function in illness prevention and nutrition (Ahmad et al., 2021; Sahak et al., 2016). In addition, Ramdan and Morsel (2002) discovered β-sitosterol and stigmasterol in the seed of the black cumin, estimating that they make up around 65% of the total sterols. Lower concentrations of campesterol, lanosterol, and D7-avenasterol were found in the seed (Ramdan and Morsel, 2002).

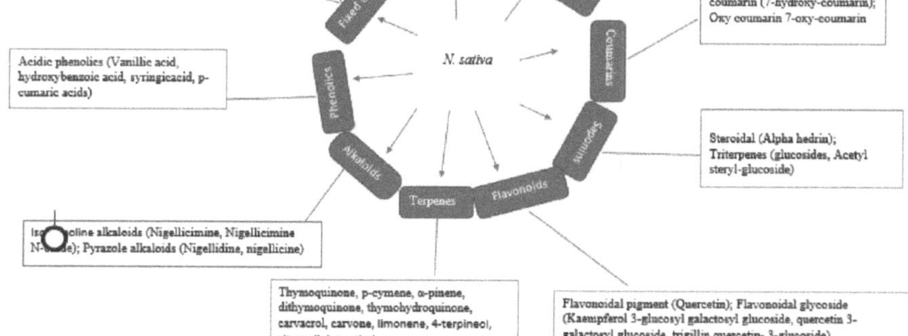

Figure 5.2a Some key chemical composition of *Nigella sativa*.

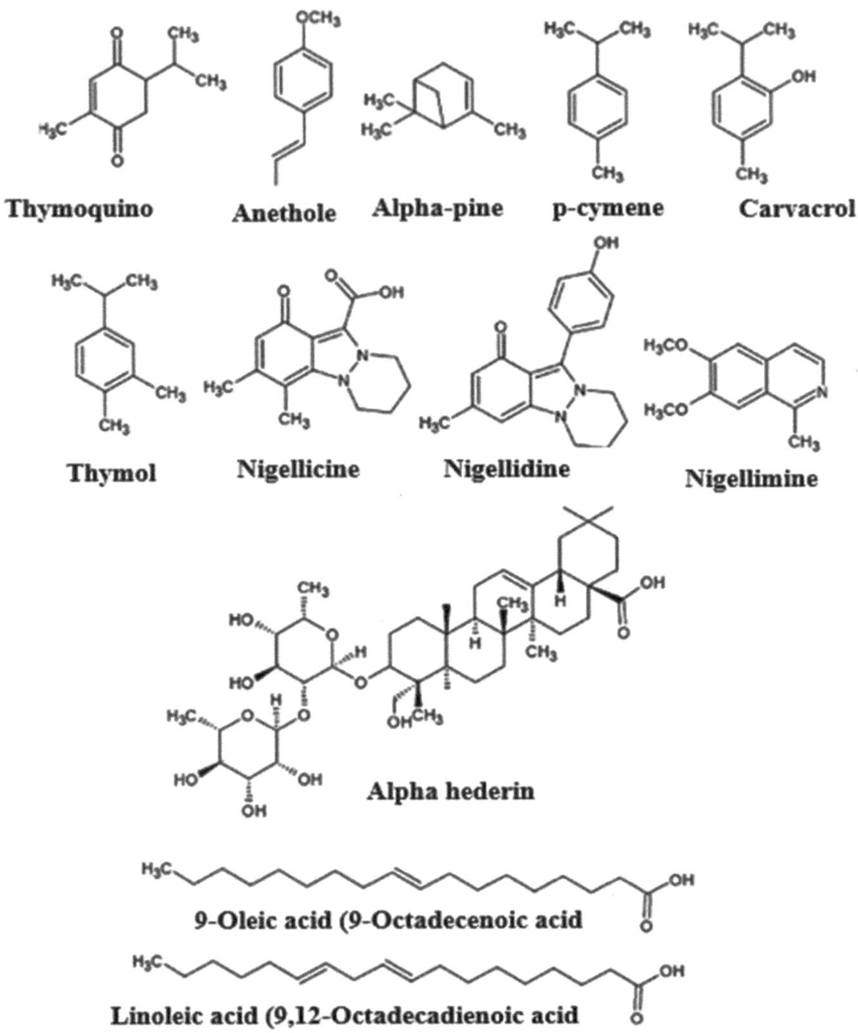

Figure 5.2b Black cumin seeds bioactive ingredients (Ahmad et al., 2021).

5.3 Nutritional value of black cumin seeds

In a separate study by Kabir et al. (2019), proximate analysis (Fig. 5.3) was utilized to determine the nutritional characteristics of BCSs. The total carbs were in the range of 20.0%–25.5%, the protein was in the range of 20.3%–23.0%, and crude fat was in the range of 32.0%–45.4%. Water

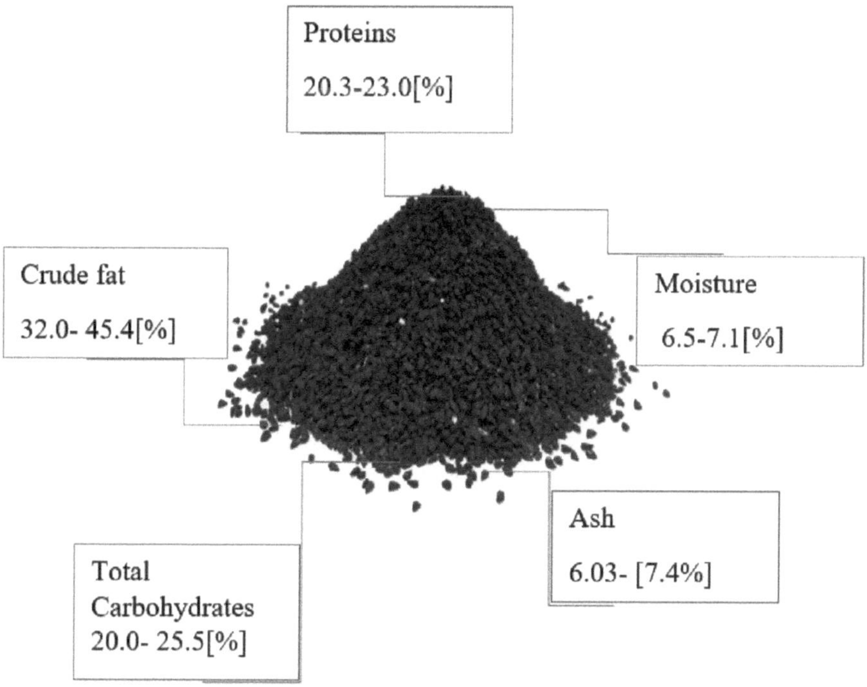

Figure 5.3 Black cumin seeds nutritional components.

content was 6.5%—7.1%, ash content was 6.0%—7.4%, and dry matter content was 9.3% in BCSs (Kabir et al., 2019). Other component analyses found that protein (20.8%), carbs (31.9%), lipids (38.2%), moisture (4.64%), ash (4.37%), and crude fiber (7.94%) were present in BCS (Mamun and Absar 2018;). BCS contains a wide range of nutrient compositional features, which can be connected to geographical distribution, harvest time, and agronomic practices used during *N. sativa* growth. During secondary metabolite research (Kabir et al., 2019; Mamun and Absar, 2018), active substances such as alkaloids (10.1 mg/100 g), flavonoids (3.7 mg/100 g), saponins (7.6 mg/100 g), and tannins (2.2 mg/100 g) were identified.

5.4 Carbohydrates in black cumin seeds

Glucose, maltose, and sucrose are all abundant carbohydrate sources in BCS (Ramadan, 2007). Research on BCS metabolism revealed that the rate of arabinose, lyxose, and xylose increased at first, then fell. Additional sugars, such as galactinol and raffinose, are produced during dehydration (Song et al., 2013). Rhamnose, arabinose, xylose, galactose, glucose, and uronic acid were detected in 30 mg/g of rhamnose, arabinose, xylose, galactose, glucose, and uronic acid in BCS from India (Manjegowda et al., 2017).

5.5 Proteins in black cumin seeds

A proximate study revealed the presence of 20%—21% protein content in BCS (Al-Jassir, 1992; Babayan et al., 1978). In BCS, the amino acid profile reveals several essential and nonessential amino acids in various concentrations (Table 5.1). The concentration of nonessential amino acids in BCS is higher than that of necessary amino acids. The oil and oil-soluble components of cumin seeds are the essential features. The medical usefulness of proteins, on the other hand, should not be disregarded, and research is currently being undertaken to evaluate the bioactivity of proteins present in BCS.

Table 5.1 Amino acid composition of black cumin seed (Kabir et al., 2019; Al-Jassir, 1992; Babayan et al., 1978).

Essential amino acid	Concentration (%)
Lysine	3.9—7.6
Threonine	1.2—3.9
Valine	3.1—4.8
Leucine	6.0—10.9
Isoleucine	3.7—4.0
Histidine	2.8—3.4
Tryptophan	0.0—1.5
Methionine	1.9—6.2
Phenylalanine	3.7—7.9
Nonessential amino acids	**Concentration (%)**
Glutamic acid	13.2—24.7
Aspartic acid	5.0—9.3
Arginine	9.0—19.5
Proline	4.9—5.3
Alanine	3.7—4.4
Serine	1.9—4.5
Tyrosine	3.4—6.1
Cystine	1.9—2.5

5.6 Vitamins and minerals in black cumin seeds

Vitamins are bioactive molecules that are necessary for optimum health and growth in humans. Their small levels, however, are sufficient to support normal metabolic function. Their absence can have severe consequences on one's health. Vitamins are found in the seeds of the sativa plant (Table 5.2). Fat-soluble vitamins include vitamins A, D, E, and K. They are antioxidants because they have radical-scavenging properties. BCS is high in these fat-soluble vitamins (Vatansev et al., 2013).

The mineral content in BCS (Table 5.3) has been a major research area. Unalike minerals exist in varying amounts in the seeds of black cumin (Table 5.3). The most prevalent minerals in BCS are phosphorus, magnesium, and iron. Zinc, copper, calcium, and manganese, on the other hand, are in relatively small quantities. There have also been reports of cadmium, lead, nickel, and molybdenum traces in the BCS (Kabir et al., 2019).

5.7 Black cumin seeds contain lipids

Neutral lipids are present in greater quantity in BCS than different lipid categories. Table 5.3 lists some of the lipids discovered in BCS as a result of independent studies by Al-Jassir (1992); Kabir et al. (2019); Kaskoos (2011); and Bourgou et al. (2010), which have been aggregated and the

Table 5.2 Major vitamins and mineral in black cumin seeds.

Vitamin (μg/g)	Dinagaran et al. (2016)	Vatansev et al. (2013)	Al-Saleh et al. (2006)	Mineral (mg/100g)	Mamun & Absar (2018)	Kabir et al. (2019)
Vitamin K	nd	1.85	Nd	Phosphorus	91.5	481.5
Vitamin D	nd	1.38	Nd	Calcium	579.3	366.7
Riboflavin	nd	Nd	Nd	Sodium	100.0	44.8
Thiamin	nd	Nd	14.6 mg/kg	Magnesium	218.3	355.2
Niacin	nd	Nd	56.5 mg/kg	Iron	41.8	42.6
Folic acid	nd	Nd	614 mg/kg	Potassium	510.3	1498.3
Tocopherol	11.3	10.1	Nd	Copper	nd	1.5
Retinol	0.27	0.18	Nd	Zinc	nd	6.7
				Manganese	nd	3.1

nd, Nondetected.

Table 5.3 Lipid components in black cumin seed separate research by Kaskoos (2011); Bourgou et al. (2010); Al-Jassir (1992); Kabir et al. (2019).

Lipids	Concentration (%)
Myristic acid (14:0)	0.16–0.4
Palmitic acid (16:0)	8.5–14.8
Palmitoleic acid (16:1)	0.16–0.28
Stearic acid (18:0)	2.2–2.9
Oleic acid (18:1)	16.6–21.8
Cis-vaccenic acid (18:1)	1.2
Linoleic acid (18:2)	42.8–58.1
Linolenic acid (18:3)	0.25–0.46
Arachidic acid (20:0)	0.18–1.1
Eicosenoic acid (20:1)	0.39–0.16
Eicosadienoic acid (20:2)	1.94–2.91
Eicosatrienoic acid (20:3	47
Eicosapentaenoic acid (20:5)	5.98
Behenic acid (22:0)	2.89
Docosahexaenoic acid (22:6)	2.97
Lignoceric acid (24:0)	3.60
Nervonic acid (24:1)	2.76
Saturated acids	15.1–22
Unsaturated acids	19.5–84

concentrations shown in a range format. The primary triaglycerols found in BCS are trilinoleoyl (24.6%), oleoyldilinoleoyl (19.6%), palmitoyldilinoleoyl (17.5%), palmitoyl oleoyl linoleoyl (12.9%), dioleoyllinoleoyl (9.6%), phosphatidylethanolamine, and phosphate. Phosphatidylglycerol, lysophosphatidylcholine, and lysophosphatidylethanol amine were discovered in trace levels.

Free sterols, steryl esters, and acylated steryl glycosides can all be present in BCS oil. The fatty acid profiles of fixed oil generated from BCS contain linoleic acid, palmitic acid, oleic acid, and stearic acid. Arachidic acid, linolenic acid, myristic acid, and palmitoleic acid are also present in large concentrations. In *N. sativa* seeds, unsaturated fatty acids are detected in higher concentrations than unsaturated fatty acids. The most abundant unsaturated fatty acids in seed oil are oleic (C18:1) and linoleic (C18:2) acids, while palmitic acid (C16:0) is the most abundant saturated fatty acid (Kabir et al., 2019; Matthaus and Zcan, 2011).

Varying lipids, such as triglycerides, diglycerides, monoglycerides, free fatty acids, and sterols, have different fatty acid content ratios. Linoleic acid can be found in a variety of foods. Free acids, triglycerides, and diglycerides, whereas myristic acid is abundant in sterols and monoglycerides (Atta, 2003). The kind of seed cultivated (landrace), seed quality characteristics (i.e., maturity, storage conditions, and damage during harvesting), and oil processing procedure all influence the lipid profile of BCS oil (Cheikh-Rouhou et al., 2007).

5.8 Black cumin seed polysaccharide

Polysaccharides are carbohydrate molecules made up of monosaccharide components linked together by a glycoside bond. The majority of these chemicals are made up of several groups of biopolymers that have industrial and biological applications and derive from natural sources such as bacteria, fungi, sea organisms, and plants. Food, cosmetics, and pharmaceutical industries all use polyfunctional polysaccharides. Polysaccharides are classified into several classes based on their structure, composition, origin, and use. Polysaccharides in plants include starch, cellulose, hemicelluloses, gums, and pectins (Ruff, 2008). Polysaccharides' bioactivities and functional peculiarities can also be influenced by their structure and content (Zhao et al., 2005). Plant polysaccharides and their bioactive components have been shown to have anticarcinogenic (Chen et al., 2010), antibacterial (Slima et al., 2018; Wu et al., 2014), antiinflammatory (Slima et al., 2018; Wu et al., 2014), antioxidant (Slima et al., 2018; Wu et al., 2014), and antiviral (Song et al., 2013) properties.

BCS polysaccharides are utilized as nutraceuticals and are used to treat chronic illnesses such as stomach ulcers (Manjegowda et al., 2017). These biopolymers are also employed as functional ingredients in food items to provide a natural supply of phytochemicals and antioxidants in various food compositions (Trigui et al., 2018). Furthermore, these biomacromolecules are stable, safe, and possess broad biological abilities or properties (Tahmouzi and Ghodsi, 2014). It has been discovered that the configuration and arrangement of polysaccharides can influence their applicative characteristics and free-radical inhibiting activity (Zhao et al., 2005). The antioxidant properties and the application attributes of BCS polysaccharides have been discovered, and it is influenced by BCS polysacharride structure and content (Trigui et al., 2018). Furthermore, these biocompounds are added to food during their formulations stage (Trigui et al., 2018).

Polysaccharides extracted from plants have recently been reviewed for potential applications (Sila et al., 2014). They are utilized as foaming and emulsifying agents, as well as enhancing flavor retention in some cases. They could be used as natural additives in food and medicinal products.

5.9 Structural attributes of BCSP

Fourier-transform infrared (FTIR) and nuclear magnetic resonance (NMR) spectroscopic investigations showed peaks of BCS polysaccharides with NMR report specifying that black cumin seed polysaccharide (BCSP) had a rhamnogalacturonan frame with a galactan and arabinan side chains (Manjegowda et al., 2017; Trigui et al., 2018). Physical studies conducted by Manjegowda et al. (2017) and Trigui et al. (2018) using X-ray diffraction to describe the kind of polysaccharides in BCS showed that the crystallography of the polysaccharides of BCSP was semicrystalline polymer (Trigui et al., 2018). The findings were comparable to Jeddou et al. (2016) results on potato peels polysaccharides. The FTIR spectrum displayed the existence of uronic acids and a pyranose (Manjegowda et al., 2017; Trigui et al., 2018). Moreover, FTIR spectral investigation specified a pectic polysaccharide type for BCSP (Manjegowda et al., 2017).

5.10 Black cumin seed polysaccharide functional properties

The functional aspects are among the characteristic attributes that may influence food materials' performance and finished products during processing, safekeeping, and utilization.

5.10.1 Water and oil-holding capacity of BCSP

Texture enhancement is one of the most important functional qualities in food preparation. The water-holding capacity (WHC) and the oil-holding capacity (OHC) are outcomes from the interactions between components such as water and lipid (Trigui et al., 2018). WHC of BCSP at dissimilar temperatures of 25°C, 50°C, and 75°C was investigated, and a significant difference was seen with the uses of the diverse temperature at $P < 0.05$. A rise in WHC was noted at a higher temperature. This may be attributed to the upsurge in the solubility of the polysaccharides due to the temperature increase (Sila et al., 2014). At 25 and 75°C, the WHC values of BCWSP were ranged between 3.3 and 5.2 g of water per g of dry matter, respectively. WHC levels in BCSP were higher compared to those reported in almond polysaccharides (1.95 g H_2O/g sample) and watermelon rinds polysaccharides (2 g H_2O/g sample) (Bentham et al., 2017). With regards to the OHC evaluated, the OHC dropped considerably ($P < .05$) as temperature improved (Trigui et al., 2018).

5.10.2 Foaming capacity or ability of BCSP

The interfacial characteristics of surface-active constituents affect foaming properties (Dickinson, 2003). The foaming capacity (FC) and foam stability (FS) of BCSP at various concentrations ranging from 1% to 5% w/v showed a varied FC and FS of BCWSP at a significance of $P < 0.05$, indicating that BCSP has an excellent foaming quality (Trigui et al., 2018). This could be attributed to the polysaccharide's capacity to upturn the viscidness and interfacial adhesion behavior as a function of air—water interface concentrations. These intriguing outcomes of BCSP could be employed as a foaming ingredient in food compositions (Trigui et al., 2018).

5.10.3 Emulsifying ability

Emulsifying ability index (EAI) and emulsifying stability index (ESI) of BCSP solutions were documented by Trigui et al. (2018). However, BCSP concentrations influenced emulsifying properties (Trigui et al., 2018). BCSP's EAI and ESI quantities decrease when the polysaccharide increases in concentration (Trigui et al., 2018). The polysaccharide-emulsifying properties of BCSP were impacted by the microscopic parts of proteins (Trigui et al., 2018). The reduced polysaccharide concentrations could explain the results. A rise in protein-polysaccharide repulsion may occur, speeding up the diffusion rate (Uruakpa and Arntfield, 2006). However, at high polysaccharide concentrations, the formation of protein-polysaccharide complexes may impede their transport to the interface (Ganzevles et al., 2006). As a result, the efficiency of polysaccharide emulsions may be affected by their concentrations (Chivero et al., 2015).

5.11 The polysaccharide of black cumin extraction

Polysaccharides and other macromolecules are distinguished by the chemistry and organizational arrangement of the plant cell wall. As a result, to acquire a larger yield, precise scientific methods must be followed to extract its macromolecules (Fig. 5.4). The extracted polysaccharides must

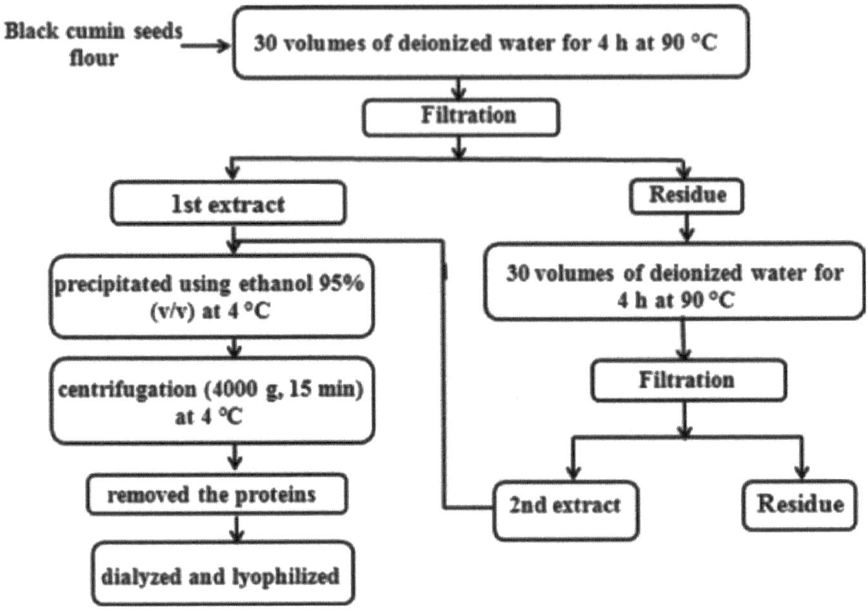

Figure 5.4 Black cumin seeds extraction diagram by Trigui et al. (2018).

be intact, contaminant-free, and homogeneous. This is determined by the extraction method, which is impacted by the plant material utilized, desired attributes, and the polysaccharide's solubility in an aqueous solution at different pH levels. There have been numerous publications on polysaccharide extraction strategies.

Furthermore, adopting an extraction method with a lower pH, higher temperature, and a longer extraction time may result in a higher extraction yield. Such conditions may result in deterioration, which hurts the extraction output and the polysaccharide's efficiency. This may encourage the extraction of harmful and undesired substances.

Polysaccharide extraction techniques are generally carried out at higher temperatures (80°C and 90°C) employing various reagents. Some of the current polysaccharide extraction procedures used are hot-water extraction, microwave-assisted extraction, and supercritical fluid extraction. These methods are remarkable for their effectiveness and scalability. as well as the polysaccharide structure and their ability to sustain the environment (Le Normand et al., 2014; Song et al., 2013). Hot water, enzymatic, ultrasonic, and microwave extraction of bioactive polysaccharides were examined. The antioxidant activity of the polysaccharide synthesized by the hot-water method was higher than that of polysaccharides made by other methods. Hot-water extraction combined with microwave extraction, ultrasound, and enzymatic pretreatment can increase extraction yield, while simultaneously preserving the biological activity of polysaccharides.

5.12 Conclusion

BCSs are unusual because they are high in carbs, vitamins, minerals, fatty acids, polysaccharides, and other biochemical ingredients. Furthermore, the seeds feature outstanding functional properties such as WHC, oil-binding capacity, emulsifying ability, FC, and bioactive compounds (phenolics, saponins, terpenes alkaloids, coumarins, flavonoids, etc.). These distinguishing characteristics of the BCS have led to its adoption in the food and pharmaceutical industries. Because of this, its use in food and medication production, which may assure good nutrition and illness prevention.

References

Al-Jassir, M.S., 1992. Chemical composition and microflora of black cumin (Nigella sativa L.) seed growing in Saudi Arabia. Food Chemistry 45 (4), 239–242.

Al-Saleh, I.A., Billedo, G., El-Doush, I.I., 2006. Levels of selenium, dl-α-tocopherol, dl-γ-tocopherol, all-trans-retinol, thymoquinone, and thymol in different brands of Nigella sativa seeds. Journal of Food Composition and Analysis 19 (2–3), 167–175.

Anjum, F.M., Jamil, A., Akhtar, S., Nasir, M., Tauseef, M., Et, S., 2009. The nutritional profile of indigenous cultivar of black cumin seeds and antioxidant potential of its fixed and essential oil. Pakistan Journal of Botany vol 41 (3), 1321–1330.

Atta, M.B., 2003. Some characteristics of nigella (Nigella sativa L.) seed cultivated in Egypt and its lipid profile. Food Chemistry 83 (1), 63–68. https://doi.org/10.1016/S0308-8146(03)00038-4.

Babayan, V.K., Koottungal, D., Halaby, G.A., 1978. Proximate analysis, fatty acid and amino acid composition of Nigella sativa L. seeds. Journal of Food Science 43 (4), 1314–1315.

Bentham, J., Di Cesare, M., Bilano, V., Bixby, H., Zhou, B., Stevens, G.A., Riley, L.M., Taddei, C., Hajifathalian, K., Lu, Y., Savin, S., Cowan, M.J., Paciorek, C.J., Chirita-Emandi, A., Hayes, A.J., Katz, J., Kelishadi, R., Kengne, A.P., Khang, Y.H., Cisneros, J.Z., 2017. Worldwide trends in body-mass index, underweight, overweight, and obesity from 1975 to 2016: a pooled analysis of 2416 population-based measurement studies in 128·9 million children, adolescents, and adults. The Lancet 390 (10113), 2627–2642. https://doi.org/10.1016/S0140-6736(17)32129-3.

Bourgou, S., Pichette, A., Marzouk, B., Legault, J., 2010. Bioactivities of black cumin essential oil and its main terpenes from Tu-nisia (. South African Journal of Botany 76, 210–216.

Cheikh-Rouhou, S., Besbes, S., Hentati, B., Blecker, C., Deroanne, C., Attia, H., 2007. Nigella sativa L.: chemical composition and physicochemical characteristics of lipid fraction. Food Chemistry 101 (2), 673–681.

Chen, R., Meng, F., Liu, Z., Chen, R., Zhang, M., 2010. Antitumor activities of different fractions of polysaccharide purified from Ornithogalum caudatum Ait. Carbohydrate Polymers 80 (3), 845–851.

Chivero, P., Gohtani, S., Yoshii, H., Nakamura, A., 2015. Effect of xanthan and guar gums on the formation and stability of soy soluble polysaccharide oil-in-water emulsions. Food Research International 70, 7–14.

Dickinson, E., 2003. Hydrocolloids at interfaces and the influence on the properties of dispersed systems. Food Hydrocolloids 17 (1), 25–39.

Dinagaran, S., Sridhar, S., Eganathan, P., 2016. Chemical composition and antioxidant activities of black seed oil (Nigella sativa L.). International Journal of Pharmaceutical Sciences and Research 7 (11), 4473.

Gali-Muhtasib, H., El-Najjar, N., Schneider-Stock, R., 2006. The medicinal potential of black seed (Nigella sativa) and its components. Advances in Phytomedicine 2 (C), 133–153. https://doi.org/10.1016/S1572-557X(05)02008-8.

Ganzevles, R.A., Zinoviadou, K., van Vliet, T., Cohen Stuart, M.A., de Jongh, H.H.J., 2006. Modulating surface rheology by electrostatic protein/polysaccharide interactions. Langmuir 22 (24), 10089–10096.

Gholamnezhad, Z., Havakhah, S., Boskabady, M.H., 2016. Preclinical and clinical effects of Nigella sativa and its constituent, thymoquinone: a review. Journal of Ethnopharmacology 190, 372–3860.

Hussein, T., Dal Maso, M., Petäjä, T., Koponen, I.K., Paatero, P., Aalto, P.P., Hämeri, K., Kulmala, M., 2005. Evaluation of an automatic algorithm for fitting the particle number size distributions. Boreal Environment Research 10 (5), 337–355.

Jeddou, K.B., Chaari, F., Maktouf, S., Nouri-ellouz, O., Helbert, C.B., Ghorbel, R.E., 2016. Structural, functional, and antioxidant properties of water-soluble polysaccharides from potatoes peels. Food Chemistry 205, 97–105.

Kabir, Y., Shirakawa, H., Komai, M., 2019. Nutritional composition of the indigenous cultivar of black cumin seeds from Bangladesh. Progress in Nutrition 21, 428–434. https://doi.org/10.23751/pn.v21i1-S.655.

Kaskoos, R.A., 2011. Fatty acid composition of black cumin oil from Iraq. Research Journal of Medicinal Plants 5, 85–89.

Le Normand, M., Moriana, R., Ek, M., 2014. Isolation and characterization of cellulose nanocrystals from spruce bark in a biorefinery perspective. Carbohydrate Polymers 111, 979–987.

Mamun, M.A., Absar, N., 2018. Major nutritional compositions of black cumin seeds—cultivated in Bangladesh and the physicochemical characteristics of its oil. International Food Research Journal 25 (6), 2634–2639.

Manjegowda, S.B., Rajagopal, H.M., Dharmesh, S.M., 2017. The polysaccharide of Black cumin (Nigella sativa) modulates the molecular signaling cascade of gastric ulcer pathogenesis. International Journal of Biological Macromolecules 101, 823–836.

Matthaus, B., Özcan, M.M., 2011. Fatty acids, tocopherol, and sterol contents of some Nigella species seed oil. Czech Journal of Food Sciences 29 (2), 145–150.

Md Ahmad, F., Ahmad, F.A., Ashraf, S.A., Saad, H.H., Wahab, S., Khan, M.I., Ali, M.S.M., Hakeem, K.R., Md Athar, T., 2021. Updated knowledge of black seed (Nigella sativa Linn.): review of phytochemical constituents and pharmacological properties. Journal of Herbal Medicine 25, 100404.

Naz, H., 2011. Nigella sativa: the miraculous herb. Pakistan Journal of Biochemistry and Molecular Biology 44 (1), 44–48.

Ramadan, Fawzy, M., 2007. Nutritional value, functional properties and nutraceutical applications of black cumin (Nigella sativa L.): an overview. International Journal of Food Science and Technology 42 (10), 1208–1218. https://doi.org/10.1111/j.1365-2621.2006.01417.x.

Ramadan, M.F., Mörsel, J.T., 2002. Neutral lipid classes of black cumin (Nigella sativa L.) seed oils. European Food Research and Technology 214 (3), 202–206. https://doi.org/10.1007/s00217-001-0423-8.

Ruff, Y., 2008. Biopolymères dynamiques: Oligo-et polysaccharides. Strasbourg 1.
Sahak, M.K.A., Kabir, N., Abbas, G., Draman, S., Hashim, N.H., Hasan, Adli, D.S., 2016. The role of Nigella sativa and its active constituents in learning and memory. eCAM 2016, 6075679.
Sila, A., Bayar, N., Ghazala, I., Bougatef, A., Ellouz-Ghorbel, R., Ellouz-Chaabouni, S., 2014. Water-soluble polysaccharides from agro-industrial by-products: functional and biological properties. International Journal of Biological Macromolecules 69, 236–243.
Slima, S.B., Ktari, N., Trabelsi, I., Moussa, H., Makni, I., Salah, R.B., 2018. Purification, characterization, and antioxidant properties of a novel polysaccharide extracted from Sorghum bicolor (L.) seeds in sausage. International Journal of Biological Macromolecules 106, 168–178.
Song, T., Pranovich, A., Holmbom, B., 2013. Separation of polymeric galactoglucomannans from a hot-water extract of spruce wood. Bioresource Technology 130, 198–203.
Tahmouzi, S., Ghodsi, M., 2014. Optimum extraction of polysaccharides from motherwort leaf and its antioxidant and antimicrobial activities. Carbohydrate Polymers 112, 396–403.
Trigui, I., Yaich, H., Sila, A., Cheikh-Rouhou, S., Bougatef, A., Blecker, C., Attia, H., Ayadi, M.A., 2018. Physicochemical properties of water-soluble polysaccharides from black cumin seeds [Elsevier B.V]. International Journal of Biological Macromolecules vol 117. https://doi.org/10.1016/j.ijbiomac.2018.05.202.
Tulukcu, E.A., 2011. Comparative study on fatty acid composition of black cumin obtained from different regions of Turkey, Iran, and Syria. African Journal of Agricultural Research 6 (4), 892–895, 5.
Uruakpa, F.O., Arntfield, S.D., 2006. Surface hydrophobicity of commercial canola proteins mixed with κ-carrageenan or guar gum. Food Chemistry 95 (2), 255–263.
Valko, M., Rhodes, C.J., Moncol, J., Izakovic, M., Mazur, M., 2006. Free radicals, metal,s, and antioxidants in oxidative stress-induced cancer. Chemico-Biological Interactions 160 (1), 1–40. https://doi.org/10.1016/j.cbi.2005.12.009.
Vatansev, H., Ciftci, H., Ozkaya, A., Ozturk, B., Evliyaoglu, N., Kiyici, A., 2013. Chemical composition of Nigella sativa L. seeds used as a medical aromatic plant from East Anatolia Region, Turkey. Asian Journal of Chemistry 25 (10), 5490–5492.
Wu, H., Zhu, J., Diao, W., Wang, C., 2014. Ultrasound-assisted enzymatic extraction and antioxidant activity of polysaccharides from pumpkin (Cucurbita moschata). Carbohydrate Polymers 113, 314–324.
Zhao, G., Kan, J., Li, Z., Chen, Z., 2005. Structural features and immunological activity of a polysaccharide from Dioscorea opposite Thunb roots. Carbohydrate Polymers 61 (2), 125–131.

CHAPTER SIX

Black cumin seed oil, protein, protein concentrates, and hydrolysate

Ahmed A.M. Elnour[1,2], Mohamed E.S. Mirghani[2], Abdelgadir A. Elfadil[3], El Bashier Sallam[4], Ibrahim Bani[5,6]

[1]Department of Biology, Faculty of Science and Mathematics, Sultan Idris Education University (UPSI), Tanjung Malim, Perak, Malaysia
[2]International Institute for Halal Research and Training (INHART), International Islamic University Malaysia (IIUM), Gombak, Kuala Lumpur, Malaysia
[3]Department of Environmental Science, Faculty of Science and Technology, Al-Neelain University, Sudan
[4]Freelance Global and Public Health Expert (FGPHE), Sudan
[5]College of Medicine, Ajman University, Ajman, United Arab Emirates
[6]Centre of Medical and Bio-Allied Health Sciences Research, Ajman University, Ajman, United Arab Emirates

6.1 Introduction

Medicinal plants (MPs) have prevented and cured diseases for many centuries by using different indigenous ecosystems in medicine and folk medicines (FMs) (Masoodi et al., 2020). Furthermore, MPs are also used to prepare herbal medicines (HMs) as they are considered safe compared to contemporary allopathic drugs. Many researchers concentrate on MPs, since only inadequate plant types have been studied systematically. They have examined their therapeutic activities, potential, safety evaluation, toxicological studies, and mechanism of action (Tiwari et al., 2019). So, considering the medicinal value and its therapeutic potential, it would spotlight the biological activities of black cumin seed oil (BCSO). Among several MPs, the black seed (also known as black cumin) (Habat-ul-Sauda), since numerous types of research have revealed its broad spectrum of pharmacological potential (Family Ranunculaceae), is emerging as a phenomenon herb with a rich history and religious background. Black cumin seed (BCS), *Nigella sativa*, is a plant native to North Africa, Southwest Asia, and Southern Europe. It is grown in many countries worldwide, including South Europe, Syria, Turkey, Saudi Arabia, the Middle East Mediterranean region, India, Pakistan, and Afghanistan (Srinivasan, 2018). Thus, the black seed is economically available worldwide.

Biochemistry, Nutrition, and Therapeutics of Black Cumin Seed
ISBN: 978-0-323-90788-0
https://doi.org/10.1016/B978-0-323-90788-0.00019-6

The BCS and their oil have been massively used for decades to treat various ailments worldwide. Moreover, it is essential to drug use in Indian FMs and advanced like Ayurveda and Unani's cure (Sudhakar et al., 2021). For example, it is considered one of the most effective forms of HMs ready to be used in Muslim countries. As mentioned, BCS has potent and all-around cures for illnesses, excluding death, as reported in one of the prophetic books by Prophetic Mohammed Hadith. It has also been recognized regularly in Tibb-e-Nabawi (Prophetic Medicine) (Maideen, 2020). Hence, BCS has religious basis evidence, particularly in Muslim communities.

BCS has been extensively studied for its biological activities and therapeutic potentials. It had been shown that it is the impact of broad spectrum of activities, such as diuretic, antidiabetic, antihypertensive, immunomodulatory, anticancer, analgesic, anthelmintic, antimicrobial, antiinflammatory, bronchodilator, spasmolytic, gastroprotective, hepatoprotective, antioxidant properties, and renal protective (DemİRbolat et al., 2019). The seeds of BCS are widely used to treat bronchitis, asthma, diarrhea, rheumatism, and skin diseases as HM.

It is also used as a liver tonic, digestive aid, antidiarrheal, appetite stimulant, emmenagogue (which increases milk production in nursing mothers), parasite management, and immune system booster (Hakim et al., 2019). Most of the therapeutic properties of this plant are due to the presence of thymoquinone (TQ) in the seed oil, which is a primary active bioactive compound. As a result of its low toxicity, BCS is also employed in nutrition to enhance the flavor of pickles and bread (Srinivasan, 2018). Thus, the available evidence suggests that TQ should be developed as a novel drug in clinical trials. The shape of the plant is presented in Fig. 6.1.

Figure 6.1 Black cumin plant; (vertical view).

6.2 Black cumin and its importance

This represents an important topic to study because BCS adds to the basal diet (Khan et al., 2013). This is a topic of pivotal importance in the application of feeding using BCS powder form on growth performance (GP) of one- to two-month-old broilers (Islam et al., 2017). Therefore, BCS has a dietary effect when blended with buckwheat (*Fagopyrum esculentum*).

These findings align with other studies on the effect of BCS oil and its impact on Japanese quail diets, GP, carcass traits, and some blood parameters (Demirci et al., 2019). There is pervasive literature on determining the effects of BCS oil Tibb-e-Nabawi. Sohail et al. (Khan et al., 2012) have been involved in research for over two decades; they used oils and seeds of BCS and reported positive broiler performance. BCS oil can be extracted using maceration or shaker for 20 h at room temperature using 96% ethanol (Rusmarilin et al., 2019). The BCS oil content of Iranian samples was 40.4% compared to other BCS nutritional components (Farhan et al., 2021). Similar work has also been pursued by Farhan (Farhan et al., 2021), who examined the effect of diets supplemented with an essential oil found to be 0.1 or 1.0 g/kg for oilseed compared to a range between 10 and 50 g/kg for BCS on broiler's GP. Thus, a limited number of comparative studies are available on cumin and black cumin regarding meat yield.

Moreover, there was a significant difference between 0.75% cumin and 0.75% black cumin on breast meat yield (25.8% vs. 23.2%), respectively (Shabaninejad et al., 2020). As a result, the impact of dietary BC seed meal on broiler growth and meat yield has sparked much curiosity in the community. This is a very active and burgeoning area of research on GP, nutrient digestibility, and immunity. As a result, it can be concluded that the addition of 0.5%–1.0% BCSO into the diets of broilers may have improved the performance of chicks and therefore enriched the meat quality due to creating meaningful changes in meat fatty acid (FA) profiles, especially in terms of total polyunsaturated fatty acids (PSFAs) (Suri et al., 2020). An approach proposed by Bakhshabadi et al. (2018) was employed in work that uses different pretreatments without any significant ($P \geq .05$) differences in the refractive index of BCSO using the influence of pulsed electric fields (PEFs) and microwave techniques. Early developments in this field were also well documented in Aytac et al. (2017), which concluded that 60 kgNha-1 supplemented with potassium (K) could enhance the seed yield, oil yield, and nitrogen (N) and K content BCSO without significant changes in FA composition. Finally, the shape and color of BCSs are presented in Fig. 6.2.

Figure 6.2 Black cumin seeds' shape and color.

6.3 Black cumin seed oil

Macroscopically, BC seeds are generally tiny, dicotyledonous, trigonous, inclined, regulate-tubercular, 2 to 3.5mm, 1 to 2 mm, darkish outwardly, in addition, white-colored inside, odor slightly aromatic, and taste bitter (Mohammed et al., 2019). Microscopically, the inclined part of the seed exhibits individual layered epidermis and skin, including elliptical, thick-walled tissue, outside the body through a dark brown cuticle with papillose cuticles. There are two to four layers of heavy-walled tangentially elongated parenchymatous cellular material and a new reddish-brown pigmented layer composed of dense walled, rectangular cells that commonly follow the epidermis. The pigment layer on the inside is available.

A new layer comprises rectangular elongated or almost columnar, elongated cells with thicker walls. The endosperm comprises oil globule-filled thin-walled, rectangular, or polygonal cellular material. Typically, the powder microscopy of seed powder exhibits dark brown—colored parenchymatous cells in addition to oil globules (Islam et al., 2017). The color of BCSO is presented in Fig. 6.3.

Several active compounds (ACs) have been separated, determined, and investigated so far in several species of BC. The main ACs are TQ 30%—48%, p-cymene 7%—15%, carvacrol 6%—12%, 4-terpineol 2%—7%,

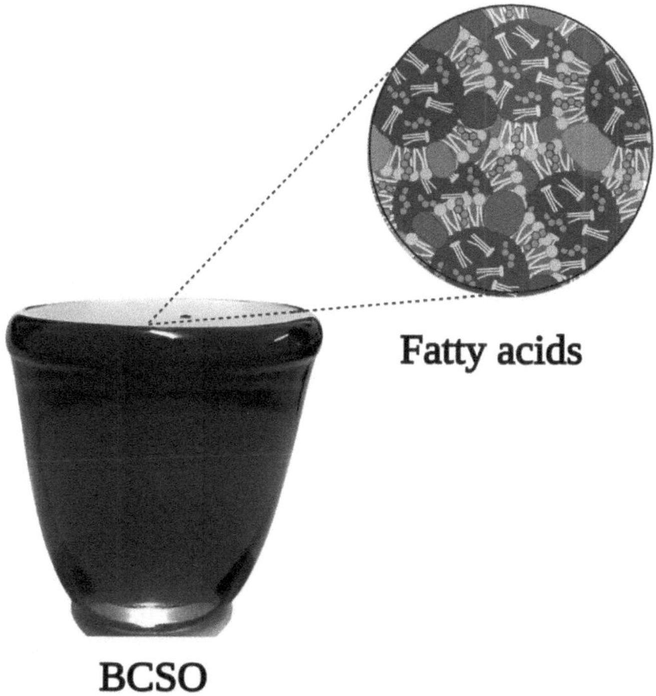

Figure 6.3 Black cumin seed oil's color.

T-anethole 1%–4%, sesquiterpene longifolene 1%–8%, and pinene, respectively (Mukhtar et al., 2021). Thus, the BCSO is rich in bioactive compounds.

In addition, BCS consists of alkaloids, such as nigellicimine-N-oxide and nigellicimine alkaloids, and indazole ring alkaloids, including nigellidine and pyrazole alkaloids. BCS contains alpha-hederin, a new standard pentacyclic water-soluble triterpene, saponin, and many anticancers (Khan et al., 2012). The BCS comprises 26.7% protein, 28.5% lipids, 24.9% sugar, 8.4% raw dietary fiber, and 4.7% total ash. The BCSO also contains good nutritional supplements like vitamins and minerals such as Cu, P, Fe, and Zn. In addition, the BCSO has carotene converted to vitamin A (Khajavi Ra et al., 2016). BCSO is rich in unsaturated fatty acids (USFAs), mainly 50%–60% linoleic acid, 20% oleic acid, 3% eicosadienoic acid, and 10% dihomolinoleic acid (Tiwari et al., 2019; Cheikh-Rouhou et al., 2007). Saturated fatty acids (SFAs) (stearic and palmitic acids) are presented to about 30% or less in BCSO; α-sitosterol is a significant sterol, which often makes up 44% and 54% reported for different BCSO samples harvested from Tunisian and Iranian, respectively (Suri et al., 2020; Farhan et al., 2021).

A review of studies undertaken on BCSO has shown several other noted chemical compounds. This has been widely observed in nigellone, avenasterol-5-ene, avenasterol-7-ene, campesterol, which is considered cholesterol-lowering agents, and cycloeucalenol as antidiabetic (Kiralan et al., 2017). This is the most relevant finding and perhaps the most significant BCSO bioactive compounds (BACPs). For example, (Seidavi et al. 2020) reported similar observations in their experiments with BACPs extracted from BCSO, including citrostadienol, gramisterol, obtusifoliol, lophenol, stigmastanol β-amyrin, and stigmasterol-7-ene. This is consistent with the findings of Kiralan et al. (2017), who found that 24-methylene-cycloartenol and cycloartenol and butyrospermol were the most BACPs of the BCS. Thus, the evaluation of BACPs of the BCSO is critical.

The evaluation of BACPs is currently focused on taraxerol (1—3)-α-L-rhamnopyranosyl, tirucallol, and 1-2-α-[L-arabino-pyranosyl]-28-O-[α-L rhamnopyranosyl] is the most important one. The vital strength of the chapter lies in the fact that BCS volatile oil was found to range from 0.5% to 1.6%, FAs 35.6% to 41.6%, in addition to oleic acid esters, which involved USFAs having C15 huge terpenoids, melanthin, hederagenin glycoside bitter basic principle, melanthigenin, and tannin, protein, glucose, and glycosidic bonds formation (Kiralan et al., 2019; Karadağoğlu et al., 2019; Kabir et al., 2019).

Much of the early work around the most widely used BC plant study is its oil (Fajar et al., 2017). The BCSO contains about 27.8% of TQ as an essential oil. Recent work by Farhan et al. (2021) reports that the Bangladeshi fat content of BCS was similar to the amount for the Iranians (40.4%). It should be noted here that according to Seidavi et al. (2020), BCS reduces cholesterol, hypoglycemia, and hypocholesterol only when provided in high doses; therefore, there is no reduction in BCSO in broiler diets.

The primary FAs of the BCSO contains myristic (C14:0), palmitic (C 16:0), and stearic (C18:0) acids as saturated fatty acids (SFAs), while the principal USFAs are oleic (C18:1), linoleic (C18:2), and linolenic (C18:3). Although some attempts have been made to address this issue, there are still changes of FAs and SFAs during the initial germination and a reduction in total monounsaturated fatty acids (Mariod et al., 2009). Both crudes of BCSO and its fractions (neutral glycolipid, fat, and phospholipids) revealed powerful quantitative radical scavenging properties associated with their total content of PSFAs, phospholipids, and unsaponifiable, in addition to the lowest hydrogen peroxide values of crude BCSO (Ozdal and Adiguzel, 2019). So, FAs are a promoting molecule in BCSO.

The yield of BCSO extracted using chloroform:methanol was much larger than that of BCSO extracted using petroleum ether and hexane in three different samples. The overall quantity of lipids extracted was more significant in the chloroform:methanol (39.2% of fresh weight) than in *n*-hexane (37.9%). The large portion of BCSO produced with chloroform:methanol was due to chloroform:methanol activity which can get polar and nonpolar macromolecules (Ramadan and Mörsel, 2002). It was supposed that the structure and organic process value (a process occurring in living organisms) were different due to country of origin, maturity, seed type, and subsequent storage conditions (Atta, 2003). It was supposed that the primary SFAs of BCSO was 11.56% stearic and 2.55% were palmitic. In addition, arachidic (C20:0), behenic (C22:0), and lignoceric (C24:0) acids have been discovered as SFA, and palmitoleic (C16:1) and erucic (C22:1n-9) acids were detected as USFAs. Previously, Cheikh et al. (2007) had shown that the BCS, which variety Bangladesh, had relatively high oleic and linoleic acids as 18.9−20.1 and 47.5−49.0, respectively, which is less than the Tunisian variety 25.0 and 50.3, respectively, and the Iranian variety 23.7 and 49.2, respectively.

Mustafa (Kiralan et al., 2014) studied the BCSO processing variables, precision of recognition, the FAs extraction, and quantitative analysis methods. They reported that the percentage of oleic acid and linoleic acid was present more than 2.5:1%. Earlier results of Houghton et al. (1995) support these outcomes. Finally, the FA profile of BCSO is presented in Table 6.1.

Table 6.1 The composition of black cumin seeds fatty acid[a].

Fatty acids (%)	Kabir et al. (2019)	Kaskoos (2011)	Bourgou et al. (2010)
Myristic acid (14:0)	0.23 ± 0.01	0.16 ± 0.02	0.40 ± 0.03
Palmitic acid (16:0)	13.1	8.51	14.8 ± 0.47
Palmitoleic acid (16:1)	0.28	0.16	0.28 ± 0.05
Stearic acid (18:0)	2.47	2.22	2.90 ± 0.01
Oleic acid (18:1)	21.8	16.6	19.2 ± 0.13
Cis-vaccenic acid (18:1)	1.18	nd	nd
Linoleic acid (18:2)	57.0	42.8	58.1 ± 0.54
Linolenic acid (18:3)	0.46	0.25	0.41 ± 0.06
Arachidic acid (20:0)	0.18	nd	1.05 ± 0.43
Eicosenoic acid (20:1)	0.39	0.16	nd
Eicosadienoic acid (20:2)	2.91	1.94	nd

(Continued)

Table 6.1 The composition of black cumin seeds fatty acid[a].—cont'd

Fatty acids (%)	Kabir et al. (2019)	Kaskoos (2011)	Bourgou et al. (2010)
Eicosatrienoic acid (20:3)	Nd	4.71	nd
Eicosapentaenoic acid (20:5)	Nd	5.98	nd
Behenic acid (22:0)	Nd	nd	2.89 ± 0.34
Docosahexaenoic acid (22:6)	Nd	2.97	nd
Lignoceric acid (24:0)	Nd	3.60	nd
Nervonic acid (24:1)	Nd	2.76	nd
Saturated acids (SFAs)	16.0	15.1	22.0 ± 0.60
Unsaturated acids (USFAs)	84.0	79.9	19.5 ± 0.20
Unsaturated/Saturated	5.25	5.29	58.5 ± 0.64

nd, not detected.
[a]Table 6.1 modified by the authors under republish license number (5085260050335).

6.4 Protein of black cumin seeds and its extraction

It has been a focus of research for more than a decade that BCS is an excellent source of plant-based proteins as they are reported to have a high index of net protein utilization, protein efficiency ratio, and net dietary protein energy percent (Sarwar and Latif, 2015). This is an important topic because BCS comprises approximately 21%–26% protein by weight (Coşkun et al., 2020; Sharma and Longvah, 2021). Moreover, the BCS chemical elements such as protein contain essential extracts, 15 amino acids (AAs) (alanine, arginine, isoleucine, lysine, tryptophan, tyrosine, threonine, asparagine, cysteine, glycine, glutamic acid, methionine, and proline), and BCS also contains niacin (vitamin B3) (Siregar et al., 2020). More recently, a reevaluation of this concept has led to the characterization of a microcapsule containing black cumin seeds protein (BCSP). Therefore, equal protein loading has been confirmed by using the protein β-actin as an internal household protein (Li et al., 2020). Another technique for addressing this topic is described by Asif et al. (2019) who conducted the expression of this protein using equalization of the control in dwarf mutants (Asif et al., 2019). The most common methods found in the literature are patterns of expression of this protein equal in the mutant plants HY1, MHY4, and LY3 compared to control (Asif et al., 2019). Having greater insight into BCSPs would yield valuable information about figure print spectroscopy. Here we see a method conceptually similar to Padhye et al. (2008), which utilizes Fourier-transform infrared data concerning the proteins' secondary structural characteristics; bands of amide I and amide II are often assessed as BCSPs.

The most recent and advanced concept proposed by Coşkun et al. (2020) offers a new approach to two-dimensional (2D) gel electrophoresis 240 μg 2D gel. BCSPs passive rehydration loaded to 11 cm, pH gradient strips (IPG) (pH 3—10). They have confirmed that BCSPs contain structural elements both α and β-secondary and, consequently, that BCSPs may have been glycosylated. The following are approximately limitations: the quality of the isolated protein was assessed by (i) separating the protect into the 2D gel system of three protein insulation methods and (ii) the pattern of expression for this protein in mutant plants HY1, MHY4, and LY3 was equal in comparison with the control (Asif et al., 2019).

Recently, Coşkun et al. (2020) reported the influence of extraction conditions on black cumin protein concentrates (BCPCs) structure based on BCSPs isoelectric point (pI) and pH. Thus, the pI values of most BCSPs AAs were found to be less than pH 7. However, various protein spots for alkaline proteins were also observed. In addition, the phenol-based approach has been superior to other methods for the amount of the extracted protein in this study (Asif et al., 2019). Moreover, Hannan et al. (2008) mentioned in the prior research that extraction considerably impacted the solubility of the BCSPs due to the surface properties variations in the BCSPs as a consequence of extraction. These proteins are the transcription factors that control various genes' functions (Najmi et al., 2008). The tris-buffer-based isolation method reported the minimum amount of protein at 2.94 mg/g leaf fresh weight (Asif et al., 2019). Therefore, BCSPs as dietary supplements need to be studied.

This well-recognized issue necessitates a better approach; as serum protein depends on dietary protein availability, the proteins of the BCS diets were more available to the birds (Kumar et al., 2017). Similar work was also carried out by Salem et al. (2010). It was shown that tryptophan is the lowest AA in BCSPs, although Cheikh et al. (2008) and Hannan et al. (2008) did not find tryptophan in BCSPs. Thus, there has been increasing research efforts and specialization in BCSPs. These findings confirm the strategy's validity in considering seed oil processed cakes as protein-based renewable sources to produce fertilizers, animal feed, culinary food, and other valuable products such as bioplastics (Sabbah et al., 2020). Thus, the nutritional composition of the BCSPs is quite crucial for study.

Once developed, this will significantly impact various areas, such as BCSPs, edible films from protein-containing defatted seeds. Moreover, this breakthrough is expected to increase the impact of PEFs and high-voltage electrical discharges on phenolic and protein extraction from sesame cake (Bakhshabadi et al., 2018). Hence, the influence of PEFs is beneficial for BCSPs edible film technology. Furthermore, these are often chosen to study BCSPs due to their development as new biomaterials from agroindustrial wastes. Recently, BCSPs edible films were recorded as a significant technological challenge, and the defatted BCS cake by-products were obtained from the oil-producing industry by Zent et al. (2021). Therefore, BCS cake was a rich, cheap source of both energy and protein. This has led to adopting advanced techniques such as a new FA ester extraction technology from BCS cake, showing potent antiprotein tyrosine phosphatase (PTP1B) activity, which has been used as a novel therapeutic target for type 2 diabetes mellitus, obesity, and related states of insulin resistance (Liu et al., 2019).

Interestingly, Alanazi et al. (2016) identified 277 metabolic and functional proteins using the UniProt database. In addition, an advanced technique was developed and utilized to determine the BCS AAs such as trypsin. For example, Zent et al. (2020) extracted trypsin and papain using BCS hydrolysates of protein for two hours. In addition, they used fast protein liquid chromatography (FPLC)-based fractionation techniques. As a result, about 20 peptide forms have been identified.

A simple setup has been employed here to study the behavior of conventional extraction techniques for extracting BCSPs. For example, Coşkun et al. (2019) have reported 57.7% and 65.8% of BCS proteins using aqueous extraction consists of alkali extraction with isoelectric precipitation (AE-IP) and then purifying using hexane. This confirms a minor angiotensin-converting enzyme inhibitory effect of protein hydrolysates. Moreover, the experimental setup given by Haas et al. (2014) has extracted BCSPs inducing apolipoprotein A-I gene expression. Therefore, BCSPs may have beneficial impacts on treating dyslipidemia and coronary cardiovascular disease.

It would be interesting to examine the nutrition and therapeutic benefits of BCSPs on living organisms. A retrospective study by Tavakoly et al. (2019) noted several promised findings. For example, BCSPs can supplement serum C-reactive protein (Tavakoly et al., 2019). Furthermore, they confirmed that BCS protein supplementation using a parallel control study significantly reduced serum levels. This is one of the earliest works on this

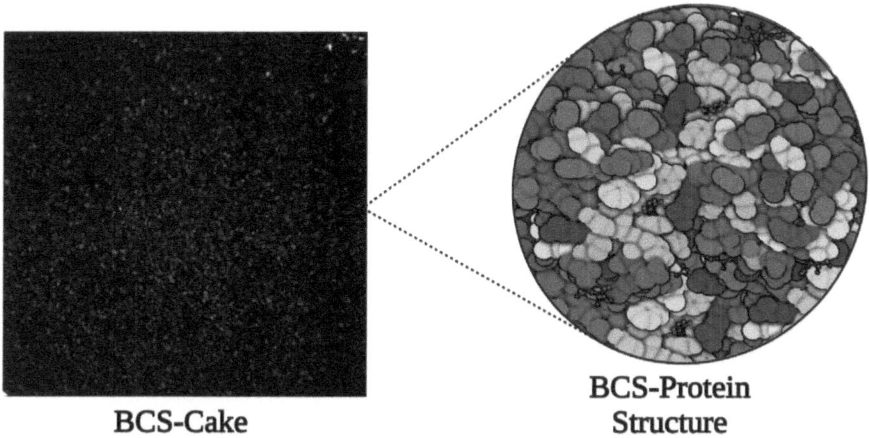

Figure 6.4 Black cumin seed cake's shape and color.

subject by Indrawan et al. (2016), who studied the impact of ethanol extract of BCSPs on proteinuria and blood pressure using the in-vivo methodology. The preeclampsia mouse model confirmed that the 2000 mg/kg/d dose was the optimum dose to reduce the ET-1 level, systolic pressure, and urine protein.

Siddiqui et al. (2020) reported several pioneering works in this area, who reported that the impact of BCS protein could target nucleocapsid when blending with the papain-like protease of SARS-CoV-2 (COVID-19) pandemic. Finally, Rahmani et al. (2019) have revealed the bioactivity of BCSPs by inducing liver injury in mice through the modulation of enzymatic antioxidant, P53, and vascular endothelial growth factor protein. They proposed that TQ can prevent CCl4-induced liver damage and be used as a therapeutic agent for liver diseases. Finally, the shape and color of BCS cake are presented in Fig. 6.4.

6.4.1 Protein concertation and amino acids profile

The BCSs are primary bioactive source compounds with potential health-promoting and antimicrobial activities. BCSs meal is rich in AAs and seems to be a promising source of foodstuffs for macro- and micronutrients. Interestingly, the BCS meal contains nonessential amino acids 67.95%, which includes: proline, glycine, alanine, glutamic acid, serine, cysteine, tyrosine, and aspartic acid, as shown in Table 6.2 (Kabir et al., 2019). Compared to the studies of Al-Jassir et al. (1992) and Babayan et al. (1978), who reported lower total essential amino acids (EAAs), ranging between 30.19% and

Table 6.2 The amino acids profile of black cumin seed proteins[a].

Amino acids %	Kabir et al. (2019)	Al-Jassir et al. (1992)	Babayan et al. (1978)
Essential amino acids (E)			
Leucine	6.00	5.82	10.88
Valine	4.77	4.61	3.06
Lysine	3.86	4.04	7.62
Threonine	3.86	3.65	1.23
Phenylalanine	3.70	3.61	7.93
Isoleucine	3.69	3.46	4.03
Histidine	2.79	3.35	00
Methionine[1]	1.88	1.65	6.16
Tryptophan[2]	1.50	00	00
Total essential amino acids	32.05	30.19	40.91
Nonessential amino acids (N)			
Glutamic acid	23.95	24.74	13.21
Aspartic acid	9.32	8.94	5.02
Arginine	8.90	9.19	19.52
Glycine	6.11	5.61	4.17
Proline	4.88	4.90	5.34
Serine	4.50	4.31	1.98
Alanine	4.39	3.73	3.77
Tyrosine	3.43	3.59	6.08
Cystine 1	2.47	1.96	00
Total nonessential amino acids	67.95	66.97	59.09
E/N	0.47	0.45	0.69

[a]Table 6.2 is modified by the authors under republish license number (5085260050335); 1AAs determination prior to oxidation treatment using performic acid (CH_2O_3) hydrolyzed by hydrochloric acid (HCL); 2AAs determined by high-performance liquid chromatography (HPLC) techniques after alkaline hydrolysis.

40.19%, respectively, which are methionine, tryptophan isoleucine, cysteine, lysine, and valine for BCSs harvested in Saudi Arabia. Kabir et al. (2019) noted that BCS meals consisted of roughly 17 AAs, ten of them are EAAs (202 g/1000 g), and lysine is about 6.41 g/kg.

Conceptually, similar work, utilizing a similar method, was reported by Kabir et al.(2019) who confirmed that BCSs meal is a rich source of leucine and glutamine (10.88% and 24.74%, respectively as described in Table 6.2). These primary acids make up around 54% of the total AAs found in the BCS protein. The overall AAs grew significantly throughout seed germination and were affected by germination conditions. Germinated BCSs showed declines in Na, K, Ca, and Fe content (Mariod et al., 2017). Thus, the results suggest that germinated BCS is an excellent food material for improving nutritional properties, as it contains glutamate (23.95%) and aspartate (9.30%).

6.5 Black cumin seed protein hydrolysate

The BCSPs have been hydrolyzed separately using pepsin, trypsin, and α-chymotrypsin. This hydrolysate was fractionalized utilizing high-performance reverse-phase liquid chromatography. As a result, arginine (8.90%), glycine (6.11%), and leucine (6.0%) are rich in BCSP. However, glutamic acid is the EAA (Kabir et al., 2019). Thus, these authors provided 433 recording chemical scores including, nine of which are EAAs recorded for Bangladeshi BCSs samples. Interestingly, Al-gaby (1998) declared the lowest amount of AAS in BCS cake proteins is tryptophan; however, Al-Jassir et al. (1992) and Babayan et al. (1978) did not report any tryptophan in BCSs.

Trigui et al. (2021) have been investigating the physical, technofunctional, and antioxidant properties of black cumin seed protein isolate (BCSPI) and its hydrolysates (BCSPH) (Trigui et al., 2021). They are characterized by a high protein content (78.4%—83.5%). Enzyme hydrolysis has been reported to improve the solubility of proteins. In addition, antioxidant activity has improved as hydrolysis has increased. The data obtained imply that the hydrolysis grade can improve protein isolate functioning and antioxidant action. The same research group of Trigui et al. (2021) reported that BCSPH generally has special technofunctional and antioxidant features that support its application as a natural dietary protein in food, cosmetics, or pharmaceuticals.

Black cumin is a seed used in traditional medicine because of the bioactive qualities of the seed oil and the solubilized components contained, as investigated by Zent et al. (2021). They used enzymatic proteolysis and FPLC-based fractionation techniques to quantify trypsin and papain hydrolysates for the first time from BCPCs were directed to serve for its dual antioxidative and acetylcholinesterase inhibitory activities (Silva et al., 2020). Using enzymatic proteolysis and FPLC-based fractionation techniques, trypsin and papain hydrolysates of BCPCs were investigated for their dual antioxidative and acetylcholinesterase inhibitory properties.

Peptides in the active fractions were identified using a liquid chromatography quadrupole time-of-flight mass spectrometry—based analytical method, which further facilitated the in-silico prediction of bioactivity for every characterized peptide sequence. The same group of researchers (Zent et al., 2021) found that the degree of suffering a usually dull persistent pain, known as "AChE" inhibitory activity, has been chiefly reduced by proteolysis. Different antioxidant activities are enhanced following proteolytic treatment. Due to its significantly increased activity, 30 min of papain-treated hydrolysates were split into four significant fractions, and their antioxidative capacities were verified in vitro. Twenty different peptide structures were identified, and their potential bioactivities were verified in silico approaches for predicting and screening promising food-derived bioactive peptides (Zent et al., 2021).

A total of 56 unknown peptides was initially found in the cumin seed protein hydrolysate (Siow et al., 2017). These peptides were then subjected to in vitro selection using the phage display technique, and three clones (CSP3, CSP4, and CSP6) showed high affinity for binding α-amylase. These clones were subjected to an inhibitory test, and only CSP4 and CSP6 exhibited high inhibitory activity. CSP4 exhibited inhibition of bacterial and human salivary α-amylases with IC_{50} values of 0.11 and 0.04 μmol, respectively, whereas CSP6 was about 0.10 and 0.15 μmol, respectively (Siow et al., 2017).

6.6 Conclusion

BCSO is widely used as an alternative medication. Many beneficial substances are generated directly from BCSO, whereas others are chemically altered natural products. This study confirmed the use of BCSO in nutritional, pharmaceutical, and medical sectors and some bioactive elements found in their oil and by-products, including AAs, FAs, and the TQ. Further research is required to understand BCSO activity, its bioactive components, and its nutritional impact. Modifying the molecular structure of alpha-hederin, TQ, and other BCSO bioactive compounds may lead to more effective and safe treatments in the future. To combat resistance issues, BCSO and its bioactive compounds such as alpha-hederin, TQ, and others can be used in appropriate combinations with existing chemotherapy drugs. Researchers should also investigate TQ's molecular and cellular targets. This book chapter is dedicated to all scientists who wish to do more preclinical and clinical research on using BCSO to treat various communicable illnesses.

Acknowledgments

This book chapter is supported by the postdoctoral scholarship merit program from the Islamic Development Bank (IsDB) (grant number ID: 2020-276278). The authors also acknowledge the International Institute for Halal Research and Training (INHART) and Biotechnology Engineering Department facilities.

Declaration of conflicts of interest

The authors declare no conflicts of interest.

References

Alanazi, I.O., et al., 2016. Proteomic analysis of the protein expression profile in the mature nigella sativa (black seed). Applied Biochemistry and Biotechnology 179 (7), 1184–1201.

Al-Gaby, A.M., 1998. Amino acid composition and biological effects of supplementing broad bean and corn proteins with Nigella sativa (black cumin) cake protein. Nahrung 42 (5), 290–294.

Al-Jassir, M.S., 1992. Chemical composition and microflora of black cumin (Nigella sativa L.) seeds growing in Saudi Arabia. Food Chemistry 45 (4), 239–242.

Asif, A., et al., 2019. Proteome profiling of the mutagen-induced morphological and yield macro-mutant lines of nigella sativa L. Plants (Basel) 8 (9).

Atta, M.B., 2003. Some characteristics of nigella (Nigella sativa L.) seed cultivated in Egypt and its lipid profile. Food Chemistry 83 (1), 63–68.

Aytac, Z., et al., 2017. Changes in N, K, and fatty acid composition of black cumin seeds affected by nitrogen doses under supplemental potassium application. Journal of Chemistry 2017, 1–7.

Babayan, V., Koottungal, D., Halaby, G., 1978. Proximate analysis, fatty acid and amino acid composition of Nigella sativa L. seeds. Journal of Food Science 43 (4), 1314–1315.

Bakhshabadi, H., et al., 2018. The influence of pulsed electric fields and microwave pretreatments on some selected physicochemical properties of oil extracted from black cumin seed. Food Sciences and Nutrition 6 (1), 111–118.

Bourgou, S., et al., 2010. Bioactivities of black cumin essential oil and its main terpenes from Tunisia. South African Journal of Botany 76 (2), 210–216.

Cheikh-Rouhou, S., et al., 2007. Nigella sativa L.: chemical composition and physicochemical characteristics of lipid fraction. Food Chemistry 101 (2), 673–681.

Cheikh-Rouhou, S., et al., 2008. Sterol composition of black cumin (Nigella sativa L.) and Aleppo pine (Pinus halepensis Mill.) seed oils. Journal of Food Composition and Analysis 21 (2), 162–168.

Coşkun, Ö., et al., 2019. Influence of extraction conditions on structural and functional characteristics of black cumin protein concentrates and ACE-inhibition in their hydrolyzates. Journal of Food Measurement and Characterization 13 (3), 2328–2338.

Coşkun, Ö., et al., 2020. Influence of Maillard reaction conditions and solvent extraction on the surface activity and foaming characteristics of black cumin protein concentrates. Journal of Food Science & Technology 58, 4323–4332.

DemİRbolat, İ., Kartal, M., Karik, Ü., 2019. Development and validation of a GC-FID method to quantify thymoquinone in black cumin seed oils. Journal of Research in Pharmacy 23 (3), 506–513.

Demirci, M., Karslı, M., Aluç, Y., 2019. Determining the effects of black cumin seed oil on performance and meat fatty acid profile of broiler chickens. South African Journal of Animal Science 49 (5), 892–899.

Fajar, D.R., Akrom, Darmawan, E., 2017. The influence of black cumin seed oil therapy with dosage of 1.5 mL/day and 3 mL/day to interleukin-21 (IL-21) expression of the patients with metabolic syndrome risk. In: International Pharmacy Conference 2017. Institute of Physics Publishing.

Farhan, N., Salih, N., Salimon, J., 2021. Physiochemical properties of Saudi Nigella sativa L.('Black cumin') seed oil. Ocl 28, 11.

Haas, M.J., et al., 2014. Induction of apolipoprotein A-I gene expression by black seed (Nigella sativa) extracts. Pharmacien Biologiste 52 (9), 1119–1127.

Hakim, A.S., et al., 2019. Assessment of immunomodulatory effects of black cumin seed (Nigella Sativa) extract on macrophage activity in vitro. International Journal of Veterinary Science 8 (4), 385–389.

Hannan, A., et al., 2008. Anti bacterial activity of Nigella sativa against clinical isolates of methicillin resistant Staphylococcus aureus. Journal of Ayub Medical College, Abbottabad 20 (3), 72–74.

Houghton, P.J., et al., 1995. Fixed oil of Nigella sativa and derived thymoquinone inhibit eicosanoid generation in leukocytes and membrane lipid peroxidation. Planta Medica 61 (01), 33–36.

Indrawan, I.W.A., et al., 2016. Effect of black cumin seeds (Nigella sativa) ethanol extracts on blood pressure, proteinuria, and ET-1 level in preeclampsia mice model. International Journal of PharmTech Research 9 (11), 133–141.

Islam, M., et al., 2017. Nigellalogy: a review on Nigella sativa. MOJ Bioequivalence & Bioavailability 3 (6), 00056.

Kabir, Y., Shirakawa, H., Komai, M., 2019. Nutritional composition of the indigenous cultivar of black cumin seeds from Bangladesh. Progress in Nutrition 21 (1-S), 428–434.

Karadağoğlu, Ö., et al., 2019. Fatty acid composition of liver and breast meat of quails fed diets containing black cumin (Nigella sativa L.) and/or coriander (Coriandrum sativum L.) seeds as unsaturated fatty acid sources. Livestock Science 223, 164–171.

Kaskoos, R., 2011. Fatty acid composition of black cumin oil from Iraq. Research Journal of Medicinal Plant 5 (1), 85–89.

Khajavi Ra, A., et al., 2016. Effects of aqueous-ethanolic extract of nigella sativa seeds (black cumin) and vitamin E on cisplatin-induced nephrotoxicity in rat. Research Journal of Medicinal Plant 10 (4), 295–302.

Khan, S.H., et al., 2012. Black cumin seeds as phytogenic product in broiler diets and its effects on performance, blood constituents, immunity and caecal microbial population. Italian Journal of Animal Science 11 (4), e77.

Khan, S.H., et al., 2013. Effects of black cumin seed (Nigella sativa L.) on performance and immune system in newly evolved crossbred laying hens. Veterinary Quarterly 33 (1), 13–19.

Kiralan, M., et al., 2014. Physicochemical properties and stability of black cumin (Nigella sativa) seed oil as affected by different extraction methods. Industrial Crops and Products 57, 52–58.

Kiralan, M., et al., 2017. Monitoring stability and volatile oxidation compounds of cold-pressed flax seed, grape seed and black cumin seed oils upon photo-oxidation. Journal of Food Measurement and Characterization 12 (1), 616–621.

Kiralan, M., et al., 2019. Stability and volatile oxidation compounds of grape seed, flax seed and black cumin seed cold-pressed oils as affected by thermal oxidation. Grasas y Aceites 70 (1), e295.

Kumar, P., et al., 2017. Effect of black cumin seeds on growth performance, nutrient utilization, immunity, gut health and nitrogen excretion in broiler chickens. Journal of the Science of Food and Agriculture 97 (11), 3742–3751.

Li, H., et al., 2020. Thymocid®, a standardized black cumin (nigella sativa) seed extract, modulates collagen cross-linking, collagenase and elastase activities, and melanogenesis in murine B16F10 melanoma cells. Nutrients 12 (7), 2146.

Liu, X., Aisa, H.A., Xin, X., 2019. A new fatty acid ester from Nigella sativa var. hispidula Boiss showing potent anti-protein tyrosine phosphatase 1B activity. Natural Product Research 33 (4), 472–476.

Maideen, N.M.P., 2020. Prophetic medicine-nigella sativa (black cumin seeds) - potential herb for COVID-19? Journal of Pharmacopuncture 23 (2), 62–70.

Mariod, A.A., et al., 2009. Antioxidant activity and phenolic content of phenolic rich fractions obtained from black cumin (Nigella sativa) seedcake. Food Chemistry 116 (1), 306–312.

Mariod, A.A., Saeed Mirghani, M.E., Hussein, I., 2017. Nigella sativa L. Black cumin. Unconventional Oilseeds and Oil Sources, pp. 73–80.

Masoodi, K.Z., et al., 2020. Botanicals from the Himalayas with anti-cancer potential: an emphasis on the Kashmir Himalayas. In: Ozturk, M., Egamberdieva, D., Pešić, M. (Eds.), Biodiversity and Biomedicine. Academic Press, pp. 189–234.

Mohammed, S.J., et al., 2019. Structural characterization, antimicrobial activity, and in vitro cytotoxicity effect of black seed oil. Evidence-based Complementary and Alternative Medicine 2019, 6515671.

Mukhtar, H., et al., 2021. Composition of nigella sativa seeds. In: Black Cumin (Nigella sativa) Seeds: Chemistry, Technology, Functionality, and Applications. Springer, pp. 45–57.

Najmi, A., et al., 2008. Effect of Nigella sativa oil on various clinical and biochemical parameters of insulin resistance syndrome. International Journal of Diabetes in Developing Countries 28 (1), 11.

Ozdal, T., Adiguzel, S., 2019. Black cumin (nigella sativa L.): bioactive compounds and health benefits. In: Medicinal Plants. CRC Press, pp. 103–114.

Padhye, S., et al., 2008. From here to eternity-the secret of Pharaohs: therapeutic potential of black cumin seeds and beyond. Cancer Therapy 6 (b), 495.

Rahmani, A.H., et al., 2019. Thymoquinone, an active constituent of black seed attenuates CCl4 induced liver injury in mice via modulation of antioxidant enzymes, PTEN, P53 and VEGF protein. Open Access Macedonian Journal of Medical Sciences 7 (3), 311–317.

Ramadan, M.F., Mörsel, J.T., 2002. Characterization of phospholipid composition of black cumin (Nigella sativa L.) seed oil. Food/Nahrung 46 (4), 240–244.

Rusmarilin, H., et al., 2019. Potential of natural antioxidants of black cumin seed (Nigella sativa) and sesame seed (Sesamum indicum) extract by microencapsulation methods. In: IOP Conference Series: Earth and Environmental Science. IOP Publishing.

Sabbah, M., et al., 2020. Black edible films from protein-containing defatted cake of nigella sativa seeds. International Journal of Molecular Sciences 21 (3), 832.

Salem, E.M., et al., 2010. Comparative study of Nigella Sativa and triple therapy in eradication of Helicobacter Pylori in patients with non-ulcer dyspepsia. Saudi Journal of Gastroenterology: Official Journal of the Saudi Gastroenterology Association 16 (3), 207.

Sarwar, A., Latif, Z., 2015. GC–MS characterisation and antibacterial activity evaluation of Nigella sativa oil against diverse strains of Salmonella. Natural Product Research 29 (5), 447–451.

Seidavi, A.R., et al., 2020. Feeding of black cumin (Nigella sativa L.) and its effects on poultry production and health. World's Poultry Science Journal 76 (2), 346–357.

Shabaninejad, Z., et al., 2020. Therapeutic potentials of curcumin in the treatment of glioblstoma. European Journal of Medicinal Chemistry 188, 112040.

Sharma, P., Longvah, T., 2021. Nigella (nigella sativa) seed. In: Oilseeds: Health Attributes and Food Applications. Springer, pp. 331–350.

Siddiqui, S., et al., 2020. Virtual screening of phytoconstituents from miracle herb nigella sativa targeting nucleocapsid protein and papain-like protease of SARS-CoV-2 for COVID-19 treatment. Journal of Biomolecular Structure and Dynamics 40, 1–21.

Silva, A.F.C., et al., 2020. Mechanism of action and the biological activities of Nigella sativa oil components. Food Bioscience 38, 100783.

Siow, H.-L., Lim, T.S., Gan, C.-Y., 2017. Development of a workflow for screening and identification of α-amylase inhibitory peptides from food source using an integrated bioinformatics-phage display approach: case study—Cumin seed. Food Chemistry 214, 67—76.

Siregar, M.H., Sinaga, H., Ridwansyah, 2020. Characterization of microcapsule containing black cumin seed (Nigella sativa) extract as preventive antioxidant. In: 3rd International Conference on Agriculture, Environment and Food Security, AEFS 2019. Institute of Physics Publishing.

Srinivasan, K., 2018. Cumin (Cuminum cyminum) and black cumin (Nigella sativa) seeds: traditional uses, chemical constituents, and nutraceutical effects. Food Quality and Safety 2 (1), 1—16.

Sudhakar, K., et al., 2021. Reverse pharmacology of phytoconstituents of food and plant in the management of diabetes: current status and perspectives. Trends in Food Science & Technology 110, 594—610.

Suri, K., et al., 2020. Influence of microwave roasting on chemical composition, oxidative stability and fatty acid composition of flaxseed (Linum usitatissimum L.) oil. Food Chemistry 326, 126974.

Tavakoly, R., et al., 2019. The effect of Nigella sativa L. supplementation on serum C-reactive protein: a systematic review and meta-analysis of randomized controlled trials. Complementary Therapies in Medicine 45, 149—155.

Tiwari, P., et al., 2019. Nigella sativa: phytochemistry, pharmacology and its therapeutic potential. Research Journal of Pharmacy and Technology 12 (7), 3111—3116.

Trigui, I., et al., 2021. Physical, techno-functional and antioxidant properties of black cumin seeds protein isolate and hydrolysates. Journal of Food Measurement and Characterization 1—10.

Zent, İ., et al., 2020. Linking collective in vitro to individual in silico peptide bioactivity through mass spectrometry (LC-Q-TOF/MS) based sequence identification: the case of black cumin protein hydrolysates. Journal of Food Measurement and Characterization 15 (1), 664—674.

Zent, İ., et al., 2021. Linking collective in vitro to individual in silico peptide bioactivity through mass spectrometry (LC-Q-TOF/MS) based sequence identification: the case of black cumin protein hydrolysates. Journal of Food Measurement and Characterization 15 (1), 664—674.

CHAPTER SEVEN

Functional properties and applications of black cumin (*Nigella sativa* L) seeds

Zanele Pamela Msibi[1,2]
[1]Food Science Nutrition and Technology, University of Eswatini, Luyengo, Eswatini
[2]AgroFood Chain Innovation and Sustainability, Ecole Nationale Supérieure de Formation de l'Enseignement Agricole, Auzeville-Tolosane, France

7.1 Introduction

Functional foods are predicted to become one of the significant dietary trends in the coming decades. Awuchi et al. (2019) define functional properties of food as those that possess essential physicochemical properties that reflect a complex interaction of molecular conformation, and structures, as their composition and physicochemical properties of food components. Some studies suggest that herbaceous plants and their derivatives are underutilized, yet they have the potential of providing several functional properties which could be used in product formulations targeted to improve health. There is an emerging need for the development of novel products from natural herbaceous resources as they prove to have the potential of being nutrient-rich; it is estimated that of 300,000 species that exist, only 15% have been explored. This justifies the importance of food value addition, exploration, and valorization (Yimer et al., 2019). Black cumin seeds' bioactive compounds are of pharmaceutical significance in improving the nutraceutical scope (Niveditha et al., 2013). From a global perspective, functional foods are projected to become widely used in food applications in a variety of dietary trends with the increase of diseases and health ailments, their responsibility will be of high benefit to address health and food safety issues, and this is an important sector for nutritionists and food scientists to pay attention to (Safvi and Yousuf, 2021). Moreover, scientists claim that plant-based foods have curative properties, hence the growth of their application in the health sector. This is mainly due to the phytochemistry composition of the plant. Plants such as herbs and spices are gaining an increase in their use to improve food composition in terms of adding to their total organoleptic qualities and nutritional qualities (Safvi and Yousuf, 2021).

Biochemistry, Nutrition, and Therapeutics of Black Cumin Seed
ISBN: 978-0-323-90788-0
https://doi.org/10.1016/B978-0-323-90788-0.00030-5

7.2 Botanical overview of *Nigella sativa* L.

Nigella sativa L. is an annual herbaceous plant of the Ranunculaceae family, which has seeds called black cumin seeds (Mazaheri et al., 2018). The plant is cultivated worldwide and grows up to 20 to 30 cm tall, it has finely divided leaves and its flowers are pale blue and white, and the flowers are said to be delicate with 5 to 10 petals in each. The fruits of this plant are capsulated and are made of 3 to 7 follicles, and in each lies numerous black cumin seeds (Farhan et al., 2021).

This herbaceous plant is an Asian angiosperm that in ancient medicinal traditions was explored for its efficacy in the culmination of the mind, body, soul, and nature. This traditional practice still exists in some parts of the world. Through research and novel academia approaches, the plant has gained much interest for its use in a broad spectrum of applications due to its native functional properties, particularly for curing disease and maintaining general health (Dabeer et al., 2022). The plant goes by several different names; in Latin, it is referred to as "Panacea," which is loosely translated to "cure-all" and in Arabic, it is popularly known as "Habbah albarakh," which means "seed of blessing." Its continued investigation has been greatly focused on its property in therapeutic applications; the main parts of the plant mostly used in functional applications are the seeds and their derivatives (Padhye et al., 2008).

7.3 Medicinal application of black cumin seeds

Herbal drugs and essential oils of cumin seeds are becoming of great dominance in medicinal applications, essential oil from these seeds has been found by researchers to treat several various diseases. Black cumin seeds are not new to existence, and they have been used over the past years mostly for religious and medicinal uses to treat several diseases. During those years, the seeds and the whole plant were used in traditional systems such as being explored for their benefit in managing postnatal depression and many other neurological disorders. Nowadays, its significance has grown to be of benefit globally, and many scientific researchers are involved in the extensive analysis of the seeds for their properties to be well-defined and characterized particularly in the scope of phytochemical and biological properties (Ahmad et al., 2020). The seeds are gaining wide research studies about their properties and characteristics in making nutritional and functional products, especially in medicinal or traditional nutrition applications, and the consumption of antioxidant-rich foods through studies is believed to improve antioxidant defense mechanism and provide protection against oxidative damage caused by free radicals (Rashida, 2019).

High considerations of its use have gained importance in a wide range of industries, small or large scale, for mainly its properties in biological activities (Ramadan, 2007). A study revealed that this plant has interesting qualities in its polysaccharides; the biomolecules that make up the seeds are stable, safe, and have different biological activities and could be a probable source for both food and pharmaceutical studies (Trigui et al., 2020). In addition, the seeds have been found to have antioxidant properties in their oil and extracts; the strong antioxidant activities these seeds possess suggest their use in product formulation and medicinal use to treat or serve as a protective role in oxidative disease (Rafiquel et al., 2012).

Antidiabetic, immunomodulation, analgesic, antimicrobial, anti-inflammatory, spasmolytic, bronchodilator, hepatoprotective, toothache, wounds, antihypertensive, neuroprotective, gastroprotective, and renal protective are some examples that are believed black cumin seeds and their extracts are responsible for (Hassanien, 2007). In a study by Ambati and Hassanien (2021), it was indicated that the main property of the *N. sativa* to its therapeutic capacity is the presence of thymoquinone bioactive compound found in the seed extract and also its oil. Black cumin seeds show to have great forms of properties for the treatment of asthma, cough, bronchitis, headaches, rheumatism, fever, and influenza eczema, as a diuretic, galactagogue, indigestion, jaundice, diarrhea, flatulence, nausea, morning sickness, dyspepsia, and vermifuge. A study done on black cumin seeds reported that the seed extracts consist of the main essential constituents that give its characteristic status in playing a major role in the therapeutic sectors (Hassanien et al., 2015); the most important compounds of black cumin seeds are thymoquinone, which accounts for 30%–48%, p-cymene (7%–15%), carvacrol (6%–12%), 4-terpineol (2%–7%), t-anethole (1%–4%), and sesquiterpene longifolene (1%–8%) (Boskabadi and Shirmohammadi, 2002).

Another study of Tunisian black cumin seeds indicated that the oil indeed has a potent biological activity, which suggests its use in medicinal applications; the results regarding the bioactivities of these main volatile compounds were an indication that the essential oil is responsible or rather connected to its chemical composition. Thymoquinone and longifolene were found to be the most active compounds, inhibiting about 95% of nitric oxide in in vitro assays, while carvacrol, terpinene-4-01, β-pinene, and p-cymene could inhibit 35%, 20.8%, 33.8%, and 32.3% respectfully (Bourgou et al., 2010). This major significance suggests that these compounds could be applied selectively or collectively in medicinal products valorization into novel food products of medicinal use, especially as an antioxidant or anti-inflammatory; this applies to carcinogenic-related diseases where studies suggest that it could serve as an effective agent to control tumor initiation, growth, and metastasis whether used independently or alongside conventional chemotherapeutic drugs (Srinivasan, 2018).

An active biomolecule that was discovered in a different study in the crude oil of black cumin, thymol proved to have the capability to lower blood pressure in invivo studies; the blood pressure was lowered through blockage of calcium channels, hence the use of the seed oil as an antihypertensive substance in food mechanisms. In addition, this compound affected multiple sites. Multiple biological activities of compounds were discovered to play a huge role as a probable disease eradicating or controlling substance; the volatile oil from the seeds and its constituents, particularly thymoquinone induced cardiovascular effects by serving as a cardiovascular depressant; significant multiple depressant effects were indicated on the heart proving that this could be of great use in medicinal systems in addressing cardiovascular complications (Gilani et al., 2004); a different study in addition to that found the richness of the essential oil with phytochemicals can be used in lifestyle disorders like hyperglycemia and hypercholesterolemia (Sultan et al., 2009). Phytosterols found in black cumin seeds are believed to be of benefit to human appetite, and they are responsible for lowering low-density lipoprotein and total cholesterol levels. The sterol component of the seeds was found to be within the range of 18%–42% of unsaponified matter, and major sterols are β-sitosterol, campesterol, stigmasterol, and 5-avenasterol (Safvi and Yousuf, 2021). These strong findings of the photochemical bio compounds need to be taken into high consideration, also food value addition and need extensive further research on how they could be including a food matrix to serve their intended therapeutic functions.

7.4 Food application of black cumin seeds

Food security has been one of the most discussed topics in many studies; nowadays, the concept is shifting more into food nutrition security, with not only the quantity taken into consideration but also the quality recieving importantance. With the rise in population, high demand for food, and increasing consumer awareness of food quality, scientists are doing whatever it takes in food evolution to meet consumer needs. Several studies indicate that black cumin seeds are nutritionally rich, providing a high amount of nutrients mainly; in fat, protein, and dietary fiber which are estimated to be 28.5%, 26.7%, and 8.4%, respectively, and total ash to be 4.8% (Boskabadi and Shirmohammadi, 2002); the seeds are rich in fatty oil, majorly unsaturated fatty acids constituting of linoleic acid about 50% −60%, oleic acid about 20%, eicosadienoic acid 3%, and dihomolinoleic acid accounting for 10% (Chehl et al., 2009). In addition, the seeds provide a considerable amount of minerals mainly; calcium: 579.33 mg/100 g, potassium: 510.30 mg/100 g, magnesium: 210.33 mg/100 g, sodium: 100 mg/100 g, phosphorus: 91.5 mg/100 g, and iron: 41.80 mg/100 g (Al-Mamun and Absar, 2019).

The use of black cumin seeds (both as whole and ground state) has been studied as an essential ingredient in product design and formulation. The seeds can be used in the concept of food value addition, food revitalization, and enhancement by providing essential biomolecules. The seeds have been widely used in most countries in applications such as spices, as an antimicrobial agent, and as an effective food preservative (Bashir et al., 2021). The cuisine sector in different parts of the world has majorly sought importance in the seed application to improve organoleptic and nutritional properties of food, majorly in condiments, as a flavoring agent, in confectionery, curries, vegetables, pulses, etc. The oil coming from the seeds has been proven to be of functional importance in its application as a flavoring agent in the liquor industry (Safvi and Yousuf, 2021). In a study where butter was infused with black cumin seeds where the essential oil was analyzed, the results indicated that the oil showed higher antioxidant activity than the controls used. Also, an indication from the laboratory analysis showed that coliform bacteria, yeast, and mold count were lower in the oil samples than in the control samples. A conclusion from these results was drawn implying that essentials oil could be of use in the food industry in stabilizing products in terms of improving the shelf life as well as the protection of nutritional properties of the food products. This strongly suggests that the oil should be taken into consideration in its significance to its utilization as a natural source of natural antioxidant properties (Cakmakci et al., 2014).

The essential oil which forms part of secondary metabolites is believed to be of benefit in food application and is generally recognized as safe by the food safety standard measures (Hassanien et al., 2015). Research has found the oil to be linked to the inhabitation of antimicrobial activities, such as the inhabitation against spoilage bacterial activity, aspergillus flavors, fusarium graminearum, fusarium moniliform, and penicillium viridicatum. The use of bioactive compounds is emerging considering consumers growing concerns over the safety of food products manufactured with synthetic antimicrobial agents in attempts to increase shelf life. Bioactive compounds such as essential oils which are believed to be derivatives of phenolic compounds can therefore provide an alternative solution to preserve food by protecting it from degradation from oxidation or microbial activity (Alzoreky and Nakahara, 2003). Such a novel initiative can be of use in the food processing industries which can also be a tool in the achievement of cost-effective outsourcing of ingredients while contributing to high-value food products.

7.5 Functional application of black cumin seed press waste material

Black cumin is initially grown for its seed production mainly in oil recovery (Cherif et al., 2017). However, press waste material from oil extraction has gained major importance in its application as a functional food additive. Cumin seeds cake is found to be of benefit, yet poorly valorized, "the cake is nutrient-dense owing to its richness in crude protein 330 g/kg, fiber, carbohydrates, vitamins, phenolics, all essential amino acids, fat estimated to be 127 g/kg, energy, and antioxidants, and these nutrients play a significant role in human and animal health. Vegetable-sourced proteins and its component, particularly bioactive peptides have been proven to be of nutritional importance in novel food products" (Zaky et al., 2021). Water holding capacity such as the swelling capacity brought about by the proteins is responsible for flavor preservation and mouthfeel (Eltayeb et al., 2011). In a study done by Różyło et al. (2021), starch bread enriched with black cumin seeds and black cumin waste was found to have a higher content of 2-hydroxybenzoic acid, 2-hydroxyphenyl acetic acid, and 4-hydroxyphenyl acetic acid; these acids play a major role in human nutrition by "inhibiting α-amylase and α-glucosidase enzymes which break down complex carbohydrates by keeping blood sugar levels low" (Różyło et al., 2015). Moreover, the addition of black cumin seeds decreased the hardness, elasticity, and chewiness of starch bread crumbs. The property of the seeds to contain volatile compounds has a significant effect on the taste and aroma of food. Valorization of waste material plays a major role in the nutritional industry due to the high nutrient-dense property of the waste material, particularly its richness in dietary fiber, minerals, and antioxidants.

The black cumin cake has been reported through several studies to its major significance in animal feed production. According to El-Ayek et al. (1999), black cumin seed press cake is believed to be a cost-effective way to replace about 50% of the protein in fodder formulas; this not only contributes to animal growth but also to the meat quality. The cake can also be used as a nutrient-dense energy supplement in feed. One study indicated that "rabbits that were fed on 12% black cumin cake gained more overall weight than the control group with 3.7% increase in feed conversion rate," hence the property of the seed press cake being nutrient-dense makes it an excellent source for animal feed (Mahmoud and Bendary, 2014), which in turn reduces the high cost in convectional feeding schemes, in that way this also provides an economic benefit. Moreover, there has been a major concern about the compromise of human health, particularly the rise of chronic illnesses which have been linked to the chemical treatments used in animals as means to promote growth such as the use of antibiotics (Kebede and Roba, 2021). It is estimated that "about 50% of the world's antibiotics are used as growth promoters in animals" (Awaisheh et al., 2019), hence the valorization of black cumin seed press cake opens a novel alternative to replace chemical growth promoters while creating explorative ideas into a paradigm shift to economic cycle whose focus is on waste minimization in the industrial production streams.

7.6 Conclusion

The properties of black cumin seeds have been indicative through studies that they can exhibit functional attributes that are of benefit in medicinal, nutraceutical, and food applications. Its waste material from seed press such as the cake can be effectively used as food additives and feed formulations owing to its nutrient-rich properties. There is an emergency need for the protection and cultivation of herbaceous species especially because such species majorly contribute to the ecosystem services to humans, particularly in the nutrition field by ensuring food nutrition security, hence the need for its diversity and sustainability.

References

Ahmad, F., Ahmad, F., Ashraf, S.A., Saad, H., Wahab, S., Khan, I., Ali, M., Mohan, S., Hakeem, K., Athar, T., 2020. Updated knowledge of black seed (*Nigella sativa* Linn): review of phytochemical constituents and pharmacological properties. Journal of Herbal Medicine.

Al-Mamun, M., Absar, N., 2019. Major nutritional compositions of black cumin seeds cultivated in Bangladesh and the physicochemical characteristics of its oil. International Food Research Journal 25, 2634–2639.

Alzoreky, N., Nakahara, K., 2003. Antimicrobial activity of extracts from some edible plants commonly consumed in Asia. International Journal of Food Microbiology 223–230. https://doi.org/10.1016/S0168-1605(02)00169-1.

Ambati, D.R.R., Hassanien, M., 2021. *Nigella sativa* seed extracts in functional foods and nutraceutical applications, pp. 501–520. https://doi.org/10.1007/978-3-030-48798-0_31.

Awaisheh, S.S., Khalifeh, M.S., Rahahleh, R.J., Al-Khaza'leh, J.M., Algroom, R.M., 2019. Sulfamethazine contamination level and exposure assessment in domestic and imported poultry meats in Jordan. Veterinary World 12 (12), 1992–1997. https://doi.org/10.14202/vetworld.2019.1992-1997.

Awuchi, C., Victory, I., Echeta, C., 2019. The Functional Properties of Foods and Flours, 5, pp. 139–160.

Bashir, O., Jan, N., Gani, G., Naik, H., Hussain, S., Reshi, M., Amin, T., 2021. Food Applications of *Nigella sativa* seeds, pp. 191–207. https://doi.org/10.1007/978-3-030-48798-0_13.

Boskabadi, M., Shirmohammadi, B., 2002. Effect of *Nigella sativa* on Isolated Guinea Pig Trachea. https://www.sid.ir/en/journal/ViewPaper.aspx?ID=13035.

Bourgou, S., Pichette, A., Marzouk, B., Legault, J., 2010. Bioactives of black cumin essential oil and its main terpenes from Tunisia. South African Journal of Botany 76, 210–216. https://doi.org/10.1016/j.sajb.2009.10.009.

Cakmakci, S., Gündoğdu, E., Dagdemir, E., Erdoğan, Ü., 2014. Investigation of the possible use of black cumin (*Nigella sativa* L.) essential oil on butter stability. Kafkas Universitesi Veteriner Fakultesi Dergisi 20. https://doi.org/10.9775/kvfd.2013.10550.

Chehl, N., Chipitsyna, G., Gong, Q., Yeo, C.J., Arafat, H.A., 2009. Anti-inflammatory effects of the *Nigella sativa* seed extract, thymoquinone, in pancreatic cancer cells. HPB 11 (5), 373–381. https://doi.org/10.1111/j.1477-2574.2009.00059.x.

Cherif, M., Salem, H., Abidi, S., 2017. Effect of the addition of *Nigella sativa* seeds to low or high concentrate diets on intake, digestion, blood metabolites, growth, and carcass traits of Barbarian lamb. Small Ruminant Research 158. https://doi.org/10.1016/j.smallrumres.2017.11.008.

Dabeer, S., Rather, M.A., Rasool, S., Rehman, M.U., Alshahrani, S., Jahan, S., Rashid, H., Halawi, M., Khan, A., 2022. Chapter 1—History and traditional uses of black seeds (*Nigella sativa*). In: Khan, A., Rehman, M. (Eds.), Black Seeds (*Nigella Sativa)*. Elsevier, pp. 1–28. https://doi.org/10.1016/B978-0-12-824462-3.00016-0.

El-Ayek, M., Gabr, A., Mehrez, A., 1999. Influence of substituting concentrate feed mixture by *Nigella sativa* meal on animal performance and carcass traits of growing lambs. Egypt J Nutr Feeds 2, 265–277.

Eltayeb, A., Ali, A.O., Abou-Arab, A., Abu-Salem, F., 2011. Chemical composition and functional properties of flour and protein isolate extracted from bambara groundnut (Vigna subterranean). https://www.semanticscholar.org/paper/Chemical-composition-and-functional-properties-of-Eltayeb-Ali/9db2d6c5076a7c87a31d2315eff161abecf48e64.

Farhan, N., Salih, N., Salimon, J., 2021. Physiochemical properties of Saudi *Nigella sativa* L. ('Black cumin') seed oil. OCL 28, 11. https://doi.org/10.1051/ocl/2020075.

Gilani, A., Jabeen, Q., Khan, M., 2004. A review of medicinal uses and pharmacological activities of *Nigella sativa*. Pakistan Journal of Biological Sciences 7. https://doi.org/10.3923/pjbs.2004.441.451.

Hassanien, 2007. Nutritional value, functional properties and nutraceutical applications of black cumin (*Nigella sativa* L.): an overview. International Journal of Food Science & Technology 42, 1208–1218. https://doi.org/10.1111/j.1365-2621.2006.01417.x.

Hassanien, M.F.R., Assiri, A.M.A., Alzohairy, A.M., Oraby, H.F., 2015. Health-promoting value and food applications of black cumin essential oil: an overview. Journal of Food Science and Technology 52 (10), 6136–6142. https://doi.org/10.1007/s13197-015-1785-4.

Kebede, A., Roba, M., 2021. Review on antibiotics residue in beef: its risk factors and public health impacts. International Journal of Veterinary Health Science & Research 260–265. https://doi.org/10.19070/2332-2748-2000049.
Mahmoud, A.E.M., Bendary, M.M., 2014. Effect of the whole substitution of a protein source by *Nigella sativa* meal and sesame seed meal in ration on the performance of growing lambs and calves. Glob. Vet. 1.
Mazaheri, Y., Torbati, M., Azadmard-Damirchi, S., April 25, 2018. Nutritional Value and Functional Properties of Black Cumin Oil (*Nigella sativa* L.): An Overview.
Niveditha, V.R., Sridhar, K., Balasubramanian, D., 2013. Physical and mechanical properties of seeds and kernels of Canavalia of coastal dunes, 20, pp. 1547–1554.
Padhye, S., Banerjee, S., Ahmad, A., Mohammad, R., Sarkar, F.H., 2008. From here to eternity - the secret of Pharaohs: therapeutic potential of black cumin seeds and beyond. Cancer Therapy 6 (b), 495–510.
Rafiquel, I., Hasan, N., Shiddique, S., Rashid, M., Mahmud, S.K., Rahman, M., Rahman, A., 2012. The black seed *Nigella sativa* Linnaeus: a study of the antioxidant activity of the essential oil and extracts. Journal of Nature Science and Sustainable Technology 7.
Ramadan, M.F., 2007. Nutritional value, functional properties and nutraceutical applications of black cumin (*Nigella sativa* L.): an overview. International Journal of Food Science & Technology 42 (10), 1208–1218. https://doi.org/10.1111/j.1365-2621.2006.01417.x.
Rashida, P., 2019. Therapeutic effects of black cumin (*Nigella sativa*). A systematic review. Progress in Nutrition 21. https://doi.org/10.23751/pn.v21i1-S.5796.
Różyło, R., Dziki, D., Gawlik-Dziki, U., Cacak-Pietrzak, G., Miś, A., Rudy, S., 2015. The physical properties of gluten-free bread are caused by water addition. International Agrophysics 29, 353–364. https://doi.org/10.1515/intag-2015-0042.
Różyło, R., Piekut, J., Wójcik, M., Kozłowicz, K., Smolewska, M., Krajewska, M., Szmigielski, M., Bourekoua, H., 2021. Black cumin pressing waste material as a functional additive for starch bread. Materials 14 (16), 4560. https://doi.org/10.3390/ma14164560.
Safvi, A.F., Yousuf, O., 2021. Application of Black Cumin (*Nigella sativa*) Seed as Functional Food. Agrospheres.
Srinivasan, K., 2018. Cumin (*Cuminum cyminum*) and black cumin (*Nigella sativa*) seeds have Traditional uses, chemical constituents, and nutraceutical effects. Food Quality and Safety 2 (1), 1–16. https://doi.org/10.1093/fqsafe/fyx031.
Sultan, M., Butt, M., Anjum, F., Jamil, A., Akhtar, S., Nasir, M., 2009. The nutritional profile of the indigenous cultivar of Black cumin seeds and the antioxidant potential of its fixed and essential oil. Pakistan Journal of Botany 41, 1321–1330.
Trigui, I., Cheikh-Rouhou, S., Attia, H., Ayadi, M., 2020. Black Cumin Polysaccharides. Research Gate 67–80. https://doi.org/10.1007/978-3-030-48798-0_7.
Yimer, E.M., Beshir, K., Karim, A., Rehman, N.U., Anwar, F., 2019. *Nigella sativa* L. (black cumin): a promising natural remedy for wide range of illnesses. Evidence-Based Complementary and Alternative Medicine 2019, 1–16. https://doi.org/10.1155/2019/1528635.
Zaky, A.A., Shim, J.-H., Abd El-Aty, A.M., 2021. A review on extraction, characterization, and applications of bioactive peptides from pressed black cumin seed cake. Frontiers in Nutrition 8. https://www.frontiersin.org/article/10.3389/fnut.2021.743909.

CHAPTER EIGHT

Black cumin (*Nigella sativa* L.) essential oil and aroma quality

Islam Md Rizwanul Fattah[1], Rahman S.M. Ashrafur[2], Arslan Ahmed[3]

[1]Centre for Green Technology (CGT), School of Civil and Environmental Engineering, Faculty of Engineering and IT, University of Technology Sydney, Ultimo, NSW, Australia
[2]Biofuel Engine Research Facility, Queensland University of Technology, Brisbane, QLD, Australia
[3]Department of Mechanical Engineering, Comsats University Islamabad, Wah Campus, Rawalpindi, Punjab, Pakistan

8.1 Introduction

The annual flowering plant of *Nigella sativa* L., commonly known as black seed and black cumin, is a small annual herb. The term "Nigella" originates from "Niger," meaning black in Latin, referring to its seeds' color (Kalidasu et al., 2017). It is one of the most important medicinal plants all over the world, particularly in regions like India, Africa, and Asia because of its multifarious uses (Khan and Abourashed, 2011; Burits and Bucar, 2000; Shabana et al., 2013). Proximate analysis of seeds reveals that they contain a composition of 39.7%–52.5% oil and 0.5%–1.5% essential oil (EO) (Al-Jassir, 1992; Takruri and Dameh, 1998). Nigella seed-fixed oil is high in linoleic, oleic, and palmitic acids. Linoleic acid, an important fatty acid that accounts for more than half of all fatty acids, is the dominant fatty acid (Bassim Atta, 2003). Nigella seed EO is mostly used in food (as flavorings), fragrances, and medications (Burt, 2004; Hassanien et al., 2015). *N. sativa* also has a long historical and religious history. Seeds and their oils have been claimed for medical purposes since the ancient Egyptians, Greeks, and Romans (Benkaci–Ali et al., 2007). *N. sativa* is regarded as one of the most powerful types of healing medicine in Islamic literature. The most significant active compounds in black cumin seed EO are thymoquinone, α-thujene, and p-cymene (Ghahramanloo et al., 2017; Kabir et al., 2020). The first one has been reported to possess anticonvulsant as well as anticancer effects (Okoye et al., 2014). Its antiinflammatory, analgesic, antipyretic, antibacterial, antiinflammatory, immunological, antihypertensive, respiratory-stimulating, antihypertensive, hematinic, and trypanocidal

Biochemistry, Nutrition, and Therapeutics of Black Cumin Seed
ISBN: 978-0-323-90788-0
https://doi.org/10.1016/B978-0-323-90788-0.00010-X

properties have also been documented (Ali and Blunden, 2003; Ekanem and Yusuf, 2008). The EO of black cumin was found to possess antibacterial properties (Kokoska et al., 2008; Shaaban et al., 2015), antioxidant activity (Pop, Bocsan et al., 2020; Bordoni et al., 2019), and antiinflammatory properties (Majdalawieh and Fayyad, 2015; Pop, Sabin et al. 2020). Although the dietary value of black cumin has received less attention in the scientific literature, it is far from insignificant, as it includes a sufficient amount of protein and fat and a significant number of important fatty acids, amino acids, vitamins, and minerals (Hannan et al., 2021).

8.1.1 Botanical description of *Nigella sativa* L.

N. sativa (Ranunculaceae family) is a native annual herbaceous plant of northern Africa, the eastern Mediterranean, Southwest Asia, and the Indian subcontinent (Hannan et al., 2021; Gharby et al., 2015). It is, nevertheless, extensively grown in several countries, including Egypt, Greece, Iran, Syria, Saudi Arabia, Albania, Turkey, India, and Pakistan (Srinivasan, 2018). The black cumin plant grows to a height of 30–60 cm, and its seeds are black in color (Ashraf et al., 2006).

8.1.2 Origin and distribution

Black cumin has a long history of crop production across the globe. The exact location of the origin of black cumin is uncertain (Yarnell and Abascal, 2011). Nevertheless, it is assumed to be originated in the Mediterranean. Several experts have voiced different opinions on the origins of black cumin (Kifelew et al., 2017; Kulloli, 2016; Sultana et al., 2015). For example, according to Sultana et al. (Sultana et al., 2015), it is indigenous to North Africa, Southwest Asia, and Southern Europe. According to Kulloli (2016) and Lal (2018), it is native to South and Southwest Asia. Moreover, Mahr (2022) reported that around 15 species of the genus Nigella (love-in-a-mist) are native to Northern Africa and Southern Europe.

Black cumin farming in olden history spanned at least North Africa, the Middle East, and South Asia, where the seed has been used in the traditional system of medicine for centuries or more (Diwakar et al., 2018). Black seed has been employed since ancient Egyptian civilizations. Although there is no archaeological evidence for the early cultivation of black cumin, black cumin seeds have been found at a number of ancient Egyptian sites, including Tutankhamun's tomb (Kulloli, 2016). Over time, it spread to many parts of northern Africa, Eastern Asia, and Southern Europe. Black seed has made its way into Eastern Europe and North America in recent decades. Over the course of ages, the crop spread throughout Europe as a significant spice used in the making of bread and cakes (Heiss and Oeggl, 2005). Its consumption is widely spread in Europe, North Africa, Asia Minor, and the Mediterranean region, whereas it is on a minor scale in Russia, Egypt, Turkey, and France (Aftab et al., 2018).

8.1.3 Taxonomic classification

Black cumin is a genus of annual plants in the Ranunculaceae family with approximately 14 species. This genus is also known as devil-in-a-bush or love-in-a-mist (Kulloli, 2016). English common names of *N. sativa* L. are: black cumin, fennel flower, nutmeg flower, black seed, black caraway, Roman coriander, Damascena, devil-in-a-bush, and wild onion seed (Kulloli, 2016; Sultana et al., 2015). The hierarchal classifications of black cumin are plants (kingdom), vascular plants (subkingdom), seed plants (superdivision), flowering plants (division), Dicotyledons (class), Magnoliidae (subclass), Ranunculales (order), Ranunculaceae (family), Nigella (genus), and *N. sativa* L. (species) (United States Department of Agriculture (USDA), 2022).

8.1.4 Morphology

Black cumin is an erect annual plant that grows up to 20—60 cm tall with a more or less branching stem (Sultana et al., 2015; Hailemichael et al., 2015). The leaves are divided, and the roots are straight taproots (Dubey et al., 2016). The flower is hermaphrodite, often cross-pollinated, and solitary on the central axis (Fig. 8.1) (Shariq et al., 2015). In the first bloom, the flower shows the pale green coloration of petaloid sepals. After full blooming, the color changes to pigeon blue (Perveen and Qaiser, 2006).

Figure 8.1 Morphology of *Nigella Sativa* L.: (a) flower (b) seeds.

The flower is without involucres of bracts, and the peduncle is long and erect. Petaloid sepals are broad and ovate in a single whorl of 4–6. On average, just 10% of flowers are nonfertile (Mukherjee et al., 2013). The stamen is in 3–4 whorls which are numerous, usually 32 to 66. The ripe fruit (capsule: 3–7 united follicles) contains numerous tiny seeds (Diwakar et al., 2018).

Black cumin seeds are triangular and have a rough surface (El-Morsy and Osman, 2021). The seed is flat on one side and convex on the other, with tapered curved ends (Fig. 8.1). The seeds are black on the outside and white on the interior, with a mild odor and a bitter taste (Perveen and Qaiser, 2006; Jansen, 1981). In general, there is considerable diversity among different cultivars of black cumin, either morphologically or physiologically (Shariq et al., 2015).

8.2 Essential oil and its extraction

EOs, usually referred to as volatile or ethereal oils, are fragrant oily liquids derived from plant material (flowers, buds, seeds, leaves, twigs, bark, herbs, wood, fruits, and roots) (Burt, 2004; Ashrafur Rahman et al., 2019). ESs have a strong odor and are generated as secondary metabolites by medicinal or fragrant plants (Bakkali et al., 2008). The word "essential oil" is likely to stem from a terminology coined in the 16th century by the Swiss reformer of medicine, Paracelsus von Hohenheim, who dubbed the active component of a medication *Quinta essentia* (Guenther, 2014). An estimated 3000 EOs are known, of which around 300 are economically important—mostly for the flavor and scent markets.

Hydrodistillation (HD), steam distillation (SD), expression, solvent extraction (SE), and supercritical fluid extraction (SFE) are the five basic extraction procedures used to acquire EOs. In the HD process, the plant material is soaked in water before being boiled in a still. The oil molecules vaporize and rise with the steam as the plant material softens. To turn the oil vapors into liquid EO, they are delivered to a condenser. The oil may then be separated using a separator. SD is identical to HD, except that instead of water, steam is created from an external source. Extraction with higher pressures and temperatures of steam is possible. The advantage is that the steam supply (amount, temperature, and pressure) may be modified, and the thermal degradation of oil components is minimal. However, this method demands a lot more capital. In other words, the plant ingredients are physically pressed. This method is used for plant materials that possess a high concentration of EOs, such as citrous fruit EOs (e.g., orange, lemon, and lime oils) (Rahman and Fattah, 2021). The source of the crop is a significant factor with pressed oils since citrus fruits are constantly treated with insecticides, which might get concentrated in the oil. A hydrocarbon solvent is employed to dissolve EOs and plant material in SE. After filtering, the solution is concentrated by distillation. After distillation, only wax and EO remain. The EO is then extracted using pure alcohol. The resulting residue after the evaporation of alcohol is the final product. In the industry, SFE under pressure is commonly used to isolate volatile oils and aroma components. This modern technique uses liquefied CO_2 under pressure to extract EOs. Once the liquid is depressurized, the CO_2 returns to a gaseous state, and only pure EO, resins, or alkaloids remain. It clearly outperforms traditional methods of SD and SE, since it does not modify the labile character of aroma components or volatile oils found in fruits and aromatic plants.

8.2.1 *Nigella sativa* essential oil extraction and its composition

In the literature on *N. sativa*, researchers typically use HD and SE to extract EO from *N. sativa* L. For example, Burits & Bucar (Burits and Bucar, 2000) used an SE process using a Soxhlet apparatus and light petroleum with a boiling point of 40–60°C. The solvent was removed under vacuum conditions, and the remaining brownish residue was then steam distilled. The EO was obtained by *n*-hexane extraction of the aqueous distillate (in the SD process) and removal of the solvent. Kabir et al. (2020) used methanol as a solvent. Seed powder is first soaked in methanol and then filtered. The supernatant was diluted with purified water and adjusted to 10% methanol. After eliminating the water-soluble compounds, the adsorbed compounds were eluted. The elute was dried using anhydrous Na_2SO_4, and the solvent was evaporated at 39.5°C under atmospheric pressure to obtain the EO. Singh et al. (2005) used a Clevenger-type apparatus for 3 h. The seeds were ground to 750 mesh and hydrodistilled using the method described in the European Pharmacopoeia. Kokoska et al. (2008) studied four different extraction methods to compare the performance, which were HD, dry SD, SE to obtain crude followed by SD (SE-SD), and SFE followed by SD (SFE-SD). For HD, they used a Clevenger-type apparatus for 2 h, whereas for SD, a glass column interposed between flask and condenser was used for 2 h. SE was performed in Soxhlet apparatus, using *n*-hexane as a solvent for 120 h. The residue was steam distilled with 200 mL of water for 1 h. The EO was isolated by extraction with *n*-hexane. To perform SFE, 67 g of milled seeds were placed in a 150-mL stainless steel extractor and operated at 25 MPa and 40°C. The specific solvent flow rate was less than 0.08 g/g/min.

The obtained saturated solution of seed oil had a high content of fatty acids. Thus, SD was carried out using the previously mentioned process to extract the EO. El-Ghorab (2003) utilized SFE to extract the EO. They placed 1 g of ground nigella seeds to Isco SEX 220 extractor cell followed by the passage of pure CO_2 (99.995%) as supercritical fluid using an Isco Model 260 D syringe pump at a rate of 2 mL/min (200 bar, 50°C) for 2 h. Table 8.1 summarizes some of the prominent researches on EO extraction in the field.

Table 8.1 Extraction of *Nigella sativa* essential oil from literature.

Extraction process	Reaction conditions	Ref.
Solvent extraction (SE)	Soxhlet apparatus and light petroleum having a boiling point 40–60 °C as the solvent	Burits & Bucar (2000)
SE	Methanol as solvent.	Kabir et al. (2020)
Hydrodistillation (HD) and microwave distillation (MD)	HD: Clevenger-type apparatus. Time 2.5 h. Water amount: 3L. MD: At 101.325 kPa in a microwave laboratory oven. A fixed power of 850 W for 10 min with the addition of 50 mL distilled water was used.	Benkaci–Ali et al. (2007)
HD	Clevenger-type apparatus for 3 h.	Singh et al. (2005)
HD	Clevenger-type apparatus for 4 h.	Wajs et al. (2008)
HD Dry steam distillation (SD) SE to obtain crude followed by SD (SE-SD) Supercritical fluid extraction (SFE) followed by SD (SFE-SD)	HD: Clevenger-type apparatus for 2 h. SD: Glass column interposed between flask and condenser for 2 h. SE-SD: Soxhlet apparatus. *n*-hexane solvent. 120 h. The residue was steam distilled for 1 h.	Kokoska et al. (2008), Sovová et al. (1994)

Table 8.1 Extraction of *Nigella sativa* essential oil from literature.—cont'd

Extraction process	Reaction conditions	Ref.
	SFE-SD: Stainless steel extractor and operated at 25 MPa and 40 °C and solvent flow rate less than 0.08 g/g/min. The obtained product was steam distilled using the above-mentioned process.	
SFE	Isco SEX™ 220 extractor. Isco Model 260 D syringe pump for supercritical CO_2 at a rate of 2 mL/min (200 bar, 50°C) for 2 h.	El-Ghorab (2003)

Generally, the volatile compounds of *N. sativa* L. belong to four chemical classes: (a) monoterpenic hydrocarbons, (b) monoterpenic alcohols, (c) terpenic ethers, and (d) monoterpenic phenols (Hamrouni-Sellami et al., 2008). For example, Hamrouni-Sellami et al. (2008) found that for *N. sativa* of Tunisian origin, p-cymene was the major compound (53.1%), followed by ocimene (18.5%) in the monoterpenic hydrocarbons category, which formed 86.3% of all volatiles. This is followed by monoterpenic alcohols, representing only 7.1%, with octen-3-ol as the major compound (6.5%). In addition, 1,8-cineole (1.9%) represented terpenic ethers, and thymol (1.8%) represented monoterpenic phenols which were 1.9% and 1.8% of total volatiles, respectively. D'Antuono et al. (2002) characterized the EO of black cumin, which originated from Morocco. They reported p-cymene is the major compound with 33.75% of the total volatiles.

Kabir et al. (2020) reported that Bangladeshi black cumin EO contained *p*-cymene (36.35%), followed by thymoquinone (29.77%), α-thujene (12.40%), carvacrol (2.85%) and β-pinene (2.41%). They also carried out the same analysis on EOs of Indian origin seeds and found that the majority compound was again *p*-cymene (41.80%), which was followed by α-thujene

(13.93%), thymoquinone (10.27%), methyl linoleate (4.02%), carvacrol (3.65%), and β-pinene (2.96%). Thus, *p*-cymene, thymoquinone, and α-thujene were the major components in the seeds of both origins. It should be noted that seeds from Bangladesh had about three times the amount of thymoquinone as seeds from India. Wajs et al. (2008) found monoterpenes (87.7%) and their oxygenated derivatives in the EO of Polish black cumin seeds (9.9%). Only a small fraction was composed of sesquiterpenes and their oxygenated derivatives (0.7% and 0.3%, respectively). The primary components of the EO were p-cymene (60.2%), γ-terpinene (12.9%), α-thujene (7.2%), carvacrol (3.0%), α-pinene (2.0%), and β-pinene (2.1%). These findings demonstrate that *N. sativa* EO contains bioactive components such as p-cymene, limonene, α-pinene, linalool, and thymol, all potent bactericides (Chouhan et al., 2017). Table 8.2 displays the chemical composition of EOs from various black cumin seeds of different origins.

Table 8.2 Chemical composition of essential oil obtained from black cumin seed.

Volatile compound	% Peak area (Singh et al., 2005)	% Total aroma compounds (Hamrouni-Sellami et al., 2008)	% Peak area (Kokoska et al., 2008)	% Mass (Singh et al., 2014)	% Area (D'Antuono et al., 2002)
α-Thujene	10.03	7.2	15.1	5.6	3.27
α-Pinene	3.33	1.4	3.3	1.4	0.7
2,4(10)-Thujadiene	0.06	—	—	—	—
Camphene	0.06	—	—	—	—
Sabinene	1.34	0.7	2	0.8	0.53
β-Pinene	3.78	1.8	4	1.7	1.12
Myrcene	—	2.1	—	—	0.29
α-Phellandrene	—	0.1	Trace	0.1	—
α-Terpinene	0.47	—	0.7	0.2	0.64
o-Cymene					3.26
p-Cymene	36.2	53.1	56.2	31.4	33.75
Limonene	1.76	0.1	2.9	1.0	1.13
Ocimene	—	18.5	—		—
1,8-Cineole	Trace	1.9	—	0.1	—
γ-Terpinene	0.16	1.2	1.2	0.2	2.4
cis-Sabinene hydrate	0.22	—	—		—
Terpinolene	0.06	0.1	—		—
trans-Sabinene hydrate	0.19		—	0.1	1.05
Linalool	0.1	0.1	Trace	—	—
Octen-3-ol	—	6.5	—	—	—
Terpinen-1-ol	0.19	—		—	—
Camphor	0.11	—	—	—	—

Table 8.2 Chemical composition of essential oil obtained from black cumin seed.—cont'd

Volatile compound	% Peak area (Singh et al., 2005)	% Total aroma compounds (Hamrouni-Sellami et al., 2008)	% Peak area (Kokoska et al., 2008)	% Mass (Singh et al., 2014)	% Area (D'Antuono et al., 2002)
Ethanone-1-(1,4-dimethyl-3-cyclohexen-1-yl)	0.46	–	–	–	–
Terpinen-4-ol	2.37	0.4	0.9	1.0	–
p-Cymen-8-ol	0.09	–	–	Trace	–
α-Terpineol	0.08	0.1	–	Trace	–
4,5-Epoxy-1-isopropyl-4-methyl-1-Cyclohexene	0.85	–	–	–	–
Cuminal	–	–	–	Trace	–
Carvone	0.16	–	1.1	Trace	–
Thujol	–	–	–	–	7.43
Thymoquinone	11.27	–	0.5	37.6	3.80
trans-Sabinene hydrate acetate	–	–	–	0.1	–
Linalool acetate	–	–	Trace	–	–
Bornyl acetate	0.44	–	0.2	0.1	–
trans-Anethole	0.61	–	–	–	–
2-Undecanone	0.09	–	–	–	–
Myrtenol	–	–	–	–	2.44
Thymol	0.13	1.8	–	0.2	26.78
Carvacrol	2.12	–	0.8	1.4	–
α-Longipinene	1.54	–	0.7	0.5	–
Longicyclene	0.06	–	–	–	–
β-Elemene	–	–	–	–	5.47
Longifolene	6.32	–	2.6	2.0	3.11
Thymohydroquinone	–	–	Trace	3.4	–
β-Caryophyllene	0.07	–	0.1	–	–
2-Tridecanone	0.1	–	–	–	–
β-Bisabolene	0.07	–	Trace	–	–
γ-Cadinene	–	–	Trace	–	–
Epizonaren	0.1	–	–	–	–
Sesquiterpene alcohol acetate	0.19	–	–	–	–
Pimara-8(14), 15-diene	0.2	–	–	–	–

(Continued)

Table 8.2 Chemical composition of essential oil obtained from black cumin seed.—cont'd

Volatile compound	% Peak area (Singh et al., 2005)	% Total aroma compounds (Hamrouni-Sellami et al., 2008)	% Peak area (Kokoska et al., 2008)	% Mass (Singh et al., 2014)	% Area (D'Antuono et al., 2002)
10-epi-γ-Eudesmol	–	–	–	0.3	–
β-Eudesmol	–	–	–	0.5	–
α-Eudesmol	–	–	–	0.4	–
Unidentified	13.32	–	–	–	–

8.3 Aroma composition of *Nigella sativa* L.

A food product's aroma is a complicated mix of a wide range of volatiles. These volatile molecules, which include groups of alcohols, aldehydes, terpenes, esters, and so on, combine to form a food product's odor (Selli et al., 2014). They are one of the most significant agents since they influence food quality and customer preferences. Aroma is the primary factor that clarifies the organoleptic qualities of foods among various chemical components. It is to be noted that only a small percentage of the vast number of scent components contribute to the overall aroma. An olfactometric approach can detect fragrance-active molecules in the complex structures of hundreds of aroma compounds. The gas chromatography-mass spectrometry-olfactometry (GC-MS-O) technique has made it possible to categorize volatile substances as aroma-active or nonaroma-active based on their concentration in the samples studied (García-González et al., 2007; Kesen et al., 2013). For example, Kesen et al. (2018) used the purge and trap extraction method to extract the volatile compounds. GC-MS-O determined the aroma-active components of black cumin seed extract. They reported obtaining a total of 13 odor-active chemicals (Table 8.3). There were two alcohols, four carboxylic acids, and two phenols among the 13 odor-active chemicals. There was one each of ketone, hydrocarbon, and terpene, and two were unknown (perceived by GC-MS-O only). The odor activity values varied from 1.0 to 170.8, which were computed by dividing the concentration of the chemicals in water by their threshold. Odor thresholds in water were obtained from the literature (Morales et al., 2005; Buttery et al., 1971; Alexander et al., 1982; Larsen and Poll, 1992).

Table 8.3 Aroma-active compounds of black cumin seed extract.

Aroma-active compounds	Odor descriptions (perceived by panellists during olfactometry)	Odor thresholds in water (ppm)	Odor activity values (OAVs)
Unknown I	Burnt plastic	–	–
3-Penten-2-ol	Green	0.4	1
Limonene	Citrusy	0.01	17.6
Styrene	Balsamic	0.0036	18.2
Acetoin	Buttery	0.055	170.8
Unknown II	Putrid fruit	–	
Propanoic acid	Fatty	0.1	11.9
Isobutanoic acid	Cheesy	0.01	22
Pentanoic acid	Cheesy	0.07	2.5
Hexanoic acid	Cheesy	0.093	3.3
Guaiacol	Smoky	0.003	3.8
Phenethyl alcohol	Honey, floral	0.086	1.5
Eugenol	Spicy-smoky	0.006	14.8

8.4 Benefits and use of *Nigella sativa* L.

The importance of black cumin has been attributed a long history in the civilization of human beings as it was used as a spice, traditional herbs, and food preservative and enhanced the flavor of bakeries (Forouzanfar et al., 2014; Ermumcu and Şanlıer, 2017). Black cumin's historical and traditional uses have been written about and well documented in ancient texts and historical documents (Botnick et al., 2012). Among several uses of black cumin, its medicinal value is of great importance. Scholars have reported its value as a natural remedy for several illnesses: treating animal and human ailments, effective antimicrobial drugs and antioxidants, and improving learning and memory ability (Mohammed et al., 2019; Shakeri et al., 2016; Hadi et al., 2021; Salehi et al., 2021). The seeds of *N. sativa* are the most valuable and medicinally used parts of the plant and are known as "Panacea (in old Latin meaning 'cure-all'), Kalonji (India), Hak Jung Chou (China), Habbatu Sawda, Habatul Baraka (in Arabic meaning 'the blessed seed'), black cumin, black seed, black caraway (English), chernushka (Russian), çörek out (Turkish), and Cyah-daneh (Iran)" (Aggarwal et al., 2008; Amin and Hosseinzadeh, 2016).

The active ingredients in black cumin oil, such as tocopherol, phenolic compounds, and thymoquinone, are thought to be responsible for the oil's therapeutic benefits on human health (Rohman et al., 2019). Its seed, for example, has been extensively used in folk medicine to cure a variety of ailments, including diarrhea, jaundice, amenorrhea, helminthiasis, ophthalmia, paralysis, and asthma (Yimer et al., 2019). Yimer et al. (2019) reported that black cumin is an important natural remedy for many illnesses. It contains properties such as antioxidant activity, antidiabetic activity, antihypertensive, neuroprotective effects, antimicrobial activity (antibacterial, antifungal, antiviral, and antiparasitic), anticancer, and effects on male infertility. According to Islam et al. (2017), black cumin has great importance in immune systems and pulmonary diseases, including diabetes mellitus, breast cancer, dermatological complications, dyspepsia, osmotic balance, dehydration, etc. Moreover, according to the experiment on the gastroprotective effect of black cumin seed, it can be mainly exerted through the antioxidant activity of the extract as well as its role in stimulating gastric mucus secretion and increasing total hexose in the gastric mucosa (Paseban et al., 2019).

Aside from its flavor, the nutritional value of black cumin is well-known. According to the research results (Dubey et al., 2016), an investigation into the nutritional profile of black cumin EO indicated significant amounts of carbohydrates, proteins, and lipids (Dubey et al., 2016). The fatty acids found in fixed oil include myristic acid, myristoleic acid, palmitic acid, stearic acid, oleic acid, linoleic acid, linolenic acid, eicosenoic acid, arachidonic acid, behenic acid, and dihomolinolenic acid, as well as saturated and monounsaturated fatty acids and polyunsaturated fatty acids (Piras et al., 2013). Magnesium, phosphorus, calcium, and potassium are the most abundant nutrients in seed oil. Other minerals and elements found in small concentrations include sodium, manganese, zinc, iron, and copper (Sultan et al., 2009).

8.5 Conclusion

N. sativa and its volatile oil are unique among herbaceous plants. The amount and composition of *N. sativa* volatile oil vary depending on environmental and extraction conditions. Many variables govern variance, including seed origin, culture period, and regional circumstances. The volatile components of *N. sativa* L. are classified as monoterpenic hydrocarbons, monoterpenic alcohols, terpenic ethers, and monoterpenic phenols. The active ingredients in black cumin oil, such as tocopherol, phenolic compounds, and thymoquinone, are thought to be responsible for the oil's therapeutic benefits on human health. An olfactometric method can identify fragrance-active molecules in hundreds of aroma compounds. The GC-MS-O approach has enabled the classification of volatile compounds as aroma-active or nonaroma-active based on their concentration in samples. *N. sativa* L. contains a total of 13 odor-active chemicals, among which there were two alcohols, four carboxylic acids, and two phenols.

Acknowledgment

The authors are grateful to Fikadu Wako for his contribution to this work.

References

Aftab, A., Yousaf, Z., Javaid, A., Rabbani, A., Ahmed, S., Khan, F., 2018. *Nigella sativa L.* from traditional to contemporary medicine: a review. International Journal of Biology and Biotechnology 15 (2), 237–254.

Aggarwal, B.B., Kunnumakkara, A.B., Harikumar, K.B., Tharakan, S.T., Sung, B., Anand, P., 2008. Potential of spice-derived phytochemicals for cancer prevention. Planta Medica 74 (13), 1560–1569.

Alexander, H.C., McCarty, W.M., Bartlett, E.A., Syverud, A.N., 1982. Aqueous odor and taste threshold values of industrial chemicals. Journal AWWA 74 (11), 595–599. https://doi.org/10.1002/j.1551-8833.1982.tb05017.x.

Ali, B.H., Blunden, G., 2003. Pharmacological and toxicological properties of *Nigella sativa*. Phytotherapy Research 17 (4), 299–305. https://doi.org/10.1002/ptr.1309.

Al-Jassir, M.S., 1992. Chemical composition and microflora of black cumin *(Nigella sativa L.)* seeds growing in Saudi Arabia. Food Chemistry 45 (4), 239–242. https://doi.org/10.1016/0308-8146(92)90153-S.

Amin, B., Hosseinzadeh, H., 2016. Black cumin *(Nigella sativa)* and its active constituent, thymoquinone: an overview on the analgesic and anti-inflammatory effects. Planta Medica 82 (1–2), 8–16.

Ashraf, M., Ali, Q., Iqbal, Z., 2006. Effect of nitrogen application rate on the content and composition of oil, essential oil and minerals in black cumin *(Nigella sativa L.)* seeds. Journal of the Science of Food and Agriculture 86 (6), 871–876.

Ashrafur Rahman, S.M., Rainey, T.J., Ristovski, Z.D., Dowell, A., Islam, M.A., Nabi, M.N., Brown, R.J., 2019. Review on the use of essential oils in compression ignition engines. In: Agarwal, A.K., et al. (Eds.), Methanol and the Alternate Fuel Economy. Springer Singapore, Singapore, pp. 157–182.

Bakkali, F., Averbeck, S., Averbeck, D., Idaomar, M., 2008. Biological effects of essential oils–a review. Food and Chemical Toxicology 46 (2), 446–475.

Bassim Atta, M., 2003. Some characteristics of nigella *(Nigella sativa L.)* seed cultivated in Egypt and its lipid profile. Food Chemistry 83 (1), 63–68. https://doi.org/10.1016/S0308-8146(03)00038-4.

Benkaci–Ali, F., Baaliouamer, A., Meklati, B.Y., Chemat, F., 2007. Chemical composition of seed essential oils from Algerian *Nigella sativa* extracted by microwave and hydrodistillation. Flavour and Fragrance Journal 22 (2), 148–153. https://doi.org/10.1016/S0308-8146(03)00038-4.

Bordoni, L., Fedeli, D., Nasuti, C., Maggi, F., Papa, F., Wabitsch, M., De Caterina, R., Gabbianelli, R., 2019. Antioxidant and anti-inflammatory properties of *Nigella sativa* oil in human pre-adipocytes. Antioxidants (Basel) 8 (2).

Botnick, I., Xue, W., Bar, E., Ibdah, M., Schwartz, A., Joel, D.M., Lev, E., Fait, A., Lewinsohn, E., 2012. Distribution of primary and specialised metabolites in *Nigella sativa* seeds, a spice with vast traditional and historical uses. Molecules 17 (9), 10159–10177.

Burits, M., Bucar, F., 2000. Antioxidant activity of *Nigella sativa* essential oil. Phytotherapy Research 14 (5), 323–328. https://doi.org/10.1002/1099-1573(200008)14:5<323::AID-PTR621>3.0.CO;2-Q.

Burt, S., 2004. Essential oils: their antibacterial properties and potential applications in foods—a review. International Journal of Food Microbiology 94 (3), 223–253. https://doi.org/10.1016/j.ijfoodmicro.2004.03.022.

Buttery, R.G., Seifert, R.M., Guadagni, D.G., Ling, L.C., 1971. Characterisation of additional volatile components of tomato. Journal of Agricultural and Food Chemistry 19 (3), 524–529.

Chouhan, S., Sharma, K., Guleria, S., 2017. Antimicrobial activity of some essential oils-present status and future perspectives. Medicines (Basel, Switzerland) 4 (3), 58.

D'Antuono, L.F., Moretti, A., Lovato, A.F.S., 2002. Seed yield, yield components, oil content and essential oil content and composition of *Nigella sativa L.* and *Nigella damascena* L. Industrial Crops and Products 15 (1), 59–69.

Diwakar, Y., Harisha, C., Singh, B., Kakani, R., Saxena, S., 2018. Floral biology and reproductive behaviour of *Nigella sativa L.* var. Ajmer Nigella-1. Journal of Pharmacognosy and Phytochemistry 7 (Sp 3), 53–58.

Dubey, P.N., Singh, B., Mishra, B.K., Kant, K., Solanki, R.K., 2016. Nigella *(Nigella sativa)*: a high value seed spice with immense medicinal potential. Indian Journal of Agricultural Sciences 86 (8), 967–979.

Ekanem, J., Yusuf, O., 2008. Some biochemical and haematological effects of black seed *(Nigella sativa)* oil on Trypanosoma brucei-infected rats. African Journal of Biotechnology 7 (2), 153–157.

El-Ghorab, A.H., 2003. Supercritical fluid extraction of the Egyptian Rosemary (Rosmarinus officinalis) leaves and *Nigella sativa L.* seeds volatile oils and their antioxidant activities. Journal of Essential Oil Bearing Plants 6 (2), 67–77.

El-Morsy, M.H., Osman, H.E.-S., 2021. Morphological characters of *Nigella sativa.* In: Fawzy Ramadan, M. (Ed.), Black Cumin *(Nigella sativa)* Seeds: Chemistry, Technology, Functionality, and Applications. Springer International Publishing, Cham, pp. 23–29.

Ermumcu, M.Ş.K., Şanlıer, N., 2017. Black cumin *(Nigella sativa)* and its active aomponent of thymoquinone: effects on health. Journal of Food and Health Science 3 (4), 170–183.

Forouzanfar, F., Bazzaz, B.S.F., Hosseinzadeh, H., 2014. Black cumin *(Nigella sativa)* and its constituent (thymoquinone): a review on antimicrobial effects. Iranian Journal of Basic Medical Sciences 17 (12), 929–938.

García-González, D.L., Tena, N., Aparicio, R., 2007. Characterization of olive paste volatiles to predict the sensory quality of virgin olive oil. European Journal of Lipid Science and Technology 109 (7), 663–672. https://doi.org/10.1002/ejlt.200700056.

Ghahramanloo, K.H., Kamalidehghan, B., Akbari Javar, H., Teguh Widodo, R., Majidzadeh, K., Noordin, M.I., 2017. Comparative analysis of essential oil composition of Iranian and Indian *Nigella sativa L.* extracted using supercritical fluid extraction and solvent extraction. Drug Design, Development and Therapy 11, 2221–2226. https://doi.org/10.2147/DDDT.S87251.

Gharby, S., Harhar, H., Guillaume, D., Roudani, A., Boulbaroud, S., Ibrahimi, M., Ahmad, M., Sultana, S., Hadda, T.B., Chafchaouni-Moussaoui, I., Charrouf, Z., 2015. Chemical investigation of *Nigella sativa L.* seed oil produced in Morocco. Journal of the Saudi Society of Agricultural Sciences 14 (2), 172–177.

Guenther, E., 2014. The Essential Oils History — Origin in Plants — Production —Analysis, vol 1. Read Books Limited, New York.

Hadi, V., Pahlavani, N., Malekahmadi, M., Nattagh-Eshtivani, E., Navashenaq, J.G., Hadi, S., Ferns, G.A., Ghayour-Mobarhan, M., Askari, G., Norouzy, A., 2021. *Nigella sativa* in controlling Type 2 diabetes, cardiovascular, and rheumatoid arthritis diseases: molecular aspects. Journal of Research in Medical Sciences: The Official Journal of Isfahan University of Medical Sciences 26 p20.

Hailemichael, G., Kifelew, H., Mitiku, H., 2015. Spices research achievements, challenges and future prospects in Ethiopia. Academic Research Journal of Agricultural Science and Research 4 (1), 9–17.

Hamrouni-Sellami, I., Elyes Kchouk, M., Marzouk, B., 2008. Lipid and aroma composition of black cumin (*Nigella sativa L.*) seeds from Tunisia. Journal of Food Biochemistry 32 (3), 335–352.

Hannan, M.A., Rahman, M.A., Sohag, A.A.M., Uddin, M.J., Dash, R., Sikder, M.H., Rahman, M.S., Timalsina, B., Munni, Y.A., Sarker, P.P., Alam, M., Mohibbullah, M., Haque, M.N., Jahan, I., Hossain, M.T., Afrin, T., Rahman, M.M., Tahjib-Ul-Arif, M., Mitra, S., Oktaviani, D.F., Khan, M.K., Choi, H.J., Moon, I.S., Kim, B., 2021. Black cumin *(Nigella sativa L.)*: a comprehensive review on phytochemistry, health benefits, molecular pharmacology, and safety. Nutrients 13 (6), 1784.

Hassanien, M.F., Assiri, A.M., Alzohairy, A.M., Oraby, H.F., 2015. Health-promoting value and food applications of black cumin essential oil: an overview. Journal of Food Science and Technology 52 (10), 6136–6142. https://doi.org/10.1007/s13197-015-1785-4.

Heiss, A.G., Oeggl, K., 2005. The oldest evidence of *Nigella damascena L.* (Ranunculaceae) and its possible introduction to central Europe. Vegetation History and Archaeobotany 14 (4), 562–570.

Islam, M.T., Guha, B., Hosen, S., Riaz, T.A., Shahadat, S., Sousa, L.d.R., Santos, J.V.d.O., da Silva Júnior, J.J., de Lima, R.M.T., Braga, A.L., dos Reis, A.C., de Alencar, M.V.O.B., Melo-Cavalcante, A.A.d.C., 2017. Nigellalogy: a review on *Nigella sativa*. MOJ Bioequivalence & Bioavailability 3 (6), 167–181.

Jansen, P.C.M., 1981. Spices, condiments and medicinal plants in Ethiopia, their taxonomy and agricultural significance. In: C.f.A.P.a. Documentation, vol 3. Pudoc, Wageningen, p. 327.

Kabir, Y., Akasaka-Hashimoto, Y., Kubota, K., Komai, M., 2020. Volatile compounds of black cumin *(Nigella sativa L.)* seeds cultivated in Bangladesh and India. Heliyon 6 (10), e05343. https://doi.org/10.1016/j.heliyon.2020.e05343.

Kalidasu, G., Reddy, G.S., Kumari, S.S., Kumari, A.L., Sivasankar, A., 2017. Secondary volatiles and metabolites from *Nigella sativa L.* seed. Indian Journal of Natural Products and Resources (IJNPR) 8 (2), 151–158.

Kesen, S., Kelebek, H., Sen, K., Ulas, M., Selli, S., 2013. GC–MS–olfactometric characterisation of the key aroma compounds in Turkish olive oils by application of the aroma extract dilution analysis. Food Research International 54 (2), 1987–1994. https://doi.org/10.1016/j.foodres.2013.09.005.

Kesen, S., Amanpour, A., Tsouli Sarhir, S., Sevindik, O., Guclu, G., Kelebek, H., Selli, S., 2018. Characterisation of aroma-active compounds in seed extract of black cumin *(Nigella sativa L.)* by aroma extract dilution analysis. Foods 7 (7), 98.

Khan, I.A., Abourashed, E.A., 2011. Leung's Encyclopedia of Common Natural Ingredients Used in Food, Drugs, and Cosmetics, 3 ed. John Wiley & Sons, Inc, Hoboken, New Jersey.

Kifelew, H., Getachew, W., Luleseged, T., Mitiku, H., Bekele, D., Fikere, D., 2017. Seed Spices Production Guideline. Ethiopian Institute of Agricultural Research. Available from: http://197.156.72.153:8080/xmlui/bitstream/handle/123456789/1659/Seed%20Spices%20Manual%20English.pdf?sequence=1&isAllowed=y.

Kokoska, L., Havlik, J., Valterova, I., Sovova, H., Sajfrtova, M., Jankovska, I., 2008. Comparison of chemical composition and antibacterial activity of *Nigella sativa* seed essential oils obtained by different extraction methods. Journal of Food Protection 71 (12), 2475–2480. https://doi.org/10.4315/0362-028x-71.12.2475.

Kulloli, S.K., 2016. Redefining black cumins. In: Spice India. Spices Board India, Kochi, India, pp. 22–28. Available from: http://www.indianspices.com/magazines/english/si-monthly.html?language=English.

Lal, G., 2018. Scenario, importance and prospects of seed spices: a review. Current Investigations in Agriculture and Current Research 4 (2), 491–498.

Larsen, M., Poll, L., 1992. Odour thresholds of some important aroma compounds in strawberries. Zeitschrift für Lebensmittel-Untersuchung und Forschung 195 (2), 120–123.

Mahr, S., 2022. Love-in-a-Mist, *Nigella Damascena*. Available from: https://hort.extension.wisc.edu/articles/love-in-a-mist-nigella-damascena/. (Accessed 20 February 2022).

Majdalawieh, A.F., Fayyad, M.W., 2015. Immunomodulatory and anti-inflammatory action of *Nigella sativa* and thymoquinone: a comprehensive review. International Immunopharmacology 28 (1), 295–304. https://doi.org/10.1016/j.intimp.2015.06.023.

Mohammed, S.J., Amin, H.H.H., Aziz, S.B., Sha, A.M., Hassan, S., Abdul Aziz, J.M., Rahman, H.S., 2019. Structural characterization, antimicrobial activity, and in vitro cytotoxicity effect of black seed oil. Evidence-Based Complementary and Alternative Medicine 2019, 6515671.

Morales, M.T., Luna, G., Aparicio, R., 2005. Comparative study of virgin olive oil sensory defects. Food Chemistry 91 (2), 293–301.

Mukherjee, S., Mandal, A., Gupta, S., Datta, A.K., 2013. Pollination events in *Nigella sativa L.* (Black cumin). International Journal of Research in Ayurveda and Pharmacy 4 (3), 342–344.

Okoye, T.C., Uzor, P.F., Onyeto, C.A., Okereke, E.K., 2014. 18 - safe African medicinal plants for clinical studies. In: Kuete, V. (Ed.), Toxicological Survey of African Medicinal Plants. Elsevier, pp. 535–555. https://doi.org/10.1016/B978-0-12-800018-2.00018-2.

Paseban, M., Niazmand, S., Soukhtanloo, M., Meybodi, N.T., 2019. The preventive effect of *Nigella sativa* seed on gastric ulcer induced by indomethacin in rat. Journal of Herbmed Pharmacology 9 (1), 12–19.

Perveen, A., Qaiser, M., 2006. Pollen flora of Pakistan – L. Ranunculaceae. Pakistan Journal of Botany 38 (3), 409–509.

Piras, A., Rosa, A., Marongiu, B., Porcedda, S., Falconieri, D., Dessì, M.A., Ozcelik, B., Koca, U., 2013. Chemical composition and in vitro bioactivity of the volatile and fixed oils of *Nigella sativa L.* extracted by supercritical carbon dioxide. Industrial Crops and Products 46, 317–323. https://doi.org/10.1016/j.indcrop.2013.02.013.

Pop, R.M., Bocsan, I.C., Buzoianu, A.D., Chedea, V.S., Socaci, S.A., Pecoraro, M., Popolo, A., et al., 2020. Evaluation of the antioxidant activity of *Nigella sativa L.* and Allium ursinum extracts in a cellular model of doxorubicin-induced cardiotoxicity. Molecules 25 (22).

Pop, R.M., Sabin, O., Suciu, Ş., Vesa, S.C., Socaci, S.A., Chedea, V.S., Bocsan, I.C., Buzoianu, A.D., 2020. *Nigella sativa's* anti-inflammatory and antioxidative effects in experimental inflammation. Antioxidants (Basel) 9 (10), 921. https://doi.org/10.3390/antiox9100921.

Rahman, S.M.A., Fattah, I.M.R., 2021. Evaluation of a compression ignition engine performance and emission characteristics using diesel-essential oil blends of high orange oil content. Australian Journal of Mechanical Engineering 1–8. https://doi.org/10.1080/14484846.2021.1913870.

Rohman, A., Lukitaningsih, E., Rafi, M., Nurrulhidayah, A.F., Windarsih, A., 2019. *Nigella sativa* oil: physico-chemical properties, authentication analysis and its antioxidant activity. Food Research 3 (6), 628–634.

Salehi, B., Quispe, C., Imran, M., Ul-Haq, I., Živković, J., Abu-Reidah, I.M., Sen, S., Taheri, Y., Acharya, K., Azadi, H., Del Mar Contreras, M., Segura-Carretero, A., Mnayer, D., Sethi, G., Martorell, M., Abdull Razis, A., Sunusi, U., Kamal, R.M., Rasul Suleria, H.A., Sharifi-Rad, J., 2021. Nigella plants - traditional uses, bioactive phytoconstituents, preclinical and clinical studies. Frontiers in Pharmacology 12, 625386.

Selli, S., Kelebek, H., Ayseli, M.T., Tokbas, H., 2014. Characterisation of the most aroma-active compounds in cherry tomato by application of the aroma extract dilution analysis. Food Chemistry 165, 540–546.

Shaaban, H.A., Sadek, Z., Edris, A.E., Saad-Hussein, A., 2015. Analysis and antibacterial activity of *Nigella sativa* essential oil formulated in microemulsion system. Journal of Oleo Science 64 (2), 223–232. https://doi.org/10.5650/jos.ess14177.

Shabana, A., El-Menyar, A., Asim, M., Al-Azzeh, H., Al Thani, H., 2013. Cardiovascular benefits of black cumin *(Nigella sativa)*. Cardiovascular Toxicology 13 (1), 9–21. https://doi.org/10.1007/s12012-012-9181-z.

Shakeri, F., Gholamnezhad, Z., Mégarbane, B., Rezaee, R., Boskabady, M.H., 2016. Gastrointestinal effects of *Nigella sativa* and its main constituent, thymoquinone: a review. Avicenna Journal of Phytomedicine 6 (1), 9–20.

Shariq, I.M., Israil, A.M., Iqbal, A., Brijesh, P., 2015. Morpho-physiological characterization of seeds and seedlings of *Nigella sativa* Linn.: study on Indian germplasm. International Research Journal of Biological Sciences 4 (4), 38–42.

Singh, G., Marimuthu, P., de Heluani, C.S., Catalan, C., 2005. Chemical constituents and antimicrobial and antioxidant potentials of essential oil and acetone extract of *Nigella sativa* seeds. Journal of the Science of Food and Agriculture 85 (13), 2297–2306.

Singh, S., Das, S.S., Singh, G., Schuff, C., de Lampasona, M.P., Catalán, C.A.N., 2014. Composition, in vitro antioxidant and antimicrobial activities of essential oil and oleoresins obtained from black cumin seeds *(Nigella sativa L.)*. BioMed Research International 2014, 918209.

Sovová, H., Komers, R., Kucera, J., Jez, J., 1994. Supercritical carbon dioxide extraction of caraway essential oil. Chemical Engineering Science 49 (15), 2499–2505.

Srinivasan, K., 2018. Cumin (Cuminum cyminum) and black cumin *(Nigella sativa)* seeds: traditional uses, chemical constituents, and nutraceutical effects. Food Quality and Safety 2 (1), 1–16.

Sultan, M.T., Butt, M.S., Anjum, F.M., Jamil, A., Akhtar, S., Nasir, M., 2009. Nutritional profile of indigenous cultivar of black cumin seeds and antioxidant potential of its fixed and essential oil. Pakistan Journal of Botany 41 (3), 1321–1330.

Sultana, S., Asif, H.M., Akhtar, N., Iqbal, A., Nazar, H., Rehman, R.U., 2015. *Nigella sativa:* monograph. Journal of Pharmacognosy and Phytochemistry 4 (4), 103–106.

Takruri, H.R.H., Dameh, M.A.F., 1998. Study of the nutritional value of black cumin seeds *(Nigella sativa L)*. Journal of the Science of Food and Agriculture 76 (3), 404–410. https://doi.org/10.1002/(SICI)1097-0010(199803)76:3<404::AID-JSFA964>3.0.CO;2-L.

United States Department of Agriculture (USDA), 2022. *Nigella Sativa L.* Available from: https://plants.sc.egov.usda.gov/home/plantProfile?symbol=NISA2. (Accessed 20 February 2022).

Wajs, A., Bonikowski, R., Kalemba, D., 2008. Composition of essential oil from seeds of *Nigella sativa L.* cultivated in Poland. Flavour and Fragrance Journal 23 (2), 126–132.

Yarnell, E., Abascal, K., 2011. Nigella sativa: holy herb of the Middle East. Alternative and Complementary Therapies 17 (2), 99–105.

Yimer, E.M., Tuem, K.B., Karim, A., Ur-Rehman, N., Anwar, F., 2019. *Nigella sativa L.* (black cumin): a promising natural remedy for wide range of illnesses. Evidence-based Complementary and Alternative Medicine 2019, 1528635. https://doi.org/10.1155/2019/1528635.

CHAPTER NINE

Effect of different food processing techniques on the composition of black cumin seed and seed oil

Richard Osae[1], Maurice Tibiru Apaliya[1], Emmanuel Kwaw[1], Newlove Akowuah Afoakwah[4], Patrick Owusu-Ansah[3], Raphael N. Alolga[2]

[1]Cape Coast Technical University, School of Applied Science and Technology, Department of Food and Postharvest Technology, Cape Coast, Ghana
[2]State Key Laboratory of Natural Medicines, Department of Pharmacognosy, China Pharmaceutical University, Nanjing, China
[3]University of Development Studies, Department of Food Processing Technology, Tamale, Ghana
[4]Department of Food Science and Technology, Faculty of Agriculture, Food and Consumer Sciences, University for Development Studies, Tamale, Ghana

9.1 Introduction

The black cumin seed (*Nigella sativa* L.) is a perennial plant that shares same family (Zingiberaceae) with other species such as *Curcuma raktakanta* Mangaly & M.Sabu, *Curcuma longa* L., *Curcuma aromatica* Salisb., *Curcuma zedoaria* Rosc., and *Curcuma amada* Roxb (Fig. 9.1) (Soleimanifar et al., 2019). Among these species, the black cumin seed has been established to contain various medicinal and pharmacological properties including antiinflammatory, anticancer, antifungal, antiviral, antiangiogenetic, antibacterial, and immunomodulatory effects (Jiang et al., 2021). The black cumin seed

Figure 9.1 Various varieties of cumin seeds (Srinivasan and safety 2018).

Biochemistry, Nutrition, and Therapeutics of Black Cumin Seed
ISBN: 978-0-323-90788-0
https://doi.org/10.1016/B978-0-323-90788-0.00008-1

and seed oil composition is generally valued in terms of its medicinal use and applicability. Its inclusion in different traditional medicines is due to its broad range of pharmacological and biological activities such as antidiabetic, anti-oxidative, and antitumor properties (Fig. 9.2). According to Hidayati et al. (2020), the high consumption of black cumin seed oil has no adverse effect or any toxic impact on the immune system.

The composition of the black cumin seed and its oil has received much attention in recent times and found to contain bioactive compounds and possess antioxidant properties and other numerous health benefits. These comprise treatment and/or management of skin conditions, asthma, high cholesterol level, weight gain, high blood sugar level, and maintenance of mental health (Hidayati et al., 2020). The black cumin seed and seed oil are active in the treatment and prevention of cardiovascular diseases (oxidative heart wound, cardiomyopathy, cardiac hypertrophy, and myocardial infarction) and vascular dysfunctions. They also offer protection to the nervous system (brain) and found to possess activity against Parkinson's and Alzheimer's diseases by virtue of their antioxidants and antiinflammatory activities (Rafiee et al., 2019). The black cumin seed and seed oil have also been established to advance wound healing (Espanany et al., 2016).

In the food industry and sector, black cumin seeds and seed oil are extensively used as natural food coloring additives. Other investigators (Hidayati et al., 2020; Soleimanifar et al., 2019) have reported that natural cumin has an attractive (orange or bright yellow) color and can be a substitute for artificial food coloring agent. According to Jiang et al. (2021), the black cumin seed and oil are usually used in enhancing the quality of food including canned fish, rice, pastries, meat, dairy products, and mustard. Due to its hot spicy and strong taste, it constitutes part of the ingredients for bread, tea, casseroles, and coffee. Additionally, black cumin can also serve as a natural preservative to extend the storage life of food. Its preservatory potential

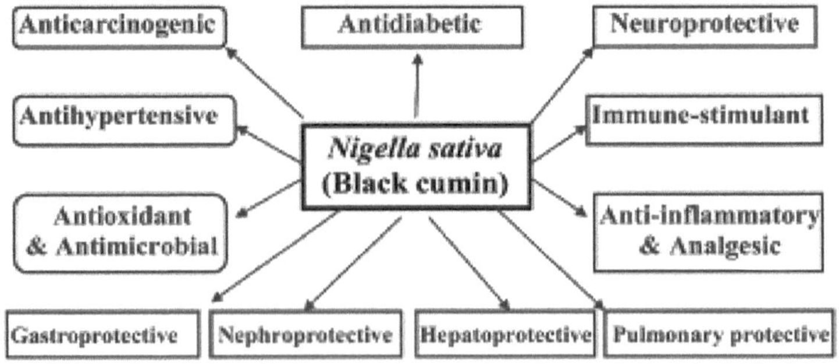

Figure 9.2 Medicinal properties of black cumin seed oils (Srinivasan and safety 2018).

for food items such as minced meat, cooked mutton, bean curd, and bread has been ascribed to its high antimicrobial activity against several microbes such as *Salmonella typhimurium*, *Escherichia coli*, *Listeria monocytogenes*, and *Staphylococcus aureus* (Altunatmaz et al., 2016).

According to Friedrich et al. (2015), the black cumin seed and oil are used in the cosmetic industry for decades due to their antiinflammatory and antioxidant properties. It has shown promise in a wide range of beauty treatments for the skin, nails, face, lips, hair, and offers protection against ultraviolet light, inflammation, hair loss, aging, nail care, and lip care. The black cumin seed also possesses antiaging, sunscreen, antiwrinkle, and moisture preservatory properties that are crucial in the cosmetics industry.

The effect of various food processing techniques on the composition of black cumin seed and seed oil is very paramount, as a bad choice will have adverse effects on the value and the consistency of same. Processing (extraction and purification techniques) of black cumin seed is an essential step that could preserve and prolong the storage life of these commodity as this will minimize the deterioration and moisture content resulting in the enhancement of the product value (oil composition) (Jiang et al., 2021). The first and most important step in recovering cumin from plant materials is extraction. All extraction methods have the following primary goals: (a) recover targeted compounds from plant materials, (b) enhance extraction selectivity, (c) increase extraction performance, and (d) provide a reliable and repeatable extraction process. The black cumin seed is usually subjected to various traditional or conventional food processing methods such as Soxhlet extraction, mechanical oil extraction, hydro/steam distillation, maceration, or solvent extraction. These methods are simple, but are usually nonselective, time-consuming, and lead to the degradation of heat-sensitive components (Zhang et al., 2019). However, the conventional food processing methods are reported to possess various merits in terms of enhancing the quality, nutritional properties, and functional compounds of the finished product (Espanany et al., 2016). Hence, the need for effective alternative food processing techniques that fast and conserve the product quality has become paramount in this regard.

Novel extraction methods, such as ultrasound-assisted extraction (UAE), microwave-assisted extraction (MAE), enzyme-assisted extraction (EAE), pulsed electric field—assisted extraction, cold pressing and supercritical liquid extraction, have emerged as effective alternatives to traditional extraction methods. These cutting-edge methods, which are environmentally friendly green innovations, have the ability to extract cumin from plant materials.

This chapter reviews: (1) the merits and demerits of various traditional or conventional food processing techniques on the composition of black cumin seed and seed oil, (2) the benefits and drawbacks of various novel or advanced food processing methods on the composition of black cumin seed and seed oil, and (3) the various purification processes of black cumin seed and seed oil.

9.2 Traditional or conventional processing techniques of black cumin seed and seed oil

The processes of oil extraction and processing are engineered to produce high-quality oil with minimal undesirable ingredients, maximum extraction yields, and high-value meal. Oil can be extracted from oilseeds, particularly black cumin seeds using a variety of traditional processing methods. Solvent extraction, Soxhlet extraction technology, hydro/steam distillation, and mechanical extraction with a screw press are the most popular oilseed food processing techniques and extraction methods.

9.2.1 Solvent extraction technology

Solvent extraction is a method of separating compounds based on their relative solubility. This method entails the use of a solvent, which is a liquid capable of dissolving another material (Chan et al., 2020). Hazardous pollutants are separated from sludge and sediments using solvent extraction technology. The method of solvent extraction of solid samples, also known as "solid-liquid extraction" (also known as "maceration" or "soaking"), is well-known and widely used. The black cumin seed has been extracted from plants using a variety of solvents, including nonpolar organic solvents and a mixture of organic solvents and water (Paulucci et al., 2013; Jiang et al., 2021). This is the most effective method for extracting oil from oilseeds. The efficiency of the process is determined by the oilseed preparation before extraction, the temperature, and the mode of operation as well as the design of the equipment (Williams 2013). Savoire et al. (2013b) revealed that after industrial solvent extraction, residual oil in the meal is predicted to be less than 1%. The choice of the solvent used is determined by the solubility of the oil in same, as well as cost and protection. Pentane (boiling point 88−97°F), hexane (boiling point 146−156°F), heptanes (boiling point 194−210°F), and octane (boiling point 215−264°C) are light paraffinic petroleum fractions that can be used for oil extraction. For the isolation of curcumin from *C. longa* L., Popuri and Pagala (2013) compared extraction solvents (isopropanol, ethanol, methanol, ethyl acetone, and acetone). When extraction with ethanol was performed at 30°C for 1 h with a solid to solvent ratio of 1:8, the maximum yield (0.26 mg/10 g) was obtained. In agreement with this outcome, Shirsath et al. (2017) established that ethanol was the most favorable organic solvent for extracting black cumin seed oil. Hexane is commonly utilized in the extraction of vegetable oils. The ethanol content in the solvent combination, on the other hand, is critical in determining extraction efficiency. The schematic diagram of solvent extraction is presented in Fig. 9.3A.

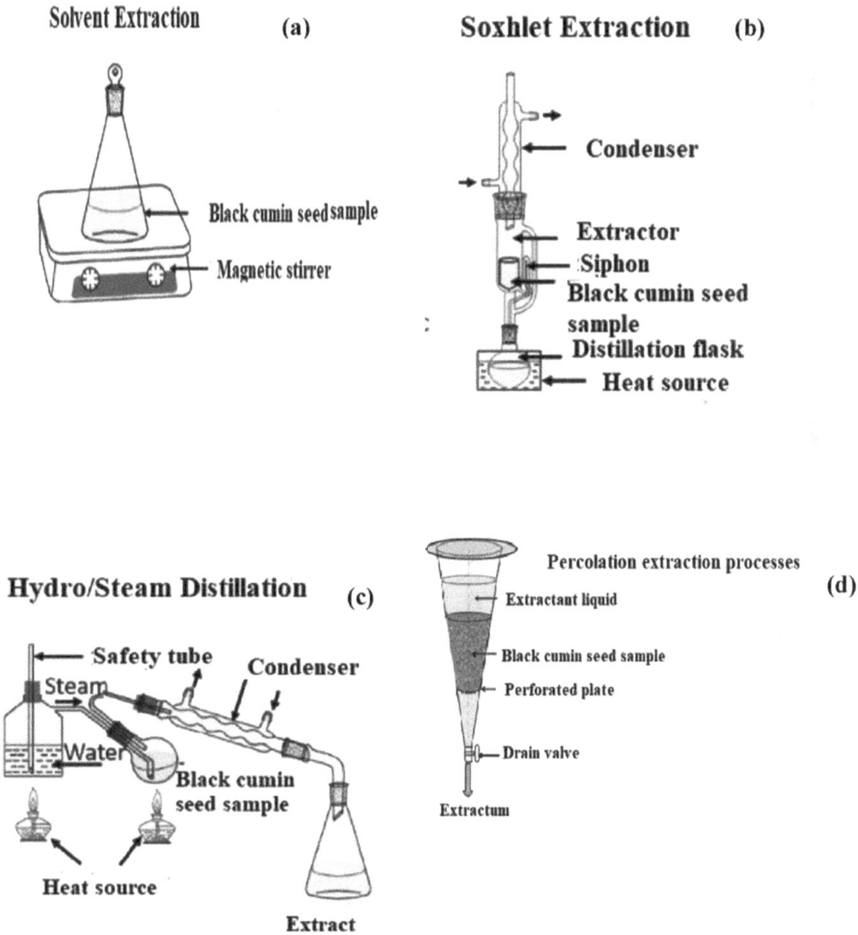

Figure 9.3 Some conventional food processing techniques on the composition of black cumin seed oil (Jiang et al., 2021).

9.2.1.1 Advantages of solvent extraction technology

1. Low energy usage
2. Capacity of large production
3. It is very fast
4. Efficient and targeted
5. Low-cost and straightforward.
6. Produces a high-quality extract
7. Product deterioration is minimized, and solvent residues are eliminated
8. Continuous operation and automation are simple (Crini and Lichtfouse 2019).

9.2.1.2 Disadvantages of solvent extraction technology

1. It is time-consuming
2. The high temperatures utilized may cause compounds to change
3. It is very expensive
4. Requires technological expertise
5. The flammability of these solvents is a significant disadvantage
6. In order to avoid fires and reduce the risk of explosions in the extraction plants, strict procedures must be followed (Crini and Lichtfouse 2019).

9.2.2 Soxhlet extraction technology

Soxhlet extraction is a time-consuming extraction method used to extract analytes that are thermally stable. The extraction solvent is cycled by virtue of the continual boiling and condensation of the same (Azwanida, 2015). This method was established by a German scientist in 1879 and remains the gold standard and reference method for solid—liquid extraction of bioactive compounds from plants. Several studies have reported on the extraction of black cumin seeds oils from plants using a Soxhlet extractor (Shirsath et al., 2017; Azwanida, 2015). During Soxhlet extraction, the sample is placed in a thimble that is gradually filled with condensed fresh extractant (the term used to refer to the solvent used for extraction) from a distillation flask as presented in Fig. 9.3B (López-Bascón and De Castro, 2020). A siphon aspirates the liquid from the thimble and returns it to the distillation flask, delivering the extracted analytes back into the bulk liquid. The procedure is repeated until the extraction is complete. Soxhlet is a hybrid continuous—discontinuous technology as a result of this performance. The assembly can be considered a batch system because the extractant acts in steps; however, because the extractant is recirculated through the sample. For example, Shirsath et al. (2017) examined the yield of black cumin seed oil by Soxhlet extraction, maceration extraction, and UAE and discovered that Soxhlet extraction produced more seed oil than the other two methods. Compared to UAE, Soxhlet extraction was carried out at a much higher temperature and for a relatively prolonged time. High-pressure Soxhlet extraction, automated Soxhlet extraction, ultrasound-assisted Soxhlet extraction, and microwave-assisted Soxhlet extraction have all been developed to increase the performance of the primary technology (i.e., Soxhlet extraction) (Vinatoru et al., 2017).

9.2.2.1 Advantages of soxhlet extraction technology

1. The sample is brought in connection with fresh sections of the extract on a regular basis, which helps to shift the mass transfer equilibrium.
2. Because some of the heat provided to the distillation flask enters the extraction cavity, the system's temperature remains rather high.
3. After the leaching step, there is no need for filtration.
4. Because the fundamental equipment is affordable, simultaneous parallel extraction can boost sample throughput.

9.2.2.2 Disadvantages of soxhlet extraction technology

1. The lengthy extraction process and the large amount of extractant waste generated, which is not only costly to dispose of but can also cause additional environmental issues.
2. Samples are often extracted at the extractant's boiling point for an extended length of time, and the risk of thermolabile compounds thermal breakdown must be overlooked.
3. The traditional Soxhlet extractor lacks agitation, which would speed up the process.
4. Due to the huge volume of solvent utilized, an evaporation/concentration phase is required after extraction.
5. The technology is limited to solvent selectivity and is difficult to automate.

9.2.3 Hydro/steam distillation

Hydro/steam distillation is another prominent approach for extracting bioactive compounds and essential oils. It is an old method of extracting plant components without the use of organic solvents. It has been used to obtain flavorless curcumin from crude cumin powder or cumin oleoresin (Jiang et al., 2021). According to Azmir et al. (2013), this method constitutes one of the popular traditional methods for extracting bioactive chemicals and seed oil from plants. Plant materials are packed in a still compartment, then an appropriate amount of water is added and the mixture is brought to its boiling point. Direct steam can also be injected directly into the plant sample. As shown in Fig. 9.3C, plant materials are packed in a still compartment and enough water is added and the temperature of the set-up increased to the boiling point of the water. For the separation of essential oils from plant materials such as black cumin seed, forms of hydrodistillation are applicable.

1. *Water distillation*

   The material is totally immersed in water, which is then boiled using direct heat, a steam jacket, a closed steam jacket, a closed steam coil, or an open steam coil. The fundamental feature of this method is that boiling water comes into direct contact with the plant material (Srivastava et al., 2021).

- *Water and steam distillation*

  Steam can be created in a satellite boiler or within water and steam distillation, but it must be separated from the plant material. Water and steam distillation are extensively employed in rural regions. The equipment is also identical to that used in water distillation, with the exception that the plant material is held above the boiling water on a perforated grid. It is not uncommon for people who start with water distillation to move on to water and steam distillation (Yadav et al., 2016).

- *Direct steam distillation*

  The process of distilling plant material with steam generated outside steam generator, sometimes known as a boiler, is known as direct steam distillation. Like in water and steam distillation, the plant material is supported on a perforated grid above the steam input. The amount of steam produced through satellite steam generation may be easily adjusted and constitutes a significant advantage (Rasul, 2018). Because steam is created in a satellite boiler, the plant material is heated to no more than 100°C and, as a result, thermal deterioration should ideally not occur. The most widely established method for producing black cumin seed oil on a large scale is steam distillation. This is a widely used method in the taste and fragrance industry (Gari et al., 2020).

9.2.3.1 Advantages of hydro/steam distillation

1. Increased oil yield.
2. The volatile oil's components are generally less vulnerable to hydrolysis and polymerization (the control of wetness on the bottom of the still affects hydrolysis, whereas the thermal conductivity of the still walls affects polymerization).

3. The loss of polar molecules is minimized when refluxing is controlled.
4. The oil quality obtained through steam and water distillation is more consistent.
5. Because no organic solvent is required, this technique is both cost-effective and environmentally friendly (Rasul, 2018).

9.2.3.2 Disadvantages of hydro/steam distillation

1. There is no way to extract all of the information.
2. The plant material near the bottom of the still may scorch when it comes into direct contact with the fire from the furnace, imparting an unpleasant odor to the essential oil.
3. The prolonged action of hot water might result in the hydrolysis of some essential oil compounds, such as esters.
4. Heat management is challenging and could result in varying distillation rates.
5. This technique necessitates a large number of stills, more room, and fuel. As a result, the procedure becomes unprofitable (López-Bascón and De Castro 2020; Rasul, 2018).

9.2.4 Maceration, infusion, percolation, and decoction technology

Maceration is a winemaking technique that has been widely adopted and employed in medicinal plant research. It is a traditional method of preparing medicines. It is a frequently used and low-cost method of obtaining natural compounds from plant material. Plant material (coarse or powdered) was soaked in a stoppered container with a solvent and left to stand at room temperature for at least 3 days with frequent agitation (Rasul 2018). The technique was designed to soften and shatter the plant's cell wall, allowing the soluble phytochemicals and oils to be released. The mixture is pressed or strained by filtration after 3 days. Heat is conveyed through convection and conduction in this traditional approach, and the solvents used depend on the type of substance recovered from the samples. Infusion and decoction are similar to maceration in that in both cases the plant material is steeped in cold or boiling water. The sample is cooked in a certain volume of water (e.g., 1:4 or 1:16) for a defined time in the case of decoction, while the maceration period for infusion is shorter (Mao et al., 2019). When compared to maceration and infusion, decoction is only suitable for extracting heat-stable phytochemicals from hard plant materials (e.g., roots and barks) and usually results in the production of more oil-soluble compounds.

Percolation is another technique with a similar core concept that uses a specialized equipment known as a percolator. The percolator is filled with dried powdered samples, which are then added to boiling water and macerated for 2 h. To achieve a concentrated extract, the percolation procedure is normally done at a modest rate (e.g., 6 drops/min) until the extraction is complete (Azwanida 2015). The percolation extraction process is depicted in Fig. 9.3D.

9.2.4.1 Advantages of maceration, infusion, percolation, and decoction processing technology

1. Maceration is a simple procedure that requires no special utensils or equipment.
2. No need for a skilled operator.
3. Its conserves energy.
4. It is suitable for specific chemicals that are not highly soluble in solvent and just need to be in contact with it for a long time.
5. Appropriate approach for less powerful and inexpensive medicines.

9.2.4.2 Disadvantages of maceration, infusion, percolation, and decoction processing technology

1. Unfortunately, the extraction processes take a long time and can take up to a week.
2. It does not fully remove the medication.
3. It is a time-consuming and slow process.
4. A lot of solvent required is required for the processes.

9.2.5 Mechanical oil extraction and processing technology

The use of mechanical presses/expellers is the most conventional method of oil processing and extraction. The forceful extraction of oil from oleaginous material under applied pressure is referred to as mechanical expression (hydraulic or screw presses) (Savoire et al., 2013a). The first mechanical oil expression equipment utilized for oilseed processing was lever and wedge press. A hydraulic press was used in the first cottonseed oil mill built in the United States in the 1920s. Between the head block and the ram of the press, seeds in filter bags were manually placed into perforated, horizontal boxes. By adding hydraulic pressure to the ram, the boxes were forced together and the oil squeezed out of the filter bag. The used cake filter bag was then manually removed from the hydraulic press. Cages were utilized instead of filter bags in later hydraulic presses. Until the 1950s, hydraulic presses were in use (Brännström et al., 2018). Continuous screw presses and continuous solvent extraction plants, which need less labor, are currently replacing them. Hydraulic presses are still used in the olive oil industry today. There is a vertical feeder and a horizontal screw with increasing body diameter advance along the length of the press to exert pressure on the oilseeds. The screw's barrel includes slots along its length, allowing for the rising internal pressure to first release air and subsequently drain the oil through the barrel (Matthäus, 2012; Di Giovacchino, 2013). The oil is collected in a trough beneath the screw, and the deoiled cake is ejected at the screw's end. In general, there are two primary phases in oil extraction: pressing (solid—liquid expression) and solvent extraction. These operations are preceded by mechanical and thermal pretreatments, which help to improve their performance. The oil recovery procedures differ depending on the seed species. Figs. 9.4 and 9.5 depict the device for mechanical oil processing and some solid—liquid expression changes in oilseed processing. Continuous screw presses are used in the industrial pressing of oilseeds. Seeds are put into the presses either in their raw forms or pretreated. Different seed species require different types of pretreatments (thermal pretreatment, size reduction, mechanical sifting, and so on) (Rasul 2018; Savoire et al., 2013a).

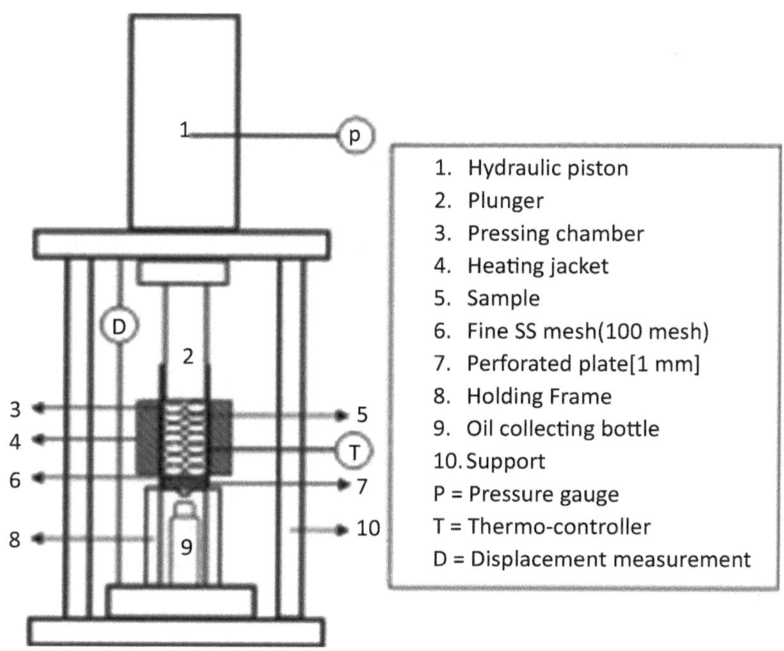

Figure 9.4 Mechanical oil extraction device (Subroto et al., 2015).

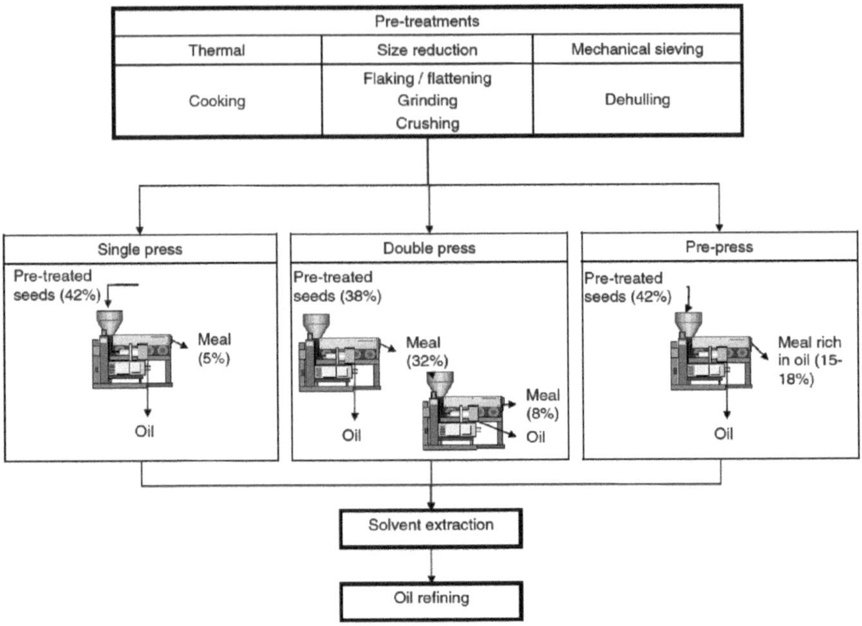

Figure 9.5 Different oil seed processes (Savoire et al., 2013b).

9.2.5.1 Advantages of mechanical oil extraction and processing technology

The advantages of pure mechanical extraction and processing technology are:

1. Lower initial investment costs
2. Higher-value edible oil and meal
3. Safer and less complex operation

9.2.5.2 Disadvantages of mechanical oil extraction and processing technology

1. High power requirements
2. Wear and tear on the equipment that requires maintenance, and Lower yield due to residual oil in the meal (Subroto et al., 2015)

9.3 Current/advanced food processing techniques on the composition of black cumin seed oil, their advantages and disadvantages

9.3.1 Ultrasound-assisted extraction in the processing of black cumin seed oil

Ultrasound is based on the idea of tandem compression and expansion, which is generated by sound waves with frequencies ranging from 20 kHz to 100 MHz. Ultrasonic waves produce microbubbles, which burst, causing cavitation, resulting in severe shear stresses, shock waves, macroturbulences, micromixing, and acoustic streaming (Zhang et al., 2019). During the sonication process, motion energy is transformed to heat, which serves as an activation energy for numerous physicochemical reactions that aid mass transfer (Alolga et al., 2021).

UAE was created based on this principle to release favorable and desirable chemicals from plant materials by enhancing mass transfer and speeding up the interaction of the solvent with the target compounds. Temperature and time, as well as the properties of the extraction solvent and the characteristics of the plant material, all have an impact on cumin seed oil yield. To this end, it is a good idea to fine-tune the UAE method before using it to recover cumin oils (Chemat et al., 2017). Previous investigation by Shirsath et al. (2017) revealed that the effect of operating parameters on black cumin seeds and cumin extraction yield using UAE was investigated, and the optimum parameters for such extraction were found to be 35°C, 1: 25 (g/mL) solid/solvent ratio, particle size of 0.09 mm, 1 h extraction time, 250 W ultrasonic power, and 22 kHz ultrasound frequency with ethanol as the solvent. The yield of 72% was significantly higher than that produced via solvent extraction or maceration/soaking, which only resulted in a yield of about 62% after 8 h. The results demonstrated that it could be a speedier and more sustainable alternative to standard procedures.

Furthermore, pulsed UAE has been employed for black cumin seed oil extraction and has proven to be an excellent alternative for obtaining oil with a high extraction rate and less energy use. The duration and interval time of pulsed UAE have a significant impact on the extraction process (Li et al., 2014). Furthermore, it was discovered that the oil yield in the pulsed UAE (1.03 0.02 g/100 g dry weight) was higher compared to UAE (0.82 0.01 g/100 g dry weight). These findings therefore suggested that a reduction in the pulsed length or interval time can lead to a high yield of seed oil under ideal conditions. Fig. 9.6A depicts a synopsis of the UAE oil extraction technique.

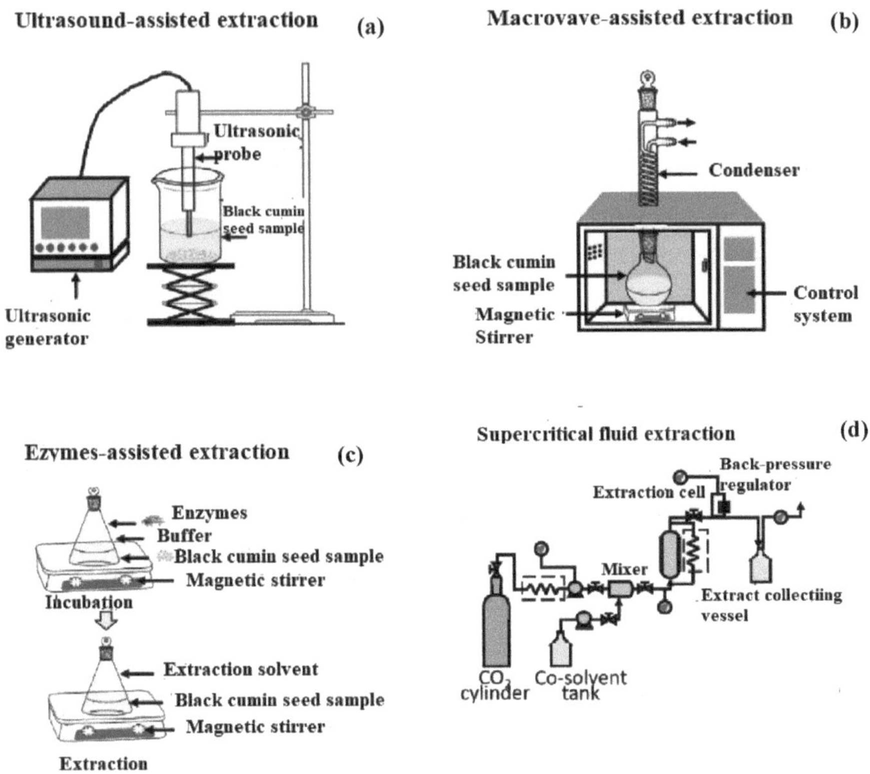

Figure 9.6 Schematic diagram of novel extraction techniques: (A) ultrasound-assisted extraction (UAE), (B) microwave-assisted extraction (MAE), (C) enzyme-assisted extraction technology, and (D) supercritical fluid extraction (SFE) (Mushtaq et al., 2020; Vinatoru et al., 2017).

9.3.1.1 Advantages of ultrasound-assisted extraction of black cumin seed oil

UAE is a more efficient process that enables the use of alternative green solvents and the manufacturing of high-quality products. Ultrasound technology has a number of features that make it a viable method for obtaining high oil extraction yields. This approach benefits both consumers and oil millers since it enables the production of high-quality oils. The advantages of UAE are mostly related to the reduction in the extraction time and solvent use (Mushtaq et al., 2020; Azwanida 2015).

9.3.1.2 Disadvantages of ultrasound-assisted extraction of black cumin seed oil

The use of ultrasound energy more than 20 kHz may have a negative effect on the active phytochemicals through the formation of free radicals. The main disadvantages of UAE include long extraction time and losses of volatile compounds (Osae et al., 2020).

9.3.2 Microwave-assisted extraction of black cumin seed oil

Microwave energy is used in MAE to help investigate partition from the sample matrix into the solvent. When microwave radiation interacts with the dipoles of polar and polarizable materials (such as solvents and samples), heat is generated near the surface of the materials, which is then transported by conduction (Destandau et al., 2013).

In the last few years, black cumin seeds and its oil have been extracted utilizing the MAE method under a range of working conditions. Microwave power, time intervals (irradiation time and cooling time), temperature, plant material size, and solvent type are among the important variables investigated by researchers for black cumin seed oil extraction. To improve extraction rate and efficiency, dry microwave irradiation (which promotes cell wall disintegration) and powdered sample preparation (which makes the sample more receptive to microwave heating) are sometimes combined with the MAE technique (Mandal et al., 2008; Wakte et al., 2011).

Bakhshabadi et al. (2017) investigated the impact of MAE on the physicochemical characteristics and bioactivity of black cumin seed oil. Different processing time of 90, 180, and 270 s, as well as microwave power of 180, 540, and 900 W were used to optimize the oil extraction process. They showed that increasing the microwave power and processing time boosted the extraction efficiency and the oil's oxidative stability value. Furthermore, the MAE approach proved more efficient than ultrasonic and supercritical carbon dioxide—assisted extraction (in terms of percent yield and required extraction time) (Li et al., 2014). A schematic summary of MAE process is shown in Fig. 9.6B.

9.3.2.1 Advantages of microwave-assisted extraction of black cumin seed oil

MAE exhibits certain advantages over traditional procedures, such as a combined reduction in extraction time, solvent and energy consumption, and a greater yield. The MAE improves analytical recovery and repeatability; however, it was advised to use suitable settings to minimize heat degradation (Mena-García et al., 2019).

9.3.2.2 Disadvantages of microwave-assisted extraction

Phenolic compounds such as the quercetin, isoflavin, and trans-resveratrol are destroyed during processing with microwave heating conditions of up to 100°C for 20 min. Additionally, MAE cycles could lead to a significant reduction in the levels of phenolics and flavanones, owing to their oxidation (Azwanida 2015). Because tannins and anthocyanins are prone to temperature degradation, they may not also be suitable for MAE.

9.3.3 Enzyme-assisted extraction technology of black cumin seed oil

According to Sahne et al. (2017), plant cell walls are made up of a series of complex structural polysaccharides that give them stability and make them resistant to the extraction of their internal components. Enzymes with unique hydrolytic capabilities are used to break down the matrix of the plant cell wall in order to gain access to the bioactive components within the cytosolic regions and even those bound to the cellular walls. In EAE, enzymes such as amylase, glucoamylase, and amyloglucosidase are used to extract black cumin seed oil.

This method is gaining popularity since it is effective, simple, long-lasting, and environmentally friendly. It is vital to remember that temperature, time, pH, and enzyme concentration all have an impact on enzyme specificity and selectivity. As a result, optimum enzyme reaction conditions are required to achieve maximum hydrolytic activity (Kujawa et al., 2021). Kurmudle et al. (2013) found that using 3% amylase at pH 5.0 after five hours of incubation and eight hours of extraction using acetone as the extraction solvent increased black cumin seed oil yield by 26.04%. Under same conditions using 2% glucoamylase at pH 4.5, they found a 31.83% increase in the yield of black cumin seed oil using the single factor approach. Sahne et al. (2016) who improved upon the process to produce the highest black cumin seed oil yield discovered that by treating black cumin seed with both amylase and amyloglucosidase resulted in a 4.1% cumin yield. EAE technology is depicted in Fig. 9.6C.

9.3.3.1 Advantages of enzyme-assisted extraction technology

Increased extraction yield, enhanced extract quality, green extraction (due to natural water and enzyme applications), the ability to scale up the process, and reduced extract filtering and purification are all advantages of EAE (Gerschenson et al., 2019).

9.3.3.2 Disadvantages of enzyme-assisted extraction technology

1. For large-scale industrial production, enzymes are relatively expensive.
2. The enzymes that are currently available are unable to completely break down the plant cell walls.
3. Because enzymes are rigorously limited by environmental conditions, EAE might not always be applicable on an industrial scale.

9.3.4 Supercritical fluid extraction of black cumin seed oil

Supercritical fluid extraction (SFE) is the most common nonconventional technique used industrially on a large scale that utilizes supercritical fluids as extraction solvents. Supercritical carbon dioxide is one of the most commonly used green extraction solvents due to lack of toxicity and side effects (Garavand et al., 2019; Li et al., 2019). The efficiency of supercritical extraction comes from its physical and thermal properties, which incorporates the features of both liquid and gaseous states (Belwal et al., 2020). A supercritical fluid can be obtained when the temperature and pressure of a gas or a liquid exceed its critical points (Guiochon and Tarafder, 2011). In SFE process, the mild processing temperature required makes it suitable for the extraction of thermally unstable and easily oxidizable compounds. SFE is generally viewed as a greener alternative to conventional extraction, since it uses less solvent, takes less time, is easier to automate, and enhances selectivity (Garavand et al., 2019). Diagram of SFE technique is presented in Fig. 9.6D.

9.3.4.1 Advantages of supercritical fluid extraction

Compared to traditional approaches, there are various advantages of employing SFE:

1. Better selectivity since the fluid's solvation power may be modified by changing temperature and pressure.
2. Supercritical fluids' lower viscosity and higher diffusivity allow for faster mass transfer of solutes.
3. The capacity to regulate extraction parameters such as temperature and pressure, which have a direct impact on fluid density, improved transport properties, higher extraction yield, and shorter extraction time are all major advantages of SFE.
4. The possibility of using a lower temperature during extraction avoids thermal degradation of the labile compounds and makes this method attractive.

9.3.4.2 Disadvantages of supercritical fluid extraction

Although, the main disadvantage of SFE is the expensive equipment and the analysis.

9.4 Purification techniques for black cumin seed and seed oil

The components of black cumin seed oils are separated and purified once they have been extracted successfully using diverse techniques. Column chromatography (CC), semipreparative high-performance liquid chromatography (HPLC), high-speed countercurrent chromatography, preparative supercritical fluid chromatography (SFC), and crystallization have all been developed for the isolation and purification of cumin based on factors such as its polarity, ligand interaction, and solubility. The next sections discuss the procedures for separating and purifying black cumin seed oils. Schematic diagram of various purification techniques is presented in Fig. 9.7.

9.4.1 Column chromatography

The most popular approach for purifying phytochemicals on a preparative scale is CC. Preparing the column, putting the sample onto the column,

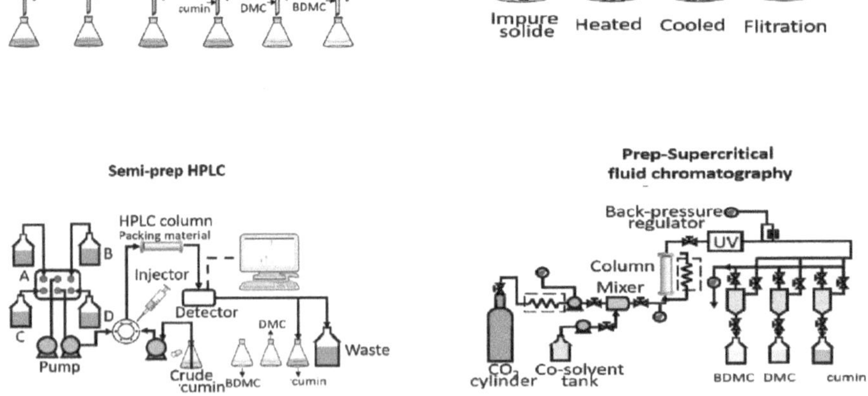

Figure 9.7 Schematic diagram of various purification techniques.

eluting the column with the mobile phase, and recovering the components constitute the main steps of CC (Cvetković et al., 2020). Adsorption chromatography, ion-exchange chromatography, affinity chromatography, and size exclusion chromatography are all examples of CC based on different stationary phases. The type of column to choose is determined by the attributes of the components to be separated. The black cumin seed and its oil are often separated via adsorption chromatography, whereby the separation is based on the adsorption capabilities of the various components on the stationary phase. The aromatic ring of black cumin seed has a distinct methoxy substitution pattern, resulting in polarity differences. The most frequently used stationary phases to separate oils from black cumin seed by CC are silica gel and resin. Existing researches suggest that column adsorption chromatography is a viable technology for producing high-purity oils on a big scale (Sahne et al., 2016; Yang et al., 2015).

9.4.2 Semipreparative high-performance liquid chromatography

Semipreparative HPLC is one of the most effective and commonly accepted methods for isolating and purifying phytochemicals, medicines, and other items. It is however more expensive and complicated for operators than traditional liquid chromatography. High-performance separation, online detection, autonomous control, high recovery, and outstanding reproducibility are all advantages of this technology (Feng et al., 2011). Because of the varied degrees of interaction (hydrophobic and ionic) with the absorbent medium, the chemicals in the sample mixture are separated from one another. The most commonly utilized column for isolating black cumin seed oils and its derivatives is a C18-based column, which is based on hydrophobic interaction. The ideal conditions (stationary phase, mobile phase, flow rate, injection volume, and sample concentration) for good resolution in semipreparative HPLC must be optimized in order to get seed oils with high purity (Mollayi et al., 2015; Owczarek et al., 2017).

9.4.3 Preparative supercritical fluid chromatography

SFC has recently gained a lot of attention because of its great efficiency and environmentally friendly nature. The concepts are comparable to those of HPLC, and the system is based on adsorption and partition chromatography, with columns that are identical to those used in HPLC systems. SFC, on the other hand, usually uses supercritical fluid as the mobile phase (Eisath et al., 2018). CO_2 is the most commonly utilized gas in SFC because it can be quickly compressed and heated to reach critical temperatures (31°C and 74 bar). It is also nontoxic, inert, readily available, and inexpensive (Yang et al., 2015). SFC has been used to prepare natural chemicals with similar structures on both an analytical and preparative scale. It has been shown to be a green and potent tool for separating oils from black cumin seeds, as high purity oil, dimethoxy curcumin (DMC), and bisdemethoxy curcumin (BDMC) can be achieved quickly and efficiently following just one separation step (Song et al., 2015).

9.4.4 High-speed countercurrent chromatography

High-speed countercurrent chromatography (HSCCC) is a liquid–liquid partition chromatographic method frequently used to extract active chemicals from natural materials. It works by separating compounds into two liquid phases that are incompatible with one another. The liquid stationary phase is immobilized by centrifugal force in the absence of a supporting matrix, preventing sample adsorption onto the solid support (Wang et al., 2020). During the liquid–liquid extraction and centrifugal distribution process, each component can be separated based on its distribution coefficients (Pan et al., 2020). HSCCC is a practical and effective process for separating oils from black cumin seeds and has the potential to be used in commercial manufacturing.

9.4.5 Crystallization

In the chemical processing industries, such as the pharmaceutical and food industries, crystallization is one of the most significant separation and purification processes. Largely, crystallization can be accomplished in four ways: antisolvent addition, evaporation, precipitation, and cooling; two of these methods can be combined for better results (Gomes et al., 2020; Tatsukawa et al., 2020). These tactics are designed to reach supersaturation in the solution, allowing the liquid phase to shift to the solid phase (Kumar et al., 2021).

9.5 Conclusion

Due to its diverse biological activities and nontoxicity, black cumin seed is widely employed in a variety of industries, including the food, cosmetics, and pharmaceutical industries. Solvent extraction, Soxhlet extraction, and hydro/steam distillation are time-consuming, environmentally unfriendly, and inefficient. To maximize extraction efficiency, limit the usage of toxic solvents, and reduce processing time and energy consumption, appropriate extraction pathways based on novel technologies such as UAE, MAE, EAE, pressurized liquid extraction, etc., have been investigated. These cutting-edge technologies could be classified as green. Purification is a crucial part of the black cumin seed oil production process. Traditional (CC, semipreparative HPLC) and supplementary modern (high-speed countercurrent chromatography, SFC, etc.) techniques have been examined. More cumin products will be produced in the future as processing techniques for obtaining high purity black cumin seed oil and a better understanding of cumin features improved.

References

Alolga, R.N., Osae, R., Essilfie, G., Saalia, F.K., Akaba, S., Chikari, F.J.F.C., 2021. Sonication, osmosonication and vacuum-assisted osmosonication pretreatment of Ghanaian garlic slices: effect on physicochemical properties and quality characteristics. Food Chemistry 343, 128535.

Altunatmaz, S.S., Aksu, F.Y., Issa, G., Kahraman, B.B., Altiner, D.D., Buyukunal, S.J.V.M., 2016. Antimicrobial effects of curcumin against L. monocytogenes, S. aureus, S. typhimurium and E. coli O157: H7 pathogens in minced meat. Veterinarni Medicina 61, 256–262.

Azmir, J., Zaidul, I.S.M., Rahman, M.M., Sharif, K.M., Mohamed, A., Sahena, F., Jahurul, M.H.A., Ghafoor, K., Norulaini, N.A.N., Omar, A.K.M., 2013. Techniques for extraction of bioactive compounds from plant materials: a review. Journal of Food Engineering 117, 426–436.

Azwanida, N.J.M.A.P., 2015. A review on the extraction methods use in medicinal plants, principle, strength and limitation. Med Aromat Plants 4, 2167-0412.

Bakhshabadi, H., Mirzaei, H., Ghodsvali, A., Jafari, S.M., Ziaiifar, A.M., Farzaneh, V., 2017. The effect of microwave pretreatment on some physico-chemical properties and bioactivity of Black cumin seeds' oil. Industrial Crops and Products 97, 1–9.

Belwal, T., Chemat, F., Venskutonis, P.R., Cravotto, G., Jaiswal, D.K., Bhatt, I.D., Devkota, H.P., Luo, Z., 2020. Recent advances in scaling-up of non-conventional extraction techniques: Learning from successes and failures. TrAC Trends in Analytical Chemistry 127, 115895.

Brännström, H., Kumar, H., Alén, R.J.B.R., 2018. Current and potential biofuel production from plant oils. BioEnergy Research 11, 592–613.

Chan, Y.H., Loh, S.K., Chin, B.L.F., Yiin, C.L., How, B.S., Cheah, K.W., Wong, M.K., Loy, A.C.M., Gwee, Y.L., Lo, S.L.Y.J.C.E.J., 2020. Fractionation and extraction of bio-oil for production of greener fuel and value-added chemicals: recent advances and future prospects. Chemical Engineering Journal 397, 125406.

Chemat, F., Rombaut, N., Sicaire, A.-G., Meullemiestre, A., Fabiano-Tixier, A.-S., Abert-Vian, M.J.U.S., 2017. Ultrasound assisted extraction of food and natural products. Mechanisms, techniques, combinations, protocols and applications. A review. Ultrasonics Sonochemistry 34, 540–560.

Crini, G., Lichtfouse, E.J.E.C.L., 2019. Advantages and disadvantages of techniques used for wastewater treatment. Environmental Chemistry Letters 17, 145–155.

Cvetković, M., Damjanović, A., Stanojković, T.P., Đorđević, I., Tešević, V., Milosavljević, S., Gođevac, D.J.T., 2020. Integration of dry-column flash chromatography with NMR and FTIR metabolomics to reveal cytotoxic metabolites from Amphoricarpos autariatus. Talanta 206, 120248.

Destandau, E., Michel, T., Elfakir, C.J.N.P.E.P., applications, 2013. Microwave-assisted Extraction, vol 113.
Di Giovacchino, L., 2013. Technological aspects. In: Handbook of Olive Oil. Springer, pp. 57–96.
Eisath, N.G., Sturm, S., Stuppner, H.J.P.M., 2018. Supercritical fluid chromatography in natural product analysis—an update. Planta Medica 84, 361–371.
Espanany, A., Fallah, S., Tadayyon, A.J.I.C., Products, 2016. Seed priming improves seed germination and reduces oxidative stress in black cumin (Nigella sativa) in presence of cadmium. Industrial Crops and Products 79, 195–204.
Feng, J., Xiao, Y., Guo, Z., Yu, D., Jin, Y., Liang, X.J.J.O.S.S., 2011. Purification of compounds from Lignum Dalbergia Odorifera using two-dimensional preparative chromatography with Click oligo (ethylene glycol) and C18 column. Journal of Separation Science 34, 299–307.
Friedrich, R.B., Kann, B., Coradini, K., Offerhaus, H.L., Beck, R.C., Windbergs, M.J.E.J.O.P.S., 2015. Skin penetration behavior of lipid-core nanocapsules for simultaneous delivery of resveratrol and curcumin. European Journal of Pharmaceutical Sciences 78, 204–213.
Garavand, F., Rahaee, S., Vahedikia, N., Jafari, S.M.J.T.I.F.S., Technology, 2019. Different techniques for extraction and micro/nanoencapsulation of saffron bioactive ingredients. Trends in Food Science & Technology vol 89, 26–44.
Gari, M.T., Admassu, S., Asfaw, B.T., Abebe, T., Jayakumar, M.J.E.T.C.E., 2020. Review on: extraction of essential oil (Gin flavor) from Juniper Berries (Juniperus communis). Emerging Trends in Chemical Engineering 7, 19–28.
Gerschenson, L.N., Rojas, A.M., Fissore, E.N., 2019. Conventional and emerging extraction technologies. In: Dietary Fiber: Properties, Recovery, and Applications. Elsevier, pp. 199–245.
Gomes, E., Rodrigues, A.J.S., Technology, P., 2020. Crystallization of vanillin from kraft lignin oxidation. Separation and Purification Technology 247, 116977.
Guiochon, G., Tarafder, A., 2011. Fundamental challenges and opportunities for preparative supercritical fluid chromatography. Journal of Chromatography A 1218 (8), 1037–1114.
Hidayati, T., Akrom, L.A., Apriani, L.J.I.J.O.P.H., 2020. The effect of physical activity on lymphocyte count in smokers who consume black cumin seed (Nigella sativa L.) oil. International Journal of Public Health Science 9, 8–14.
Jiang, T., Ghosh, R., Charcosset, C.J.T.I.F.S., Technology, 2021. Extraction, purification and applications of curcumin from plant materials-a comprehensive review. Trends in Food Science & Technology 112, 419–430.
Kujawa, J., Głodek, M., Li, G., Al-Gharabli, S., Knozowska, K., Kujawski, W.J.S.O.T.T.E., 2021. Highly effective enzymes immobilization on ceramics: requirements for supports and enzymes. Science of The Total Environment 801, 149647.
Kumar, K.V., Ramisetty, K.A., Devi, K.R., Krishna, G.R., Heffernan, C., Stewart, A.A., Guo, J., Gadipelli, S., Brett, D.J., Favvas, E.P.J.A.O., 2021. Pure curcumin spherulites from impure solutions via nonclassical crystallization. ACS Omega 6, 23884–23900.
Kurmudle, N., Kagliwal, L.D., Bankar, S.B., Singhal, R.S.J.F.B., 2013. Enzyme-assisted extraction for enhanced yields of turmeric oleoresin and its constituents. Food Bioscience 3, 36–41.
Li, M., Ngadi, M.O., Ma, Y.J.F.C., 2014. Optimisation of pulsed ultrasonic and microwave-assisted extraction for curcuminoids by response surface methodology and kinetic study. Food Chemistry 165, 29–34.
Li, X., Yu, J., Jaroniec, M., Chen, X., 2019. Cocatalysts for selective photoreduction of CO2 into solar fuels. Chemical reviews 119 (6), 3962–4179.
López-Bascón, M., De Castro, M.L., 2020. Soxhlet extraction. In: Liquid-Phase Extraction. Elsevier, pp. 327–354.

Mandal, V., Mohan, Y., Hemalatha, S.J. J.O.p., Analysis, B., 2008. Microwave assisted extraction of curcumin by sample—solvent dual heating mechanism using Taguchi L9 orthogonal design. Journal of Pharmaceutical and Biomedical Analysis 46, 322—327.

Mao, G., Wu, D., Wei, C., Tao, W., Ye, X., Linhardt, R.J., Orfila, C., Chen, S.J.T.I.F.S., Technology, 2019. Reconsidering conventional and innovative methods for pectin extraction from fruit and vegetable waste: Targeting rhamnogalacturonan I. Trends in Food Science & Technology 94, 65—78.

Matthäus, B., 2012. Oil technology. Technological Innovations in Major World Oil Crops, Vol 2. Springer, pp. 23—92.

Mena-García, A., Ruiz-Matute, A.I., Soria, A.C., Sanz, M.L.J.T.T.I.A.C., 2019. Green techniques for extraction of bioactive carbohydrates. TrAC Trends in Analytical Chemistry 119, 115612.

Mollayi, S., Tamhidi, S., Hashempour, H., Ghassempour, A.J.A.C., 2015. Recycling preparative high performance liquid chromatography for the separation of curcumin from curcuminoids in curcuma longa L. Acta Chromatographica 27, 387—398.

Mushtaq, A., Roobab, U., Denoya, G.I., Inam-Ur-Raheem, M., Gullón, B., Lorenzo, J.M., Barba, F.J., Zeng, X.A., Wali, A., Aadil, R.M.J.J.O.F.P., Preservation, 2020. Advances in green processing of seed oils using ultrasound-assisted extraction: A review. Journal of Food Processing and Preservation 44, e14740.

Osae, R., Essilfie, G., Alolga, R.N., Akaba, S., Song, X., Owusu-Ansah, P., Zhou, C.J.J.O.T.S.O.F., Agriculture, 2020. Application of non-thermal pretreatment techniques on agricultural products prior to drying: a review. Journal of the Science of Food and Agriculture 100, 2585—2599.

Owczarek, A., Magiera, A., Matczak, M., Piotrowska, D.G., Olszewska, M.A., Marchelak, A.J.P.L., 2017. Optimisation of preparative HPLC separation of four isomeric kaempferol diglycosides from Prunus spinosa L. by application of the response surface methodology. Phytochemistry Letters 20, 415—424.

Pan, Y., Ju, R., Cao, X., Pei, H., Zheng, T., Wang, W.J.J.O.S.S., 2020. Optimization extraction and purification of biological activity curcumin from curcuma longa L by high-performance counter-current chromatography. Journal of Separation Science vol 43, 1586—1592.

Paulucci, V.P., Couto, R.O., Teixeira, C.C., Freitas, L.A.P.J.R.B.D.F., 2013. Optimization of the extraction of curcumin from Curcuma longa rhizomes. Revista Brasileira de Farmacognosia 23, 94—100.

Popuri, A.K., Pagala, B.J.I.J.I.R.S., 2013. Extraction of curcumin from turmeric roots. International Journal of Innovative Research Studies 2, 289—299.

Rafiee, Z., Nejatian, M., Daeihamed, M., Jafari, S.M.J.T.I.F.S., Technology, 2019. Application of curcumin-loaded nanocarriers for food. Drug and Cosmetic Purposes 88, 445—458.

Rasul, M.G.J.I.J.B.S.A.C., 2018. Conventional extraction methods use in medicinal plants, their advantages and disadvantages. International Journal of Basic Sciences and Applied Computing vol. 2, 10—14.

Sahne, F., Mohammadi, M., Najafpour, G.D., Moghadamnia, A.A.J.P.J.O.B., 2016. Extraction of bioactive compound curcumin from turmeric (Curcuma longa L.) via different routes: A comparative study. Pakistan Journal of Biotechnology 13, 173—180.

Sahne, F., Mohammadi, M., Najafpour, G.D., Moghadamnia, A.A., 2017. Enzyme-assisted ionic liquid extraction of bioactive compound from turmeric (Curcuma longa L.): isolation, purification and analysis of curcumin. Industrial Crops and Products 95, 686—694.

Savoire, R., Lanoisellé, J.-L., Vorobiev, E., 2013a. Mechanical continuous oil expression from oilseeds: a review. Food and Bioprocess Technology 6, 1—16.

Savoire, R., Lanoisellé, J.-L., Vorobiev, E.J.F., Technology, B., 2013b. Mechanical continuous oil expression from oilseeds: a review. Food and Bioprocess Technology 6, 1—16.

Shirsath, S., Sable, S., Gaikwad, S., Sonawane, S., Saini, D., Gogate, P.J.U.S., 2017. Intensification of extraction of curcumin from Curcuma amada using ultrasound assisted approach: effect of different operating parameters. Ultrasonics Sonochemistry 38, 437–445.

Soleimanifar, M., Niazmand, R., Jafari, S.M.J.J.O.F.M., Characterization, 2019. Evaluation of oxidative stability, fatty acid profile, and antioxidant properties of black cumin seed oil and extract. Journal of Food Measurement and Characterization 13, 383–389.

Song, W., Qiao, X., Liang, W.f., Ji, S., Yang, L., Wang, Y., Xu, Y.W., Yang, Y., Guo, D.A., Ye, M.J.J.O.S.S., 2015. Efficient separation of curcumin, demethoxycurcumin, and bisdemethoxycurcumin from turmeric using supercritical fluid chromatography: from analytical to preparative scale. Journal of Separation Science 38, 3450–3453.

Srinivasan, K.J.F.Q., safety, 2018. Cumin (Cuminum cyminum) and black cumin (Nigella sativa) seeds: traditional uses, chemical constituents, and nutraceutical effects. Food Quality and Safety 2, 1–16.

Srivastava, N., Singh, A., Kumari, P., Nishad, J.H., Gautam, V.S., Yadav, M., Bharti, R., Kumar, D., Kharwar, R.N., 2021. Chapter 21 - Advances in extraction technologies: isolation and purification of bioactive compounds from biological materials. In: Sinha, R.P., Häder, D.-P. (Eds.), Natural Bioactive Compounds. Academic Press, pp. 409–433.

Subroto, E., Manurung, R., Heeres, H.J., Broekhuis, A.A., 2015. Mechanical extraction of oil from Jatropha curcas L. kernel: effect of processing parameters. Industrial Crops and Products 63, 303–310.

Tatsukawa, S., Kadota, K., Yoshida, M., Shirakawa, Y.J.J.O.M.L., 2020. Development of quantifying supersaturation to determine the effect of the anti-solvent on precipitation in liquid-liquid interfacial crystallization. Journal of Molecular Liquids 309, 113097.

Vinatoru, M., Mason, T., Calinescu, I.J. T.T.i.A.C., 2017. Ultrasonically assisted extraction (UAE) and microwave assisted extraction (MAE) of functional compounds from plant materials. TrAC Trends in Analytical Chemistry 97, 159–178.

Wakte, P.S., Sachin, B., Patil, A., Mohato, D., Band, T., Shinde, D.J.S., Technology, P., 2011. Optimization of microwave, ultra-sonic and supercritical carbon dioxide assisted extraction techniques for curcumin from Curcuma longa. Separation and Purification Technology 79, 50–55.

Wang, C.-X., Wang, L.-X., Li, C.-Y., Hu, C., Zhao, S.-H.J.E.F.R., Technology, 2020. Anti-proliferation activities of three bioactive components purified by high-speed counter-current chromatography in essential oil from ginger. European Food Research and Technology 246, 795–805.

Williams, M., 2013. Separation technologies in oilseed processing. In: Separation, Extraction and Concentration Processes in the Food, Beverage and Nutraceutical Industries. Elsevier, pp. 396–429.

Yadav, R., Yadav, N.J.I.J.O.P., Analysis, D., 2016. Review on extraction methods of natural essential oils. International Journal of Pharmaceutics and Drug Analysis 4, 266–273.

Yang, Z., Zheng, S., Rui, Z., Fang, Y., Ji, H.J.S., Technology, P., 2015. Enhanced separation and purification of curcuminoids on polyamide column via noncovalent interactions. Separation and Purification Technology 152, 155–159.

Zhang, R., Li, S., Zhu, Z., He, J.J.T.I.F.S., Technology, 2019. Recent advances in valorization of chaenomeles fruit: a review of botanical profile, phytochemistry. Advanced Extraction Technologies and Bioactivities 91, 467–482.

CHAPTER TEN

Biological activities: antiinflammatory, antihyperlipidemic, and antimicrobial of black cumin seeds

Mosab Nouraldein Mohammed Hamad
Faculty of Health Sciences, Elsheikh Abdallah Elbadri University, Berber, Sudan

10.1 Introduction

Black cumin is a well-known folk medicine in several parts of the world. It has enriched history and beliefs as an effective medicinal plant (Kulyar et al., 2021). And Muslims consider it to be a cure-all for all diseases (Al-Bukhari).

10.2 Chemical composition

Black cumin holds many useful proteins, carbohydrates, minerals, vitamins, and oils. The percentage of those vital elements depends on agricultural method, harvest period, and the geographical area. Its extracted oil has both the characteristics of fixed and volatile oil, and the fixed oil contains unsaturated fatty acids and slightly saturated fatty acids, counting A and E vitamins. In black seeds, minerals such as potassium, calcium, and iron are present in large proportions, while zinc, magnesium, manganese, selenium, and copper are present in lesser amounts. Furthermore, vital oil extract from black cumin is of useful significance owing to its wealthy volatiles, such as 18.4%–24% *thymoquinone* (TQ), 46.1% *monoterpenes* counting p-cymene, -pinene, thymol, dithymoquinone, nigellone, and thymohydroquinone (Black Cumin, 2016).

Biochemistry, Nutrition, and Therapeutics of Black Cumin Seed
ISBN: 978-0-323-90788-0
https://doi.org/10.1016/B978-0-323-90788-0.00007-X

10.3 Biological activities

It showed antioxidant, antiinflammatory, antihypertensive, antibacterial, antiviral, antiparasitic, antifungal, antianxiety, antiulcer, anticancer, hypoglycaemic activity, hepatoprotective, immunomodulatory, and muscle relaxant properties. Furthermore, it improves immunity in patients with compromised immune systems (Forouzanfar et al., 2014; Padhye et al., 2008).The bulk of the curative characteristics of this medicinal plant are owing to the existence of TQ, which is a chief bioactive part of the vital oil (Beheshti et al., 2016). Kolahdooz et al. (2014) discovered that black cumin oil remedy significantly improved sperm count, motility, morphology, sperm volume, and pH.

Black cumin oil is used by Persians and Indians to increase milk production in nursing mothers (Drugs and, 2006). On the other hand, women throughout the globe choose to use herbal sources as milk production stimulators, since herbal medicines are supposed to be safe and easily accessible in the local market (Nigella sativa, 2020). Antimicrobial resistance and the high cost of manufacturing of synthetic medicines increase attention to alternative herbs as they possess the capability to kill microbes and maintain and enhance microbial flora (Adegbeye et al., 2020).

10.4 Antiinflammatory activity

Black cumin oil has been found to show strong antiinflammatory influences on many inflammation-based models through the inhibition of prostaglandins and leukotrienes (Shabana et al., 2013). Its supplementation is used to improve the clinical and inflammatory parameters of asthma, such as pulmonary function tests, eosinophil count, and total serum immunoglobulin E. Furthermore, its oil promotes interferon-/interleukin-4 balance, acts in children with asthma, and inhibits histamine release from mast cells (Ikhsan et al., 2018). Activities of black cumin oil are attributed to its ability to suppress *eicosanoid* production in leukocytes by TQ and the fixed oil of it, by suppressing both cyclooxygenase and 5-lipoxygenase. Black cumin shows an inhibitory influence on nitric oxide release by macrophages, and TQ inhibits the action of histone deacetylase and improves inflammation-associated cancer (Chehl et al., 2009).

10.5 Antihyperlipidemic properties

Black cumin seeds, chiefly *TQ*, act as a lowering agent for cholesterol, triglycerides, and low-density lipoprotein, and they raise high-density lipoprotein. The elevated content of polyunsaturated fatty acids of the black seed synergized the hypolipidemic activity of TQ (Shabana et al., 2013).

10.6 Antimicrobial properties

Antimicrobial properties of black cumin seeds have been found to be effective against a variety of pathogenic bacteria, viruses, parasites, and fungi (Forouzanfar et al., 2014).

10.6.1 Antibacterial activities

Black cumin serves as a strong antibacterial agent against numerous bacterial strains such as *Escherichia coli and Pseudomonas aeruginosa.* Aside from its usefulness in combating antibiotic resistance (e.g., suppressing the growth of methicillin-resistant *Staphylococcus aureus*) (Shabana et al., 2013). Its usage is recommended in the food industry in order to prevent spoilage. TQ components of black cumin inhibit a wide range of gram-negative and gram-positive bacteria.

10.6.2 Antifungal properties

A reasonable inhibitory effect of black cumin seed and oil was observed against several pathogenic fungi such as *Candida albicans*, *dermatophytes*, *non-dermatophytes*, and some *aflatoxin*-producing fungi (Abdallah, 2017).

10.6.3 Antiparasitic properties

Black cumin acts actively against numerous parasitic infections and serves as an antileishmanial, anticestodal, antinematodal, and antischistosomal. Also, it has effective antioxidant properties and might be a good phytotherapeutic agent against malaria parasite infection (Mahmoudv et al., 2015).

10.6.4 Antiviral properties

The limited effectiveness of synthetic antiviral medicines guides people to search for other accessible, cheap herbal remedies like black cumin, which shows antiviral activity against influenza virus *(H9N2)*, infectious laryngotracheitis virus, *hepatitis C virus*, *human immunodeficiency virus*, and severe acute respiratory syndrome virus (SARS-COV-2) (Abdallah, 2017).

10.7 Conclusion

Black cumin represents a promising medicinal plant owing to its biological actions against pathological conditions (inflammation, hyperlipidemia) and pathogens (bacteria, viruses, fungi, parasites), which inhibit the growth of antimicrobial-resistant bacteria, are economical and accessible, and have no limited shelf life as synthetic remedies. Furthermore, it improves male fertility.

References

Abdallah, E.M., November 2017. A mini-review of black seed (Nigella sativa) as an antimicrobial drug. Appropriate Drug Development and Development 3 (2), 555603.

Adegbeye, M.J., Elghandour, M.M., Faniyi, T.O., Rivero Perez, N., Barbabosa-Pilego, A., Zaragoza-Bastida, A., Salem, A.Z., August 2020. Black cumin, pawpaw, and mustard seeds have antimicrobial and antihelminthic impacts on livestock production and health. Agroforestry Systems 94 (4), 1255–1268.

Al-Bukhari, S. Sahih Al-Bukhari 5687, Book 76, Hadith 10. Online at https://sunnah.com/bukhari/76.

Beheshti, F., Khazaei, M., Hosseini, M., 2016 . Neuropharmacological effects of Nigella sativa. Avicenna's Journal of Phytomedicine 6 (1), 104–116. PMCID: PMC4884225; PMID: 27247928.

Black cumin (Nigella sativa) and its active constituent, thymoquinone: an overview of the analgesic and anti-inflammatory effects. Planta Med 82 (1–2), January 2016, 8–16. https://doi.org/10.1055/s-0035-1557838. PMID: 26366755. Published on September 14, 2015.

Chehl, N., Chipitsyna, G., Gong, Q., Yeo, C.J., Arafat, H.A., 2009. Anti-inflammatory effects of Nigella sativa seed extract, thymoquinone, on pancreatic cancer cells. HPB (Oxford). https://doi.org/10.1111/j.1477-2574.2009.00059.x. PMCID: PMC2742606; PMID: 19768141.

Cumin. [Revised on November 15, 2021] Drugs and Lactation Database (LactMed) [Internet], 2006. National Library of Medicine (US), Bethesda (MD). Accessible at: https://www.ncbi.nlm.nih.gov/books/NBK501873/.

Forouzanfar, F., Bazzaz, B.S., Hosseinzadeh, H., December 2014. Antimicrobial effects of black cumin (Nigella sativa) and its constituent (thymoquinone). Iranian Journal of Basic Medical Sciences 17 (12), 929–938. PMCID: PMC4387228; PMID: 25859296.

Ikhsan, M., Hiedayati, N., Maeyama, K., Nurwidya, F., December 2018. Nigella sativa as an anti-inflammatory agent in asthma. BMC Research Notes 11 (1), 1–5.

Kolahdooz, M., Nasri, S., Modarres, S., Kianbakht, S., Huseini, H.F., May 15, 2014. A randomized, double-blind, placebo-controlled clinical trial of Nigella sativa L. seed oil on abnormal sperm quality in infertile men. Phytomedicine 21 (6), 901–905. https://doi.org/10.1016/j.phymed.2014.02.006. Epub 2014 Mar 25. PMID: 24680621.

Kulyar, M.F., Li, R., Mehmood, K., Waqas, M., Li, K., Li, J., May 2021. A chance to slow the COVID-19 pandemic.Epub 2020 Jul 10. Phytomedicine 85, 153277. https://doi.org/10.1016/j.phymed.2020.153277. PMCID: PMC7347483; PMID: 32773257.

Mahmoudv, H., Tavakoli, R., Sharififar, F., Minaie, K., Ezatpour, B., Jahanbakhsh, S., Sharifi, I., 2015. Leishmanicidal and cytotoxic activities of Nigella sativa and its active principle, thymoquinone. Pharmaceutical Biology 53 (7), 1052–1057.

Nigella sativa as a galactagogue: a systematic review, Aini Farzana Zulkefli, ruszymah BT HJ Idrus, and Adila A Hamid. Sains Malaysiana 49 (7), 2020, 1719–1727. https://doi.org/10.17576/jsm-2020-4907-22.

Padhye, S., Banerjee, S., Ahmad, A., Mohammad, R., Sarkar, F.H., 2008. From Here to Eternity, the Secret of the Pharaohs: The Therapeutic Potential of Black Cumin Seeds Goes beyond. Cancer Ther 6 (b), 495. PMCID: PMC2583426; PMID: 19018291.

Shabana, A., El-Menyar, A., Asim, M., Al-Azzeh, H., Al-Thani, H., March 2013. Black cumin (Nigella sativa) has cardiovascular benefits. PMID: 22911452. Cardiovasc Toxicol 13 (1), 9–21. https://doi.org/10.1007/s12012-012-9181-z.

CHAPTER ELEVEN

Black cumin seeds as animal feed

Suzy Munir Salama
Indigenous Knowledge and Heritage Center, Ghibaish College of Science and Technology, Ghibaish, Sudan

11.1 Introduction

Globally, *Nigella sativa* seeds (black cumin seeds [BCUS]) and their active constituents have been used for many years as one of the phytomedicinal herbs in treating various diseases of humans and animals (Awulachew, 2021). *N. sativa* L. is an ancient herb that belongs to the family Ranunculaceae and native to Eastern Europe and Asia (Amin and Hosseinzadeh, 2016). Traditionally, BCUS were mentioned in different ancient medicines such as Indian and Chinese medicines. Also, it was given the name of *Kitzah* in the Bible, *Habba-sauda* in Islamic medicine, and *Habba-albaraka* in Arab countries (Ahmad et al., 2021). Studies revealed that BCUS are considered as a good source of antioxidants that play a vital role in protecting tissues from many ailments including hypertension, skin disorders, gastrointestinal disorders, and immunological disorders (Mohammed and Al-Hameed, 2021). Additionally, the seeds reported anticancer and protective activities against kidney and liver diseases (Ahmad et al., 2021). Further, researches proved that phytogenic feed additives have the ability to improve growth and health of animals via their bioactivities including probiotic, prebiotic, and immune-promoting activities (Abd El-Hack, Abdel-Moneim, 2021). Essential oils isolated from BCUS that constitute an average of 1.45% were reported for their antibacterial potential against different types of pathogenic bacteria when supplemented to animals (Awulachew, 2021; Kumar and Patra, 2017). This review will focus on the beneficial effect of BCUS on animals when supplemented as feed additives in different animal models including fish, poultry, and small ruminants.

11.2 Discussion

11.2.1 Nutritional composition of BCUS

Nutritionists found that BCUS exhibit high nutritional values that can provide animals and humans with the necessary substances (Abd El-Hack et al., 2016; Sultan et al., 2009). Based on Sultan et al. (2009), the proximate

Biochemistry, Nutrition, and Therapeutics of Black Cumin Seed
ISBN: 978-0-323-90788-0
https://doi.org/10.1016/B978-0-323-90788-0.00001-9

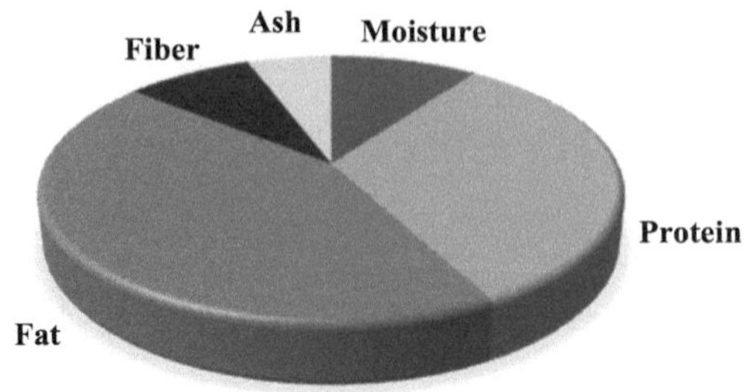

Figure 11.1 Percentage proximate composition of BCUS.

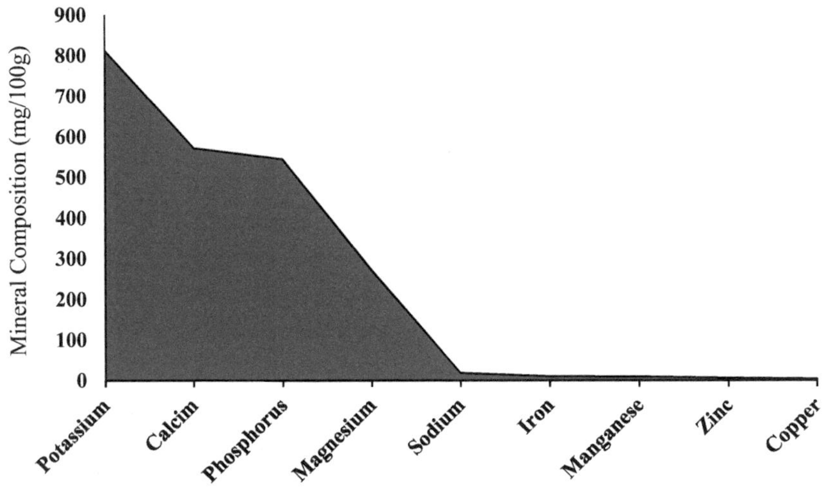

Figure 11.2 Mineral composition of BCUS measured in mg/100 g dry matter.

composition of BCUS recorded approximately moisture 7%, protein 23%, fat 31%, fiber 6%, and ash 4%. Regards mineral composition, BCUS showed high content of the necessary minerals measured per dry matter as potassium 810 mg/100 g, calcium 572 mg/100 g, phosphorus 544 mg/100 g, magnesium 270 mg/100 g, sodium 18 mg/100 g, iron 10 mg/100 g, manganese 9 mg/100 g, zinc 6 mg/100 g, and copper 3 mg/100 g (Sultan et al., 2009) as shown in Figs. 11.1 and 11.2.

11.2.2 Phytochemical profiling of BCUS

The phytochemical profile of BCUS exposed high concentration of antioxidant oil components attributed to total polyphenols as approximately 310 mg gallic acid per kg seed oil, while its content of alkaloids was about 2% (Sultan et al., 2009). In addition, the essential oil content of BCUS reported respected active contents of carotenoids and tocopherols that collectively reach about 451 mg/kg of seed oil (Abd El-Hack et al., 2016; Sultan et al., 2009). One of the active essential oil components studied by researchers was the volatile oil thymoquinone, which was proved for its protective potential against experimental diabetes, cancer, and kidney problems (Ahmad et al., 2021; Sahak et al., 2016). Further, BCUS were proved to contain approximately 55% linoleic acid that contains omega-3 fatty acids and has antineoplastic efficacy (Abd El-Hack, Abdelnour, 2021). Moreover, the protein profile of BCUS reported mainly of nine essential amino acids constituting more than 50% of the total amino acid components of the seed protein (Kabir et al., 2019).

11.2.3 BCUS as animal feed

11.2.3.1 BCUS as fish feed

Fish represents one of the most important nutritive sources for humans due to its flesh contents from healthy fatty acids such as omega-3 and vitamins such as vitamin D. Therefore, researchers put into consideration studies on health of fish. Mahboub et al. (2022) have reported that supplementing Nile tilapia with a diet containing 3%/BCUS oil/body weight for 45 days significantly improved the immunity parameters measured and protected their livers and kidneys from the oxidative damage caused by *Burkholderia cepacia* bacterial infection (Mahboub et al., 2022). A recent study published that BCUS and their oil derivatives reported growth-enhancing effects and immune-promoting activities on different fish species such as tilapia, catfish, rainbow trout, and climbing perch (Altunoglu et al., 2017; Khatun et al., 2015; Mohammed and Arias, 2016; Öz et al., 2018). In addition, many studies stated that BCUS considerably improved the innate immune system and the hematology parameters of rainbow trout fish against pathogenic infection (Altunoglu et al., 2017; Khatun et al., 2015; Mohammed and Arias, 2016; Öz et al., 2018; Bektaş et al., 2018; Fadeifard et al., 2018). Kalil et al. (2011) reported that BCUS extract and its oil thymoquinone exhibited antimutagenic activity in Tilapia and zebrafish via the improvement of histopathological and biochemical parameters (Khalil et al., 2011). Diab et al. (2018) proved that the supplementation of Tilapia with a diet containing 30g BCUS/kg could ameliorate the toxic effect of ochratoxin A (Diab et al., 2018). The effects of dietary supplementation of BCUS or its oil derivative on fish are summarized in Fig. 11.3.

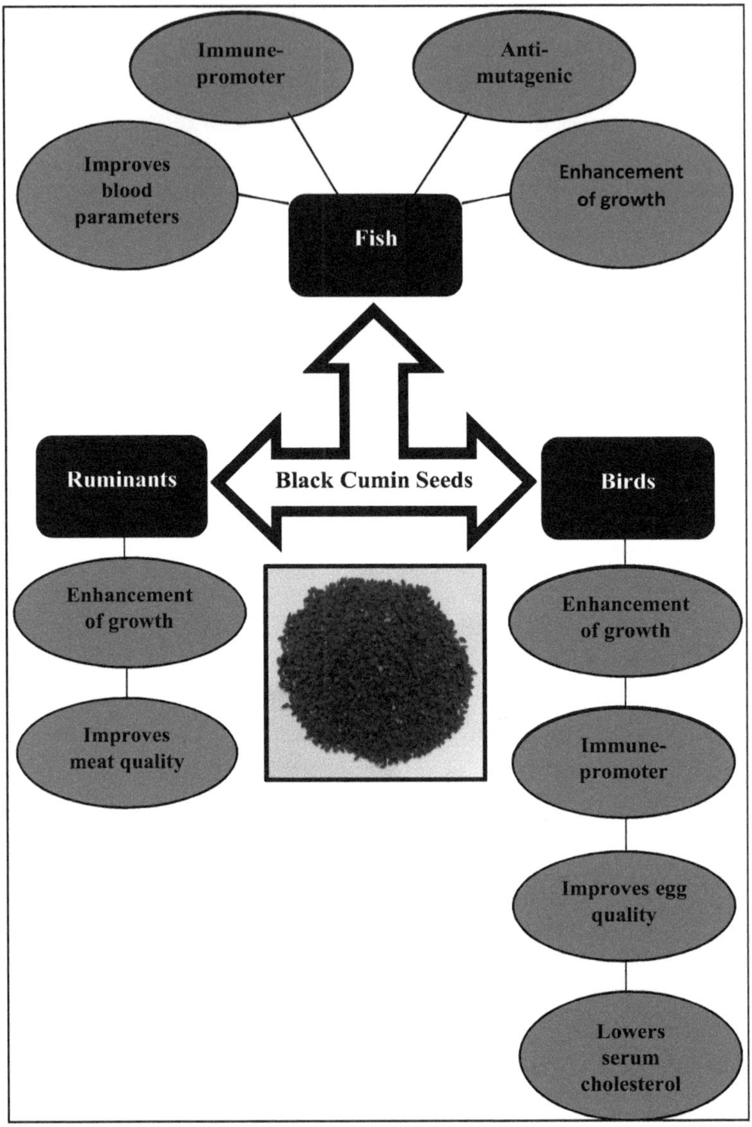

Figure 11.3 Summary of the effect of dietary supplementation of BCUS or its oil derivative on fish, poultry breeds, and small ruminants.

11.2.3.2 BCUS as poultry breeds feed

In the recent days and in comparison to livestock meat, poultry such as chicken, pigeon, quail, and duck are preferable to humans due to its low cost and healthy quality of protein (Abd El-Hack, Abdelnour, 2021). Indigenously, many houses use phytogenic additives while raising poultry breeds for human consumption (Madhupriya et al., 2018). Many researchers published that BCUS have positively promoted the growth of broilers through their constituents of antioxidant essential oils with special regards to the main oil thymoquinone (Abd El-Hack, Abdel-Moneim, 2021; Hassan, 2018; Soliman et al., 2017). In addition, the high quality of broiler's meat fed on BCUS could be attributed to the nine essential amino acid constituents of BCUS (Kabir et al., 2019). Further BCUS was found to enhance the absorption ability of broiler's intestine to the nutrients resulting in the promotion of growth (Abd El-Hack, Abdel-Moneim, 2021). Kumar et al. (2017) found that supplementing birds with a diet containing BCUS at concentration of 10 g per kg body weight improved the quality of the breast and thigh meat of the bird (Kumar and Patra, 2017). Studies showed that fertility of broiler chicks along with their hatchability was significantly improved on feeding them diet containing up to 1% BCUS or their oil isolate (Abdulkarim and Al-Sardary, 2009). Khan et al. (2013) published that supplementing Leghorn birds with diets containing different concentrations of BCUS between 3% and 5% enhanced egg hatchability and improved egg quality through decrement of serum cholesterol of the bird as well as the egg yolk's cholesterol constituent. Authors added that BCUS increased the immunity of the birds against viral infection (Khan et al., 2013). In another scenario, Saleh et al. (2019) and Kumar et al. (2017) manifested that diet rich in BCUS revealed lowered level of serum cholesterol and triglycerides when administered to birds (Kumar et al., 2019; Saleh et al., 2019). In another study, the oil isolated from BCUS significantly increased the egg shell thickness and decreased the egg albumin constituent (Saleh et al., 2019). Scientists reported paradox of results regards the effect of supplementing high concentrations of BCUS on the protein and albumin content of chicken's eggs. One study reported that high concentrations of BCUS in the diet elevated the serum level of albumen and proteins of fed chickens. Authors attributed the results to the immune-promoting activity of BCUS (Tolba and Hassan, 2003). On the contrary, Japanese quails supplemented with diet containing high concentration of BCUS reaching 15 g/kg body weight did not affect the serum level of protein and albumin in the tested birds (Shokrollahi and Sharifi, 2018).

BCUS revealed detoxifying activity against diet contaminated with aflatoxin B1 through the increment of *Lactobacillus* bacteria and the reduction of *Escherichia coli* in the gut of birds when orally administered to quails (Rasouli-Hiq et al., 2017). Diet containing 3% black cumin oil has reduced microflora of the quail's gut along with increasing their immune response against pathogens (Seidavi et al., 2020). Feeding chicks with BCUS has shown antioxidant potential through reduction of lipid peroxidation and enhancement of endogenous antioxidant enzymes production (Abd El-Hack, Abdel-Moneim, 2021). The effects of dietary supplementation of BSCE or its oil derivative on poultry breeds are summarized in Fig. 11.3.

11.2.3.3 BCUS as ruminants feed

Ruminants such as sheep, goats, and cattle are raised in cultural farms for meat, milk, and its production worldwide including the Middle East. Therefore, diets from plant sources are good choice for improving the production of these animals (Obeidat, 2020). A study showed that supplementing 30% of calves diet with BCUS remarkably enhanced their growth (Abdel-Magid et al., 2007). In another study, researchers found that BCUS could improve digestion and enhanced growth in calves and lambs (Mahmoud and Bendary, 2014). Recently, a Jordanian study proved that supplementing Awassi lambs with a meat composed of 150g BCUS per kg weight of animals showed positive effect on the growth of animals, their bodies' abilities to retain nitrogen, and improvement of their digestibility (Obeidat, 2020). The effects of dietary supplementation of BCUS or its oil derivative on ruminants are summarized in Fig. 11.3.

11.3 Conclusion

In conclusion, dietary supplementation of fish, breeding birds, and small ruminants with BCUS or their essential oil derivatives proved enhancement of their growth, improved the immune response of fish and birds, and increased the blood biochemistry parameters of fish and birds. Additionally, BCUS diet showed improvement in the egg hatchability, cholesterol content of the egg's yolk, and albumin content of the egg's white. Future studies require more focus on the potential activity of BCUS diet on other animal models.

References

Abd El-Hack, M.E., Alagawany, M., Farag, M.R., Tiwari, R., Karthik, K., Dhama, K., 2016. Nutritional, healthical and therapeutic efficacy of black cumin (*Nigella sativa*) in animals, poultry and humans. International Journal of Pharmacology 12 (3), 232–248.

Abd El-Hack, M.E., Abdelnour, S.A., Khafaga, A.F., Taha, A.E., Abdel-Latif, H.M., 2021. *Nigella sativa* Seeds and Its Derivatives in Fish Feed. Black Cumin (Nigella Sativa) Seeds: Chemistry, Technology, Functionality, and Applications. Springer, pp. 297–315.

Abd El-Hack, M.E., Abdel-Moneim, A.-M.E., Mesalam, N.M., Mahrose, K.M., Khafaga, A.F., Taha, A.E., et al., 2021. Nigella Sativa Seeds and Its Derivatives in Poultry Feed. Black Cumin (Nigella Sativa) Seeds: Chemistry, Technology, Functionality, and Applications. Springer, pp. 265–296.

Abdel-Magid, S.S., El-Kady, R., Gad, S.M., Awadalla, I., 2007. Using cheep and local non-conventional protein meal (*Nigella sativa*) as least cost rations formula on performance of crossbreed calves. International Journal of Agriculture and Biology 9 (6), 877–880.

Abdulkarim, S.M., Al-Sardary, S.Y., 2009. Effect of black seed (*Nigella sativa* L.) on some reproductive traits in Ross broiler breeder male chickens. Journal of Bombay Veterinary College 17 (1), 19–28.

Ahmad, M.F., Ahmad, F.A., Ashraf, S.A., Saad, H.H., Wahab, S., Khan, M.I., et al., 2021. An updated knowledge of Black seed (*Nigella sativa* Linn.): review of phytochemical constituents and pharmacological properties. Journal of Herbal Medicine 25. https://doi.org/10.1016/j.hermed.2020.100404.

Altunoglu, Y.C., Bilen, S., Ulu, F., Biswas, G., 2017. Immune responses to methanolic extract of black cumin (*Nigella sativa*) in rainbow trout (*Oncorhynchus mykiss*). Fish & Shellfish Immunology 67, 103–109.

Amin, B., Hosseinzadeh, H., 2016. Black cumin (*Nigella sativa*) and its active constituent, thymoquinone: an overview on the analgesic and anti-inflammatory effects. Planta Medica 82 (01/02), 8–16.

Awulachew, M.T., 2021. Black cumin (*Nigella sativa* L): a review on effect and scientific developments in animal and human ailments. International Journal of Medical Biotechnology & Genetics 8 (11), 64–72.

Bektaş, Z.H., Savaşer, S., Akçimen, U., Ceylan, M., Yener, O., Bulut, C., 2018. Using of BCUS powder (*Nigella sativa*) as immunostimulant and growth promoter in rainbow trout, *Oncorhynchus mykiss* (Walbaum). Turkish Journal of Fisheries and Aquatic Sciences 19, 987–999.

Diab, A.M., Salem, R.M., Abeer, E.-K.M., Ali, G.I., El-Habashi, N., 2018. Experimental ochratoxicosis A in Nile tilapia and its amelioration by some feed additives. International Journal of Veterinary Science and Medicine 6 (2), 149–158.

Fadeifard, F., Raissy, M., Jafarian, M., Boroujeni, H.R., Rahimi, M., Faghani, M., 2018. Effects of black seed (*Nigella sativa*), ginger (*Zingiber officinale*) and cone flower (*Echinacea angustifolia*) on the immune system of rainbow trout, *Oncorhynchus mykiss*. Arquivo Brasileiro de Medicina Veterinária e Zootecnia 70, 199–204.

Hassan, S.S., 2018. Effect of *Nigella sativa* seeds on growth performance, carcass traits and economic efficiency of broiler chicks under Egyptian condition. Egyptian Poultry Science Journal 38 (2), 331–344.

Kabir, Y., Shirakawa, H., Komai, M., 2019. Nutritional composition of the indigenous cultivar of BCUS from Bangladesh. Progress in Nutrition 21 (1-S), 428–434.

Khalil, W., Abdel-Gawad, F., Belattar, N., Senator, A., Abdel-Wahhab, M., 2011. Protective effects of *Nigella sativa* extract against chromium (VI)-induced genotoxicity in Nile tilapia (*Oreochromis niloticus*) and Zebrafish (*Danio rerio*). Global Veterinaria 7 (3), 283–293.

Khan, S.H., Anjum, M.A., Parveen, A., Khawaja, T., Ashraf, N.M., 2013. Effects of BCUS (*Nigella sativa* L.) on performance and immune system in newly evolved crossbred laying hens. Veterinary Quarterly 33 (1), 13–19.

Khatun, A., Hossain, M., Rahman, M., Alam, M., Yasmin, F., Islam, M., 2015. Effect of BCUS oil (*Nigella sativa*) on enhancement of immunity in the climbing perch, *Anabas testudineus*. Microbiology Research Journal International 6 (6), 331–339.

Kumar, P., Patra, A., 2017. Beneficial uses of black cumin (*Nigella sativa* L.) seeds as a feed additive in poultry nutrition. World's Poultry Science Journal 73 (4), 872–885.

Kumar, M., Arya, R., Dhuria, R., Nehra, R., Jain, D., 2019. Effect of supplementation of ginger root powder and BCUS powder on performance of white leghorn layers. Journal of Animal Research 9 (5), 683–688.

Madhupriya, V., Shamsudeen, P., Manohar, G.R., Senthilkumar, S., Soundarapandiyan, V., Moorthy, M., 2018. Phyto feed additives in poultry nutrition—a review. International Journal of Science, Environment and Technology 7 (3), 815–822.

Mahboub, H.H., Elsheshtawy, H.M., Sheraiba, N.I., Fahmy, E.M., Mohamed, E.A., Abdelnaeim, N.S., et al., 2022. Dietary black cumin (*Nigella sativa*) improved hemato-biochemical, oxidative stress, gene expression, and immunological response of Nile tilapia (*Oreochromis niloticus*) infected by *Burkholderia cepacia*. Aquaculture Reports 22. https://doi.org/10.1016/j.aqrep.2021.100943.

Mahmoud, A., Bendary, M., 2014. Effect of whole substitution of protein source by *Nigella sativa* meal and sesame seed meal in ration on performance of growing lambs and calves. Global Veterinaria 13 (3), 391–396.

Effect of dietary BCUS (*Nigella stavia*), garlic (*allium sativum*) and lettuce leaves (*lactuca sativa*) on performance and egg qualty traits of native layer hens. In: Mohammed, H.A., Al-Hameed, S.A. (Eds.), 2021. IOP Conference Series: Earth and Environmental Science. IOP Publishing.

Mohammed, H., Arias, C., 2016. Protective efficacy of *Nigella sativa* seeds and oil against columnaris disease in fishes. Journal of Fish Diseases 39 (6), 693–703.

Obeidat, B.S., 2020. The inclusion of black cumin meal improves growth performance of growing Awassi lambs. Veterinary Sciences 7 (2). https://doi.org/10.3390/vetsci7020040.

Öz, M., Dikel, S., Durmus, M., 2018. Effect of black cumin oil (*Nigella sativa*) on the growth performance, body composition and fatty acid profile of rainbow trout (*Oncorhynchus mykiss*). Iranian Journal of Fisheries Sciences 17 (4), 713–724.

Rasouli-Hiq, A., Bagherzadeh-Kasmani, F., Mehri, M., Karimi-Torshizi, M., 2017. *Nigella sativa* (BCUS) as a biological detoxifier in diet contaminated with aflatoxin B1. Journal of Animal Physiology and Animal Nutrition 101 (5), e77–e86.

Sahak, M.K.A., Kabir, N., Abbas, G., Draman, S., Hashim, N.H., Hasan Adli, D.S., 2016. The role of *Nigella sativa* and its active constituents in learning and memory. Evidence-Based Complementary and Alternative Medicine 2016. https://doi.org/10.1155/2016/6075679.

Saleh, A.A., Kirrella, A.A., Dawood, M.A., Ebeid, T.A., 2019. Effect of dietary inclusion of cumin seed oil on the performance, egg quality, immune response and ovarian development in laying hens under high ambient temperature. Journal of Animal Physiology and Animal Nutrition 103 (6), 1810–1817.

Seidavi, A., Laudadio, V., Khazaei, R., Puvača, N., Selvaggi, M., Tufarelli, V., 2020. Feeding of black cumin (*Nigella sativa* L.) and its effects on poultry production and health. World's Poultry Science Journal 76 (2), 346–357.

Shokrollahi, B., Sharifi, B., 2018. Effect of *Nigella sativa* seeds on growth performance, blood parameters, carcass quality and antibody production in Japanese quails. Journal of Livestock Science 9 (1), 56–64.

Soliman, E.S., Hamad, R.T., Ahmed, A., 2017. Prophylactic and immune modulatory influences of *Nigella sativa* Linn. in broilers exposed to biological challenge. Veterinary World 10 (12), 1447–1455.

Sultan, M.T., Butt, M.S., Anjum, F.M., Jamil, A., Akhtar, S., Nasir, M., 2009. Nutritional profile of indigenous cultivar of BCUS and antioxidant potential of its fixed and essential oil. Pakistan Journal of Botany 41 (3), 1321–1330.

Tolba, A., Hassan, M., 2003. Using some natural additives to improve physiological and productive performance of broiler chicks under high temperature condition. Egyptian Poultry Science 23 (2), 327–340.

CHAPTER TWELVE

Therapeutic potential of *Nigella sativa*

Nor Zaihana Abdul-Rahman and Nurul Szawani Mohd-Zubri
Department of Basic Sciences, Faculty of Dentistry, Universiti Sains Islam Malaysia, Kuala Lumpur, Selangor, Malaysia

12.1 Introduction

Mounting evidence has shown the therapeutic effect of medicinal plants. They have been used and discovered for years as part or alternative of treatment, and until today, globally it is being used for well-being (WHO, 2021). One of the most researched plants is *Nigella sativa* or commonly known as black seed or black cumin, a native plant in Southern Europe, Southwest Asia as well as North Africa. Moreover, it is also commonly found in other countries such as India, Pakistan, Turkey, and the Middle East (Khare, 2004). Historically, *N. sativa* has been traditionally been used to treat a number of diseases and widely used as a liver tonic, anti-hypertensive, antibacterial, and analgesic (Ahmad et al., 2013).

The medicinal benefit of *N. sativa* will be highlighted in this chapter. The investigations include those that are undertaken clinically, either by human volunteers or participants or in a lab setting, such as an animal study or an in vitro research.

12.2 Clinical therapeutic potential of *Nigella sativa*

This section contains a collection of studies that were conducted with human volunteers or participants. Clinical trials or clinical observations were used to perform the research. Among the studies include improving thyroid status, reduction of lipid profile and hypercholesterolemia, remedy for insulin resistance syndrome, and reducing rhinitis.

12.2.1 Improving thyroid status

Hashimoto thyroiditis (HT), a type of autoimmune disease is the common cause of hypothyroidism. It is an organ-specific T-cell—mediated illness that

Biochemistry, Nutrition, and Therapeutics of Black Cumin Seed
ISBN: 978-0-323-90788-0
https://doi.org/10.1016/B978-0-323-90788-0.00017-2

affects more females than males and is implicated in alteration in the composition and transport of lipoprotein. A randomized controlled trial was done on 40 HT patients. The patients were divided into two groups, intervention and control. The intervention group were supplied with *N. sativa* powder or placebo (control group) daily for 8 weeks. The results showed a significant positive effect after *N. sativa* supplementation on thyroid function (Farhangi et al., 2016). Comparing both groups, among the effects observed were the serum concentration of T3 was increased and body weight and body mass index in HT patients were considerably lowered consequence of treatment with *N. sativa*. This study indicates the potential therapeutic effect of *N. sativa* in improving thyroid function.

12.2.2 Reduction of lipid profile and hypercholesterolemia

Coronary heart disease is the major cause of death worldwide. It is estimated killing 17.9 lives every year. Based on a previous study, high lipoprotein level is considered as risk factors for coronary heart disease (Zhao et al., 2021). A study to investigate the influence of lipid profile in healthy people was conducted at the Institute of Cardiology in Multan, Pakistan. Eighty patients were divided into two groups: the intervention and the noninterventional group and were sampled for blood and subjected for evaluation of total cholesterol (TC), triacylglycerides (TGs), high-density lipoprotein (HDL), very low-density lipoprotein (VDL), and low-density lipoprotein (LDL). The results indicated that in the interventional group, the cholesterol, LDL, TGs, and VDL were reduced, whereas the HDL increased significantly after 6 months compared with the noninterventional group (Tasawar et al., 2011). These results suggest that *N. sativa* may have a therapeutic effect on the reduction lipid profile and subsequently be beneficial for cardiovascular health. Moreover, another randomized, double-blind, placebo-controlled trial has been conducted to investigate the effect of *N. sativa* on hypercholesterolemia patients. Administration of *N. sativa* seeds for 4 weeks has been demonstrated to reduce TC and LDL (Pelegrin et al., 2019).

Apart from that, patients with type-2 diabetes mellitus (DM) have a complex alteration in plasma lipids, which is defined by an increased level of TGs and lowered level of HDL (Krentz, 2003). A prospective study was done on 94 uncontrolled type-2 DM patients to investigate TC, TG, LDL, and HDL. The patients were divided into three groups and supplied a different dose of *N. sativa* for 12 weeks (1, 2, and 3 g per day) to see how the lipid profile changed. It is reported that the patients supplemented with 2g of *N. sativa* has shown a significant alleviation in both TC and TG levels (Kaatabi et al., 2012). It can be concluded that *N. sativa* has promising hypolipidemic effects and is effective in lowering blood lipids and is postulated to have the potential to control hypercholesterolemia.

12.2.3 Therapy for insulin resistance syndrome

A study was conducted to examine the effect of *N. sativa* in metabolic syndrome patients clinically. The patients were divided into three groups, a group were given the standard regimen, the second group were given standard regimen plus *N. sativa* oil (NSO), and the last group was the therapeutic group which was given tablet metformin and atorvastatin (Najmi et al., 2008). The outcome revealed positive effects on fasting blood glucose, TC, and LDL cholesterol.

12.2.4 Improving reproductive system disorder

Infertility impacted around 186 million individuals living worldwide, which indirectly affect families and communities. In the male reproductive system, the common causes of infertility are problems in the ejection of semen, absence or low level of sperm as well as abnormal shape and motility of the sperm (WHO, 2021).

To study the effect of *N. sativa* on infertility, 80 infertile men around 20–45 years old were recruited to a double-blind and block randomization study. The patients were grouped into either the *N. sativa* group treatment or placebo groups. After 2 months, it was discovered that consumption of 5 mL NSO daily improves sperm count, morphology, and motility, as well as volume, pH, and the number of round cells in the sperm (Kolahdooz et al., 2014). The result concurs with previous in vitro reports that showed spermatogenesis, infertility in male rats and chickens (Abdulkarim and Al-Sardary, 2009).

12.2.5 *Nigella sativa* significantly reduces rhinitis

The effectiveness of *N. sativa* in the treatment of allergic rhinitis was also demonstrated in the clinical setting. A prospective and double-blind clinical trial was conducted in which allergy rhinitis patients were supplemented with *N. sativa* seed oil (6 mg/kg daily) for 30 days. It has been observed that administration of *N. sativa* significantly reduced the respiratory symptoms including nasal congestion and itching, runny nose, sneezing as well as turbinate hypertrophy (Nikakhlagh et al., 2011). The study was in agreement with another report that investigated the effect of *N. sativa* nasal spray in a randomized clinical study. The patients in this study were grouped into two: the intervention group that received (2 puffs/day) of *N. sativa* and the other group received 2 puffs/day of sodium chloride spray. It has been observed that *N. sativa* has shown significant positive effects in the intervention group (Rezaeian and Khouzain, 2018).

12.3 Nonclinical study

Multiple studies had been done to investigate the potential of *N. sativa* to be developed as a novel drug. In this section, the fundamental, in vitro *or* in vivo experiments of *N. sativa* laboratory test are laid off. The experiments are laboratory-based, and some of the tests are conducted on animals model. The results demonstrated an opportunity for *N. sativa* seeds to be developed as an alternative treatment in clinical settings.

12.3.1 Alleviates liver fibrosis

Hepatic fibrosis, also known as liver fibrosis, is a form of chronic wound healing response caused by long-term liver injury (Böttcher and Pinzani, 2017). It is manifested by excessive accumulation of extracellular matrix (ECM) protein such as collagen Type I and Type II in the liver (Bataller and Brenner, 2020). This disease can cause the liver to lose its function and remodeling of the liver parenchyma (Tsuchida and Friedman, 2017). Mounting evidence from the research suggested that activated hepatic stellate cells (HSCs) produce an abnormal build-up of ECM protein in liver fibrosis. Thus, inhibiting HSC activation has been one of the approaches for treating liver fibrosis (Geng et al., 2021).

Thymoquinone (TQ) has been demonstrated to reduce liver fibrosis in vivo and in vitro through the regulation of miR-30a and Snai1 (Geng et al., 2020). By using carbon tetrachloride mice, TQ has been shown to ameliorate liver fibrosis and collagen accumulation. As in in vitro, TQ inhibited HSC activation in a few ways including a few ways that have been postulated as to how TQ inhibits liver fibrosis, such as reduced proliferation, α-smooth muscle actin, and collagen.

12.3.2 Antimicrobial efficacy

Nigella sativa is primarily studied for its antimicrobial properties based on the availability of multiple bioactive compounds that possess multiple functions including antimicrobial properties. The antimicrobial properties of black seeds extracts and essential oils are studied on a wide variety of bacteria including multiple drugs resistance bacteria. The availability of bioactive compounds from the compound groups of terpenes, terpenoids, flavonoids, phenolics, and also anthraquinones exhibits antibacterial properties that were reported in various studies (Abdallah, 2017).

A study found that black cumin seed oil exhibits antimicrobial activity against four types of gram-positive bacteria, which include methicillin-resistant *Staphylococcus aureus*, *S. aureus*, *Bacillus subtilis*, and *Bacillus cereus*. The antimicrobial activity was attributed to the availability of TQ and its isomeric forms such as dithymoquinone, thymohydroquinone, and thymol which were assessed by Fourier-transform infrared (FTIR) and thin-layer chromatography (TLC) (Mohammed et al., 2019).

N. sativa seed ethanolic extract was found to demonstrate antimicrobial properties against multidrug-resistant bacteria isolated from urine samples of urinary tract infection patients. Ethanolic extract of *N. sativa* shown excellent inhibition activity against urinary pathogens, *Escherichia coli, Klebsiella pneumonia, Proteus mirabilis, Pseudomonas aeruginosa,* and *Acinetobacter baumanni*. Interestingly, all of the stated pathogens were tested for antibiotic resistance in the same study and were resistant to at least five different antibiotics (Dhanasekar, 2019). This study proved that *N. sativa* extract possesses the potential to be developed as a new drug against drug-resistant bacteria.

The applicability of black cumin seeds as an antimicrobial compound is not only restricted to a specific site of the human body but also multiple locations of pathogen reservoirs. One study assessed the antimicrobial activity of *N. sativa* seeds oil extract against dental caries pathogens, *Streptococcus mutans,* and *Streptococcus sanguinis*. The extract from *N. sativa* seeds oil extracts showed inhibition activity against both pathogens. Aside from that, the study also assessed the antibiofilm formation properties of the seeds where the qualitative observation using staining and scanning electron microscope showed significant antibiofilm activity by the black cumin seeds (Gasong et al., 2017). The study has proven the versatile applicability of *N. sativa* seeds as an antimicrobial treatment at various sites of the human body. Other than that, the oil extract from black cumin seed is also effective to combat biofilm formation, which had been the foundation of various pathogens-related diseases.

The applicability of *N. sativa* seed extract against specific disease-related pathogens was also investigated in a study. The study highlighted the antimicrobial activity of *N. sativa* seed oil against pathogens related to infection in the middle and external ear. The black cumin seed oil was found to be effective against three species related to an ear infection which are *Streptococcus pneumonia, Haemophilus influenza,* and *Moraxella catarrhalis* isolated clinically from infected patients (Kocoglu et al., 2019). In addition, another study reveals the efficacy of *N. sativa* seeds oil formulated for topical application against acne inducing bacteria. The topical formulation of *N. sativa* seed oil was found to show antimicrobial activity against *S. aureus* and *Propionibacterium acne* (Nawarathne et al., 2019). These studies showed that the applicability of *N. sativa* seed oil or extract has the potential to be developed for therapeutic purposes in clinical settings.

Multiple other studies were reporting the antimicrobial properties of *N. sativa* seeds oil or extract against various bacteria. For example, TQ of black cumin seeds exhibits antimicrobial activity against anaerobic bacteria against four bacteria which are *Clostridium difficile, Clostridium perfringens, Bacteroides fragilis,* and *Bacteroides thetaiotaomicron* (Randhawa et al., 2017). In another study, NSO extract containing a high percentage of TQ showed antimicrobial and antibiofilm efficacy against reference *S. aureus* and ten *S. aureus* clinical isolates which includes two methicillin-resistant strains (Gawron et al., 2019). The antimicrobial activity of black cumin seed extract was also recorded against methicillin-resistant *Salmonella enterica* isolates (Ashraf et al., 2018).

In order to understand the antimicrobial properties of *N. sativa* seeds, the availability of bioactive compounds and their underlying action were assessed in a few studies. A few studies employ the use of gas chromatography-mass spectrometry (GC-MS) to detect the composition of black cumin seeds oil. These studies revealed that black cumin seeds oil contains TQ, Beta-pinene, O-cymene, thymol, various organic acids, terpenes, and cuminal. Black cumin seed oils were found to exhibit antimicrobial activity against multiple species of bacteria in the same studies. These studies concluded that the antimicrobial activity of black cumin seed oil might be attributed to the availability of well-known bioactive compounds (Alizadeh Behbahani et al., 2019; Badger-Emeka et al., 2021; Reddy et al., 2018; Saleh et al., 2017). In the other studies, the effect of NSO exposure on bacteria was studied by using electron microscopies. The study indicated that exposure to *N. sativa* seeds oil at minimum inhibitory concentration brought detrimental effects on the tested strains cell wall integrity and membrane structure (Alizadeh Behbahani et al., 2019; Badger-Emeka et al., 2021). These studies demonstrated that the *N. sativa* seeds oil rich in bioactive compounds possess the ability to inhibit various types of pathogens. Disruption of the bacterial cell wall might be the underlying mechanism of action for the antimicrobial activity of *N. sativa* seeds.

12.3.3 Antifungal properties of *Nigella sativa* seeds

Further investigation to its antimicrobial properties, *N. sativa* seeds were also studied for their antifungal properties against disease-inducing fungi species. The antifungal properties of *N. sativa* seed can be credited to the high percentage of bioactive compounds available in the extract (Ahmad et al., 2021; Kooti et al., 2016). Several studies had evaluated the efficacy of *N. sativa* seeds oil or extract against a few species of disease-inducing fungal species.

N. sativa seed was studied for its efficacy against *Trichophyton rubrum*, a fungi species related to dermatophyte condition. Dermatophyte refers to a condition of fungi infection on keratinized tissue on human skin, hair, and nails. In recent years, dermatophyte fungi including *T. rubrum* had become less susceptible to antifungal treatment which brought an urge to seek for novel source to counter the problems (Khurana et al., 2019). However, based on the study, *N. sativa* seeds in the form of essential oil were revealed to exhibit strong antifungal activity against 18 clinical isolates of *T. rubrum* when being compared to the current antifungal drug, fluconazole. The contributor to the antifungal activity was studied by utilizing the GC-MS technology, and it was found that NSOs contain TQ, α-thujene, anethole, carvacrol, and longifolene at different percentages (Kaya and Duran, 2018).

Similarly, research has been done to identify the antifungal activity of *N. sativa* seed extracts against clinically significant fungi strains. *Aspergillus niger, Aspergillus fumigatus, Rhizopus stolonifer, Penicillium digitatum,* and *Acremonium alternatum* isolated from infantile sepsis and gallbladder patients were investigated in the study because of the species significance in infectious disease. The ethanolic, methanolic, and water extract of *N. sativa* seeds showed significant inhibition against the stated clinically isolated fungi species (Shah et al., 2018).

Another interesting study found that isolated peptides from the thionin group of *N. sativa* termed nigellothionin were found to exhibit antifungal activity against the *Aspergillus* and *Candida* strains. Both strains were clinical isolates that were identified as the causative agents of invasive pulmonary aspergillosis and candidiasis. The results demonstrated that the nigellothionin isolated from black cumin seeds exhibit excellent antifungal activity against the stated fungi strains and were competitive with the current antifungal drugs. Despite the decrease of antifungal activity dynamic over time, the antifungal activity of Nigellothionins was remarkable and have the potential to be developed as a novel antifungal agent (Barashkova et al., 2021).

A study profiled the *N. sativa* methanolic extract by utilizing the GC-MS technology to identify the active compounds that exhibited antifungal activity. The antifungal activity was determined by observing the reduction of fungal mass after exposure to methanolic extract of *N. sativa*. *N. sativa* methanolic extract was found to reduce the fungal mass of two fungi strains, *Fusarium oxysporum* and *Macrophomina phaseolina*. *F. oxysporum* were studied as one of the strains that caused human fusariosis, a fungal infection disease that is hard to diagnose and treat (Batista et al., 2020). The GC-MS profiling of *N. sativa* methanolic extract revealed percentages of various organic acids that were deduced as the active compound in the extract (Aftab et al., 2019).

The application of *N. sativa* antifungal properties was expanded to inhibit the colonization and plaque formation by *Candida albicans* on acrylic dentures pieces. The study applied ethanolic extract of *N. sativa* onto the acrylic dentures that were contaminated purposely with *C. albicans*. The result showed that the decrease of *C. albicans* colony counts corresponded to an increase in the extract concentration. The extract exhibited lower antifungal activity compared to the positive control; however, it showed that *N. sativa* has the potential to be developed as plant-based antifungal agents (Ariamanesh et al., 2019).

12.3.4 Antidiabetic properties

Among other therapeutic potential of *N. sativa* seeds, extracts are their antidiabetic properties. Multiple studies had been done to investigate the ability of *N. sativa* to be developed as novel antidiabetic agents. Most of the studies were done in vivo on typical diabetic animals model, and the results demonstrated an opportunity for *N. sativa* seeds to be developed as an alternative treatment clinical setting.

A study assessed the antidiabetic ability of *N. sativa* methanolic extract on streptozotocin-induced diabetes in rats model. The study also investigates the effect of *N. sativa* seeds methanolic extract on diabetic neuropathy-inducing proteins. The result highlighted the positive effects of *N. sativa* seeds methanolic extract on various diabetes markers and conditions of tissue related to diabetes. The study reported a significant reduction of fasting blood glucose in the induced diabetic male rats. Other than that, the study demonstrated the restoration of lipid peroxidation level parameters to normal level after being treated with *N. sativa* methanolic extracts. The level of diabetes-related proinflammatory cytokine and immunoglobulins was also reduced with the treatment aside from the reduction of diabetic neuropathy biomarkers. The renal and liver functions of the studied rats were also positively affected with the treatment based on the tissue observation and chemical markers profiling. A high percentage of flavonoids and carotenoids from the extract indicates that the positive effects were caused by the presence of the antioxidants in the extract (El Rabey et al., 2017).

Another study assessed the same positive outcomes on the high fat diet-induced diabetic rats model. However, the study took an interesting take for the evaluation by using the *N. sativa* seeds polysaccharides (NSSPs) instead of the oils or extracts. The study demonstrated a reduction of fasting blood glucose and glycosylated proteins level compared to the nontreatment group. Furthermore, in the treatment group, there is an improvement of blood insulin dose-dependently, and the results were comparable with the metformin-treated group. In another aspect, the supplementation of NSSP caused significant positive outcomes on the mice lipid metabolism with the reduction of blood triglycerides, TC, LDL, and improvement of blood HDL. The antioxidative stress ability biomarkers and diabetes-related cytokines were also measured where the result showed positive outcomes for all the parameters studied. The histology study of mice kidney and liver tissues has shown an ameliorative effect of NSSP, which were indicated by the restoration of stated tissue to a normal condition. Generally, the result indicated that NSSP has the potential to be developed as an adjuvant anti-diabetic agent in the future clinical setting (Dong et al., 2020).

Chronic DM brought a serious nerve problem known as diabetic neuropathy, which can cause restricted motor function, pain, and even amputation in serious cases (Dewanjee et al., 2018). In a study, green synthesis of silver nanoparticles (AgNPs) by *N. sativa* extracts potential to ameliorate diabetes and diabetic neuropathy was investigated. The study revealed a new potential of *N. sativa* extract utilization for better performance. Similar parameters with previously stated studies were also observed on streptozotocin-induced diabetic neuropathy rats. The study reported significant positive outcomes in serum glucose level, serum insulin, and serum advanced glycation end product, which plays an important role in causing diabetes-related complications and oxidative stress biomarkers level. Histological study of the brain tissue after exposure to green AgNP with *N. sativa* extract showed less lesion score and tissue degeneration compared to the nontreated group. The positive impact was also reported in the reduction of TNF-α and NF-KB that might cause nerve inflammation. The exposure to AgNP with *N. sativa* extract restored the level of nerve-related biomarkers to a normal level. The results indicated that AgNPs technology combined with *N. sativa* extract might be able to target specific disease sites of diabetes-related complications (Alkhalaf et al., 2020).

12.3.5 Diabetic wound healing properties

N. sativa is also functional for diabetic wounds recovery not only due to its antimicrobial properties but also its potential to alleviate diabetic wound inflammation. Several studies had observed the applicability of *N. sativa* seeds extract or oils as for its wound healing activity.

In a study, the wound healing properties of NSO gel was observed and compared with aloe vera gel on an induced diabetes mice model. Alloxan monohydrate was injected to induce diabetes in the mice, and a 1 cm wound was made on the mice dorsal part. The physical observation of the mice wounds was made, and the result showed that mice treated with NSO gel showed wound size reduction starting from day 1 after treatment to day 7 of observation. The mice treated with NSO gel were reported to have less inflammation score. However, the treatment with aloe vera gel showed higher efficiency of wound healing compared with NSO gel based on the histological study. Regardless of the performance of NSO gel compared with aloe vera gel to promote wound healing, the study had shown the possibility of NSO gel to be developed as a diabetic wound healing agent (Sari et al., 2018).

Other than that, *N. sativa* seed extract was developed as a wound dressing in the form of polyvinyl alcohol (PVA)-Nigella nanofibrous mat (PNNM). The study combined *N. sativa* extract with PVA to form polymers that can be applied as a wound dressing. The wound observation revealed that rabbits treated with PNNM had reduced wound area and wound liquid on the second day of treatment. The study also reported that PNNM wound dressing showed inhibition area against *S. aureus* and *E. coli*, which are two common pathogens on diabetic wounds. The moisture profile study also reported that wound dressing with *N. sativa* dressing might have one-way moisture release where the wound liquid is absorbed into the material and external moisture is not permitted to pass through due to the oil extracted from *N. sativa* (Ali et al., 2020). This study showed that NSO can be incorporated into other materials to be applied as a diabetic wound dressing.

One of the underlying actions of *N. sativa* in promoting diabetic wound healing is by exhibiting antimicrobial properties against diabetic ulcers bacteria. A comparative study showed that NSO had excellent inhibition against two bacteria isolated from diabetic foot ulcers. The other essential oils were reported with no inhibition or inhibition on only one species of bacteria isolated from diabetic foot ulcers. This might be one of the underlying activities of NSO in promoting diabetic wound healing which is by eradicating the diabetic wound pathogens from the wound site (Bady et al., 2019).

12.3.6 Anticancer properties

Another interesting therapeutic potential of *N. sativa* is its anticancer properties. A number of *N. sativa* bioactive compounds especially TQ were reported to demonstrate positive effects on different types of cancer hallmarks. Furthermore, compounds from *N. sativa* were also applied in nanoparticles technology as one of the novel approaches to combat cancer cells progression (Yimer et al., 2019).

For instance, a study investigated active saponin anticancer properties extracted from *N. sativa* termed as alpha-hederin (α-hederin). The compounds were concluded to promote ovarian cancer cells (SKOV-3) apoptosis. α-hederin was reported to reduce SKOV-3 cell viability, which was observed by the cell viability system. MTT assays on the cell line also revealed that dose-dependently α-hederin reduced the viability of the cell. Other than that, staining of SKOV-3 cell nucleus and apoptosis staining showed that α-hederin exhibits nucleus fragmentation and induced apoptosis on the cell line. Mitochondrial activity of the SKOV-3 cell line was also reduced with the exposure to α-hederin besides the inactivation of antiapoptosis protein after being treated with the compound. The exposure to α-hederin also caused arrest in the cancer cell cycle (Adamska et al., 2019).

Moreover, the anticancer potential of TQ was also reported in a study. The study highlights the excellent anticancer properties on the human hepatocellular carcinoma cell line by TQ and resveratrol, both extracted from *N. sativa* seeds. The MTT assay on the cell line reported that a combination of TQ and resveratrol reduced cell viability more effectively than either TQ or resveratrol alone. The combination also increases the caspase-3 content by 166% compared to the negative control. Caspase-3 is important in the execution of cancerous cell apoptosis where the increment of caspase-3 content in the cell is important to control cancerous cell migration and invasion (Zhou et al., 2018). Other than that, a marker of oxidative stress in cells, malondialdehyde was also reduced after 72 h of exposure toward TQ-resveratrol combination, which indicates that the compounds exhibit antioxidant activity on the cell line (Ismail et al., 2018).

Another application of *N. sativa* seeds anticancer potential was utilizing its extract for green synthesis of AgNPs. Green synthesis of AgNP using the *N. sativa* extract demonstrated anticancer activity against human breast cancer cell lines (HCC712). The stable form of synthesized AgNP with *N. sativa* extract as confirmed by FTIR, X-ray diffraction, and microscopic study were tested on breast cancer cell line for its cytotoxic activity. The cytotoxic study showed dose-dependent cytotoxic activity of the synthesized nanoparticles against human breast cancer cell lines. Microscopic observation showed a reduction of breast cancer cell density which indicated that exposure to synthesized AgNP with *N. sativa* extract promoted cell apoptosis (Almatroudi et al., 2020).

Another example of *N. sativa* TQ anticancer properties was discussed in a study. The study approached multiple endpoints analyses to assess the potential of TQ from *N. sativa* anticancer activity against the human renal carcinoma cell line. Similarly, to other studies, TQ showed dose-dependent cytotoxic activity as observed by MTT assay. Other than that, the percentage of apoptotic cells increased with the exposure to TQ, and reduction of cancer cells migration and invasion were also observed in the treated cell line. Intracellular reactive oxygen species (ROS) increased moderately in the cell line treated with TQ from *N. sativa*. A moderate amount of ROS is important in healthy cell signaling to trigger programmed cell death for irregular cells, while to abrupt increment of ROS is dangerous for regular cells and might encourage the cancer cells activity (Perillo et al., 2020). The moderate increment of ROS when being exposed to TQ treatment is beneficial to eradicate cancerous cells (Costa et al., 2020).

The studies discussed the potential anticancer activity of *N. sativa* seeds in various forms and derivatives showing the capability and possibility to develop *N. sativa* seeds derivatives as novel anticancer agents.

12.4 Conclusion

Holistically, *N. sativa* seeds extracts, oils, and their various derivatives have the potential to be developed for a variety of applications. The plethora of studies that discussed the potential of *N. sativa* seeds gave insight into the ability of this plant which is rich in bioactive compounds, beneficial peptides, and also flexible for multiple novel applications. Further research which might expand into clinical settings is very beneficial in order to assess the safety, efficacy, and applicability of the *N. sativa* seeds for various therapeutic purposes.

References

Abdallah, E.M., 2017. Black seed (Nigella sativa) as antimicrobial drug: a mini-review. Novel Approach Drug Designing Development 3 (2), 6.

Abdulkarim, S.M., Al-Sardary, S.Y., 2009. Effects of black seed (Nigella sativa) on some reproductive traits in Ross broiler breeder male chickens. Journal of Bombay Veterinary College 17, 19–28.

Adamska, A., Stefanowicz-Hajduk, J., Ochocka, J.R., 2019. Alpha-hederin, the active saponin of *Nigella sativa*, as an anticancer agent inducing apoptosis in the SKOV-3 cell line. Molecules 24 (16), 2958. https://doi.org/10.3390/molecules24162958.

Aftab, A., Yousaf, Z., Javaid, A., Riaz, N., Younas, A., Rashid, M., Bushra, H., Chahel, A.A., 2019. Antifungal activity of vegetative methanolic extracts of Nigella sativa against fusarium oxysporum and macrophomina phaseolina and its phytochemical profiling by GC-MS analysis. International Journal of Agriculture and Biology 21 (3), 9.

Ahmad, M.F., Ahmad, F.A., Ashraf, S.A., Saad, H.H., Wahab, S., Khan, M.I., Ali, M., Mohan, S., Hakeem, K.R., Athar, M.T., 2021. An updated knowledge of Black seed (*Nigella sativa* Linn.): review of phytochemical constituents and pharmacological properties. Journal of Herbal Medicine 25, 100404. https://doi.org/10.1016/j.hermed.2020.100404.

Ahmad, A., Husain, A., Mujeeb, M., Khan, S.A., Abul Kalam, N., Siddique, N.A., Damanhouri, Z.A., Anwar, F., 2013. A review on therapeutic potential of Nigella sativa: A miracle herb. Asian Pacific Journal of Tropical Biomedicine 3, 337–352. https://doi.org/10.1016/S2221-1691(13)60075-1.

Ali, A., Mohebbullah, M., Shahid, M.A., Alam, S., Uddin, M.N., Miah, M.S., Jamal, M.S.I., Khan, M.S., 2020. PVA- *Nigella sativa* nanofibrous mat: antibacterial efficacy and wound healing potentiality. Journal of the Textile Institute 1–11. https://doi.org/10.1080/00405000.2020.1831168.

Alizadeh Behbahani, B., Noshad, M., Falah, F., 2019. Cumin essential oil: phytochemical analysis, antimicrobial activity and investigation of its mechanism of action through scanning electron microscopy. Microbial Pathogenesis 136, 103716. https://doi.org/10.1016/j.micpath.2019.103716.

Alkhalaf, M.I., Hussein, R.H., Hamza, A., 2020. Green synthesis of silver nanoparticles by Nigella sativa extract alleviates diabetic neuropathy through anti-inflammatory and antioxidant effects. Saudi Journal of Biological Sciences 27 (9), 2410–2419. https://doi.org/10.1016/j.sjbs.2020.05.005.

Almatroudi, A., Khadri, H., Azam, M., Rahmani, A.H., Al Khaleefah, F.K., Khateef, R., Ansari, M.A., Allemailem, K.S., 2020. Antibacterial, antibiofilm and anticancer activity of biologically synthesized silver nanoparticles using seed extract of Nigella sativa. Processes 8 (4), 388. https://doi.org/10.3390/pr8040388.

Ariamanesh, H., Tamizi, N., Yazdinezhad, A., Salah, S., Motamed, N., Amanloo, S., 2019. The effectiveness of Nigella sativa alcoholic extract on the inhibition of Candida albicans colonization and formation of plaque on acrylic denture plates: an in vitro study. Journal of Dentistry 20 (3). https://doi.org/10.30476/dentjods.2019.44911.

Ashraf, S., Anjum, A.A., Ahmad, A., Firyal, S., Sana, S., Latif, A.A., 2018. In vitro activity of Nigella sativa against antibiotic-resistant Salmonella enterica. Environmental Toxicology and Pharmacology 58, 54–58. https://doi.org/10.1016/j.etap.2017.12.017.

Badger-Emeka, L.I., Emeka, P.M., Ibrahim, H.I.M., 2021. Molecular insight into the synergistic mechanism of Nigella sativa (black cumin) with β-lactam antibiotics against clinical isolates of methicillin-resistant Staphylococcus aureus. Applied Sciences 11 (7), 3206. https://doi.org/10.3390/app11073206.

Bady, E.A., Ibrahim, A.Y., Marzoq, A.H.K., 2019. The effect of Nigella sativa and other plants on bacteria isolated from diabetic foot ulcers. Drug Invention Today 11 (10), 7.

Barashkova, A.S., Sadykova, V.S., Salo, V.A., Zavriev, S.K., Rogozhin, E.A., 2021. Nigellothionins from black cumin (Nigella sativa L.) seeds demonstrate strong antifungal and cytotoxic activity. Antibiotics 10 (2), 166. https://doi.org/10.3390/antibiotics10020166.

Bataller, R., Brenner, D.A., 2020. Liver fibrosis. Science & Medicine 66 (4), e36–e41. https://doi.org/10.1201/9780429427275-10.

Batista, B.G., Chaves, M. A. de, Reginatto, P., Saraiva, O.J., Fuentefria, A.M., 2020. Human fusariosis: an emerging infection that is difficult to treat. Revista da Sociedade Brasileira de Medicina Tropical 53, e20200013. https://doi.org/10.1590/0037-8682-0013-2020.

Böttcher, K., Pinzani, M., 2017. Pathophysiology of liver fibrosis and the methodological barriers to the development of anti-fibrogenic agents. Advanced Drug Delivery Reviews 121, 3–8. https://doi.org/10.1016/j.addr.2017.05.016.

Costa, J.G., Keser, V., Jackson, C., Saraiva, N., Guerreiro, Í., Almeida, N., Camões, S.P., Manguinhas, R., Castro, M., Miranda, J.P., Fernandes, A.S., Oliveira, N.G., 2020. A multiple endpoint approach reveals potential in vitro anticancer properties of thymoquinone in human renal carcinoma cells. Food and Chemical Toxicology 136, 111076. https://doi.org/10.1016/j.fct.2019.111076.

Dewanjee, S., Das, S., Das, A.K., Bhattacharjee, N., Dihingia, A., Dua, T.K., Kalita, J., Manna, P., 2018. Molecular mechanism of diabetic neuropathy and its pharmacotherapeutic targets. European Journal of Pharmacology 833, 472–523. https://doi.org/10.1016/j.ejphar.2018.06.034.

Dhanasekar, S., 2019. Alteration of multi-drug resistance activities by ethanolic extracts of Nigella sativa against urinary pathogens. International Journal of Pharmacology 15 (8), 962–969. https://doi.org/10.3923/ijp.2019.962.969.

Dong, J., Liang, Q., Niu, Y., Jiang, S., Zhou, L., Wang, J., Ma, C., Kang, W., 2020. Effects of Nigella sativa seed polysaccharides on type 2 diabetic mice and gut microbiota. International Journal of Biological Macromolecules 159, 725–738. https://doi.org/10.1016/j.ijbiomac.2020.05.042.

El Rabey, H.A., Al-Seeni, M.N., Bakhashwain, A.S., 2017. The antidiabetic activity of *Nigella sativa* and propolis on streptozotocin-induced diabetes and diabetic nephropathy in male rats. Evidence-based Complementary and Alternative Medicine 2017, 1–14. https://doi.org/10.1155/2017/5439645.

Farhangi, M.A., Dehghan, P., Tajmiri, S., Abbasi, M.M., 2016. The effects of Nigella sativa on thyroid function, serum Vascular Endothelial Growth Factor (VEGF) - 1, Nesfatin-1 and anthropometric features in patients with Hashimoto's thyroiditis: a randomized controlled trial. BMC Complementary and Alternative Medicine 16 (1), 1–9. https://doi.org/10.1186/s12906-016-1432-2.

Gasong, B.T., Hartanti, A.W., Tjandrawinata, R.R., 2017. Antibacterial activity of Nigella sativa L.seed oil in water emulsion against dental cariogenic bacteria. International Journal of Pharmaceutical Sciences and Research 8 (7), 8.

Gawron, G., Krzyczkowski, W., Lemke, K., Ołdak, A., Kadziński, L., Banecki, B., 2019. Nigella sativa seed extract applicability in preparations against methicillin-resistant Staphylococcus aureus and effects on human dermal fibroblasts viability. Journal of Ethnopharmacology 244, 112135. https://doi.org/10.1016/j.jep.2019.112135.

Geng, W., Li, C., Zhan, Y., Zhang, R., Zheng, J., 2020. Thymoquinone alleviates liver fibrosis via miR-30a-mediated epithelial-mesenchymal transition. Journal of Cellular Physiology. https://doi.org/10.1002/jcp.30097.

Geng, W., Li, C., Zhan, Y., Zhang, R., Zheng, J., 2021. Thymoquinone alleviates liver fibrosis via miR-30a-mediated epithelial-mesenchymal transition. Journal of Cellular Physiology 236 (5), 3629–3640. https://doi.org/10.1002/jcp.30097.

Ismail, N., Abdel–Mottaleb, Y., Eissa Ahmed, A.A., El-Maraghy, N.N., 2018. Novel combination of thymoquinone and resveratrol enhances anticancer effect on hepatocellular

carcinoma cell line. Future Journal of Pharmaceutical Sciences 4 (1), 41–46. https://doi.org/10.1016/j.fjps.2017.08.001.
Kaatabi, H., Bamosa, A.O., Lebda, F.M., Al Elq, A.H., Al-Sultan, A.I., 2012. Favorable impact of Nigella sativa seeds on lipid profile in type 2 diabetic patients. Journal of Family & Community Medicine 19 (3), 155–161. https://doi.org/10.4103/2230-8229.102311.
Kaya, D.A., Duran, N., 2018. Antifungal activity of Nigella sativa L. and thymbra spicata L. essential oils against trichophyton rubrum, pp. 439–444. https://doi.org/10.24264/icams-2018.VIII.9.
Khare, C.P., 2004. Encyclopedia of Indian Medicinal Plants. Springes-Verlag Berlin Heidelberg, NewYork.
Khurana, A., Sardana, K., Chowdhary, A., 2019. Antifungal resistance in dermatophytes: recent trends and therapeutic implications. Fungal Genetics and Biology 132, 103255. https://doi.org/10.1016/j.fgb.2019.103255.
Kocoglu, E., Kalcioglu, M.T., Uzun, L., Zengin, F., Celik, S., Serifler, S., Gulbay, H., Gonullu, N., 2019. In vitro investigation of the antibacterial activity of Nigella sativa oil on some of the most commonly isolated bacteria in otitis media and externa. The Eurasian Journal of Medicine 51 (3), 247–251. https://doi.org/10.5152/eurasianjmed.2019.18386.
Kolahdooz, M., Nasri, S., Modarres, S.Z., Kianbakht, S., Huseini, H.F., 2014. Effects of Nigella sativa L. seed oil on abnormal semen quality in infertile men: a randomized, double-blind, placebo-controlled clinical trial. Phytomedicine 21 (6), 901–905. https://doi.org/10.1016/j.phymed.2014.02.006.
Kooti, W., Hasanzadeh-Noohi, Z., Sharafi-Ahvazi, N., Asadi-Samani, M., Ashtary-Larky, D., 2016. Phytochemistry, pharmacology, and therapeutic uses of black seed (Nigella sativa). Chinese Journal of Natural Medicines 14 (10), 732–745. https://doi.org/10.1016/S1875-5364(16)30088-7.
Krentz, A.J., 2003. Lipoprotein abnormalities and their consequences for patients with Type 2 diabetes. Diabetes 5. S19-27.
Mohammed, S.J., Amin, H.H.H., Aziz, S.B., Sha, A.M., Hassan, S., Abdul Aziz, J.M., Rahman, H.S., 2019. Structural characterization, antimicrobial activity, and *in vitro* cytotoxicity effect of black seed oil. Evidence-based Complementary and Alternative Medicine 2019, 1–9. https://doi.org/10.1155/2019/6515671.
Najmi, A., Nasiruddin, M., Khan, R., Haque, S., 2008. Effect of Nigella sativa oil on various clinical and biochemical parameters of insulin resistance syndrome. International Journal of Diabetes in Developing Countries 28 (1), 11–14. https://doi.org/10.4103/0973-3930.41980.
Nawarathne, N.W., Wijesekera, K., Gaya Bandara Wijayaratne, W.M.D., Napagoda, M., 2019. Development of novel topical cosmeceutical formulations from *Nigella sativa* L. with antimicrobial activity against acne-causing microorganisms. The Scientific World Journal 2019, 1–7. https://doi.org/10.1155/2019/5985207.
Nikakhlagh, S., Rahim, F., Aryani, F.H., Syahpoush, A., Brougerdnya, M.G., Saki, N., 2011. Herbal treatment of allergic rhinitis: the use of Nigella sativa. American Journal of Otolaryngology 32, 402–407.
Pelegrin, S., Galtier, F., Chalancon, A., Gagnol, J.P., Barbanel, A.M., Pelissier, Y., et al., 2019. Effects of Nigella sativa seeds (black cumin) on insulin secretion and lipid profile: a pilot study in healthy volunteers. British Journal of Pharmacology 85, 1607–1611.
Perillo, B., Di Donato, M., Pezone, A., Di Zazzo, E., Giovannelli, P., Galasso, G., Castoria, G., Migliaccio, A., 2020. ROS in cancer therapy: the bright side of the moon. Experimental & Molecular Medicine 52 (2), 192–203. https://doi.org/10.1038/s12276-020-0384-2.
Randhawa, M., Alenazy, A., Alrowaili, M., Basha, J., 2017. An active principle of Nigella sativa L, thymoquinone, showed significant antimicrobial activity against anaerobic

bacteria. Journal of Intercultural Ethnopharmacology 6 (1), 97. https://doi.org/10.5455/jice.20161018021238.

Reddy, S.H., Al-Kalbani, A.S., Al-Rawahi, A.S., 2018. Studies on phytochemical screening—GC-MS characterization, antimicrobial and antioxidant assay of black cumin seeds (Nigella sativa) and Senna alexandria (Cassia angustifolia) solvent extracts. International Journal of Pharmaceutical Sciences and Research 9 (2), 9.

Rezaeian, A., Amoushahi Khouzain, S., 2018. Effect of Nigella sativa nasal spray on the treatment of chronic rhinosinusitis without a nasal polyp. Allergy Rhinology (Providence) 9, 2152656718800059.

Saleh, F.A., El-Darra, N., Raafat, K., Ghazzawi, I.E., 2017. Phytochemical analysis of Nigella sativa L. Utilizing GC-MS exploring its antimicrobial effects against multidrug-resistant bacteria. Pharmacognosy Journal 10 (1), 99–105. https://doi.org/10.5530/pj.2018.1.18.

Sari, Y., Purnawan, I., Kurniawan, D.W., Sutrisna, E., 2018. A comparative study of the effects of *Nigella sativa* oil gel and aloe vera gel on wound healing in diabetic rats. Journal of Evidence-Based Integrative Medicine 23. https://doi.org/10.1177/2515690X18772804, 2515690X1877280.

Shah, S.B., Shah, S.I.A., Sartaj, L., et al., 2018. Effect of Nigella sativa seeds extracts on clinically important bacterial and fungal species. MOJ Bioequivalence & Bioavailability 5 (4), 218–221. https://doi.org/10.15406/mojbb.2018.05.00105.

Tasawar, Z., Siraj, Z., Ahmad, N., Lashari, M.H., 2011. The effects of Nigella sativa on lipid profile in patients with stable coronary artery disease in Multan, Pakistan. Pakistan Journal of Nutrition 10, 162–167.

Tsuchida, T., Friedman, S.L., 2017. Mechanisms of hepatic stellate cell activation. Nature Review Gastroenterology & Hepatology 14, 397–411. https://doi.org/10.1038/nrgastro.2017.38.

WHO, 2021. Infertility. Retrieved from. https://www.who.int/news-room/fact-sheets/detail/infertility.

Yimer, E.M., Tuem, K.B., Karim, A., Ur-Rehman, N., Anwar, F., 2019. *Nigella sativa* L. (black cumin): a promising natural remedy for wide range of illnesses. Evidence-based Complementary and Alternative Medicine 2019, 1–16. https://doi.org/10.1155/2019/1528635.

Zhao, X., Wang, D., Qin, L., 2021. Lipid profile and prognosis in patients with coronary heart disease: a meta-analysis of prospective cohort studies. BMC Cardiovascular Disorders 21 (1), 1–15. https://doi.org/10.1186/s12872-020-01835-0.

Zhou, M., Liu, X., Li, Z., Huang, Q., Li, F., Li, C.-Y., 2018. Caspase-3 regulates the migration, invasion and metastasis of colon cancer cells: metastasis of colon cancer cells. International Journal of Cancer 143 (4), 921–930. https://doi.org/10.1002/ijc.31374.

Further reading

Bamosa, A., Lebda, F., Al Elq, A., Al-Sultan, A., Kaatabi, H., 2012. Favorable impact of Nigella sativa seeds on lipid profile in type 2 diabetic patients. Journal of Family and Community Medicine 19 (3), 155. https://doi.org/10.4103/2230-8229.102311.

Salehi, B., Quispe, C., Imran, M., Ul-Haq, I., Zivkovic, J., Abu-Reidah, I.M., Sen, S., Taheri, Y., Acharya, K., Azadi, H., Contretas, M.D.M., Carretero, et al., 2021. Nigella plants – traditional uses, bioactive phytoconstituents, preclinical and clinical studies. Frontiers in Pharmacology 12. https://doi.org/10.3389/fphar.2021.

CHAPTER THIRTEEN

Antihypercholesterolemic and antihyperglycemic activity of *Nigella sativa*

Siti Aisyah Abd Ghafar[1], Muhammad Fahmi Yakop[1], Rohazila Mohamad Hanafiah[1], Norsharina Ismail[2]

[1]Department of Basic Sciences, Faculty of Dentistry, Universiti Sains Islam Malaysia, Kuala Lumpur, Selangor, Malaysia

[2]Natural Medicines and Products Research Laboratory, Institute of Bioscience, Universiti Putra Malaysia, Serdang, Selangor, Malaysia

Abbreviations

2 hPg 2 hours plasma glucose
AMPK 5′-adenosine monophosphate (AMP)-activated protein kinase
ASCVD atherosclerotic cardiovascular disease
CAD coronary artery disease
CO_2 carbon dioxide
CoV coronavirus
FBG fasting blood glucose
GLUT4 glucose transporter type 4
HbA1C hemoglobin A1c
HDL high-density lipoprotein
HDL-C high-density lipoprotein cholesterol
HLP-C hyperlipidemic control
HLP-ME hyperlipidemic-methanolic extract
HLP-VO hyperlipidemic-volatile oil
HMG-COA 3-hydroxy-3-methyl-glutaryl-coenzyme A
LDL low-density lipoprotein
LDL-C low-density lipoprotein cholesterol
NC negative control
NLP-C normolipidemic control
NS *Nigella sativa*
OGTT oral Glucose Tolerance Test
PBC powdered black cumin
PON1 paraoxonase 1
PPBG postprandial blood glucose
RCT randomized control trial
STZ streptozotocin
TC total cholesterol
TG triglyceride

Biochemistry, Nutrition, and Therapeutics of Black Cumin Seed
ISBN: 978-0-323-90788-0
https://doi.org/10.1016/B978-0-323-90788-0.00003-2

TRP transient receptor potential protein
VLDL very low-density lipoprotein
WHO World Health Organization

13.1 Introduction

Blessed seed, Al Habbat Al Sawda, and black cumin seeds are a few of many great names given to *Nigella sativa* (NS), which is one of the natural drugs in the list of "Tibb-e-Nabawi" or "Medicine of the Prophet Muhammad." In the word of the Holy Prophet Muhammad saw, "This black cumin is healing for all diseases except As-Sam@Death" (sahih al-Bukhari 5687) (Al-Bukhari, 1976; Ramadan et al., 2020). Based on this statement, NS is one of many great herbs that had been studied exhaustively not only in modern medicine but also it had been studied extensively since the era of Unani medicine when one of the greatest scholars, Ibn Sina described it in the *Canon of Medicine textbook*, which he described it as seeds that stimulate body energy and cure from fatigue and depression. It is not an understatement to call it a blessed seed as it had been reported extensively for its richness in the bioactive compound such as its volatile oil that contains thymoquinone, saponin, tocopherols, carotenoids, stigmasterol, and many more (Ahmad et al., 2013).

In traditional uses, NS is used as a treatment for diarrhea, dyslipidemia, rheumatoid arthritis, dermatological disease, and digestive illnesses. Besides that, it also has been used to treat colic, cough, and as antiseptic agent for eye infections.

Researchers have reported various pharmacological properties such as anticancer, antiinflammatory, antibacterial, bronchodilator, antihypertensive, hepatoprotective, nephroprotective, antioxidant, gastroprotective, antidiabetic, antihypercholesterolemic, and hypoglycemic properties. It has been reported that NS can decrease the virus load and transient receptor potential gene expression after corona virus (CoV) infection (Ulasli et al., 2014). Since many medicinal properties of NS have been found, this chapter will try to review all the relevant studies that have been done on the antihypercholesterolemic and antihyperglycemic properties of NS.

13.2 Antihypercholesterolemic activity of *Nigella sativa* in hypercholesterolemic-induced animal model

Hypercholesterolemia is a primary root of noncommunicable disorder problems, accounting for one-third of ischemic heart disease and one-fifth stroke (Aguilar-Ballester et al., 2021; Adeloye et al., 2020).

Hypercholesterolemia is also one of the health risks for atherosclerotic cardiovascular disease. The World Health Organization has characterized hypercholesterolemia in adults (≥18 years old and above) with total cholesterol (TC) levels of more than 5.0 mmol/L or 190 mg/dL (World Health Organization, 2021). Hypercholesterolemia also known as hyperlipidemia is described as rises of fasting TC concentration which could or could not be correlated with raised triglyceride (TG) concentration. However, lipids are not solvable in plasma but are rather carried in fragments identified as lipoproteins. Hence, the classification of hyperlipidemia or hypercholesterolemia is also based on abnormalities of apolipoprotein (Nelson, 2013). There are four most important forms of lipoproteins which are very low-density lipoprotein, intermediate-density lipoprotein, low-density lipoprotein (LDL), and high-density lipoprotein (HDL). LDL is recognized as bad cholesterol, whereas HDL is known as good cholesterol (Nelson, 2013). Another disease that always interchanged or overlapped with hypercholesterolemia is dyslipidemia. Dyslipidemia is considered by an increase of serum TC, low-density lipoprotein cholesterol (LDL-C), or TGs and decreased serum high-density lipoprotein cholesterol (HDL-C) concentration (Hedayatnia et al., 2020). These parameters also called lipid profile parameters are among the common measured parameters when evaluating hypocholesterolemic properties of natural products mainly from plants.

Study done by Sultan et al. (2011) revealed the usage of powdered black cumin (*N. sativa*) (PBC) on 30 normal Sprague Dawley rats. The rats were fed with 1% and 2% PBC once daily for 56 days. Results showed that PBC lessens the possibility of hypercholesterolemia on normal Sprague Dawley rats by reducing TC, TG, and LDL with nonsignificant HDL level elevation (Sultan et al., 2011).

Another interesting research was done to evaluate the efficacy of NS extracts on normal, hypercholesterolemic, and hyperglycemic-induced Sprague Dawley rats, respectively, using conventional (ethanol [50% v/v]) and supercritical fluid extraction methods. NS extracts were formulated together with the diet given to each group of rats, respectively. The experimental period was executed for 2 months. Both groups' (conventional and supercritical fluid) lipid profile results were compared with control group. From this study, maximum reduction of serum cholesterol level was seen in hypercholesterolemic-induced rats supplemented with NS extract from supercritical fluid preparation with 12.8% followed by 10.7% reduction from conventional group. A similar pattern was seen in serum LDL level, whereby supercritical fluid reduced 17.12% followed by conventional group with 16.00%. TG reduction can be seen with 11.6% trailed by conventional group with 10.00%. Meanwhile, serum HDL level was seen to be increased but not significant ($P > .05$) with 3.1% in supercritical group and 2.7% in conventional group (Iqbal et al., 2017). This study shows supercritical fluid extraction able to increase potential of NS as an antihypercholesterolemic agent in hypercholesterolemic-induced Sprague Dawley rats. This may be due to supercritical fluid extraction method uses carbon dioxide (CO_2) as solvent able to extract more bioactive compounds as compared with the conventional method (ethanol 50% v/v) (Abd Ghafar et al., 2013).

A preclinical study was executed on Wistar rats induced with hyperlipidemia to evaluate hypolipidemic activity of thymoquinone-enriched methanolic and volatile oil extracts from NS oil. Four groups of male Wistar rats were utilized in this research which were normolipidemic control (NLP-C), hyperlipidemic control (HLP-C), methanolic extract (HLP-ME), and volatile oil (HLP-VO). HLP-C, HLP-ME, and HLP-VO groups were induced with hyperlipidemia, while NLP-C is normal untreated Wistar rats. All interventions were given before the administration of atherogenic suspension once daily for 30 days. Results shown forced-feeding of 100 mg ME or 20 mg VO per rat daily lowered plasma TG level to almost normal while significantly ($P < .05$) improved HDL-C level and its subfraction along with arylesterase level (Ahmad et al., 2013a).

Another similar study was done on hyperlipidemia-induced Wistar rats to evaluate on the efficacy of thymoquinone (bioactive compound found in NS seed) as hypolipidemic agents. Results shown thymoquinone able to reduce all parameters involved in cardiovascular risk parameters such as TG, LDL, and TC. Thymoquinone also increased HDLc/LDLc and HDLc/Tc ratios to 259% and 110%, respectively, as compared to control hyperlipidemic rats' group. These changes correspond to the reduction of HMG-CoA reductase activity alongside with upsurge of arylesterase activity in thymoquinone-fed group (Ahmad et al., 2013b).

Oxidative stress Sprague Dawley rats' model was used to evaluate the effect of NS on serum lipid profile. Study shown oxidative stress increased TC, LDL, and TG level. Treatment with either NS-fixed oil or NS essential oil reduced TC, LDL along with TG level in oxidative stress-induced rats. However, no significant changes can be seen in serum HDL level in all three groups (control oxidative stress, oxidative stress + NS-fixed oil, oxidative stress + NS essential oil) (Sultan et al., 2020).

13.3 Hypocholesterolemic effects of *Nigella sativa* in clinical trial

Hypolipidemic effects of NS were tested on 37 menopausal women. The randomized control trial (RCT) was done with intervention of 500 mg NS powder capsule twice daily for 2 months. Meanwhile, placebo group was given 100 mg wheat germ capsule. Results shown NS significantly ameliorated lipid profiles of menopausal women with reduced TC, LDL-C, and TG levels. A significant increase of HDL-C was found when compared to placebo group after 2 months of intervention. A month after termination of the intervention, lipid profiles in treatment group have a tendency to change toward the pretreatment levels (Ibrahim et al., 2014).

A randomized, placebo-controlled clinical trial was done to assess NS seeds as a treatment for hyperlipidemia. A total of 88 subjects aged more than 18 years old with a TC level more than 200 mg/dL were recruited. They were given crushed NS seeds capsules which contained 500 ± 10 mg each. Subjects were instructed to take 2 g NS crushed seeds daily (four capsules every day). Patients were instructed to take two capsules in the morning before breakfast and two more capsules in the afternoon before their meal. This experiment was done in 4 weeks. Patients were also instructed not to change their usual diets during experimental period. At the completion of investigational phase, full blood was collected, and serum lipid profile was analyzed. A significant decrease ($P < .05$) was detected in concentration of TC (4.78%), LDL (7.6%), and TG (16.65%). TGs have shown the most decreased as compared to placebo group. Consumption of 2 g/day of NS crushed seeds for 4 weeks has shown some beneficial effects in treating hyperlipidemia patients (Sabzghabaee et al., 2012).

A clinical trial was performed on healthy and hypercholesterolemic subjects were supplemented with the combination of NS seed and honey (50 mg/kg). The intervention was given daily for 3 months. Results shown significant ($P < .05$) declines in TG (20.1%), ratio of TC:HDLC reduced to 12.3%, complemented with heightened HDLC value by 15% in healthy supplemented group. Meanwhile in hypercholesterolemic supplemented group also decreased significantly ($P < .05$) in TC (6.2%), TG (13.3%), ratio of TC:HDLC (19.5%) with improved HDLC (8.4%). Additionally, significant ($P < .05$) reductions were observed in body mass index (BMI) and body fat in healthy supplemented group and hypercholesterolemic supplemented group (Mohamad et al., 2014).

NS oil was tested on patients with coronary artery disease (CAD). A randomized, double-blind, placebo-controlled clinical trial was done to assess the effects of NS oil supplementation on CAD patients. A total of 60 patients with CAD were recruited and accepted either 2 g/day of NS oil or sunflower oil as a placebo for 56 days (8 weeks). From this study, body weight, BMI, waist circumference, hip circumference, waist to hip ratio, systolic and diastolic blood pressures together with fasting blood glucose (FBG) values have been significantly ($P < .05$) reduced, whereas for TC, TG, HDL, and LDL values, no significant changes have been observed when compared with placebo group (fed with sunflower oil 2 g/day). The results show the potential of NS oil in managing CAD patients by improvement of metabolic parameters, anthropometric indices, blood pressure, and FBG (Tavakoli-Rouzbehani et al., 2021).

A randomized, double-blind, placebo-controlled, clinical trial has been done to appraise the efficacy of NS seeds oil in lessening cardiovascular risk in hypertensive patient. This study evaluated NS seeds oil as an adjunct therapy meant for hypertension, glucose control, and lipid metabolism. A total of 26 hypertensive patients acquired 2.5 mL NS seeds oil (intervention group), meanwhile 29 hypertensive patients in the placebo group accepted 2.5 mL of sunflower oil twice daily for 8 weeks. Systolic blood pressure in the intervention group was significantly reduced ($P < .001$) after contrasted with baseline values and placebo group ($P < .05$). A significant decrease was noticed in the levels of diastolic blood pressure, TC, LDL, malondialdehyde, and FBG ($P < .001$), whereas a significant rise was reflected in the HDL level and glutathione reductase ($P < .001$). Absence of adverse effects was discovered in the placebo group and intervention group by evaluating renal and hepatic-related enzymes (Shoaei-Hagh et al., 2021).

A preliminary study upon the effects of NS seeds on lipid profile had been done in healthy subjects. The authors performed a double-blind, randomized, placebo-controlled 4-week experiment on 30 healthy male volunteers, getting NS seed powder (1 g/day) or placebo orally (15 subjects/group). Lipid profile (TC, LDL, HDL and TG) was measured before and after treatment. No significant differences on serum lipids were noted following therapy neither in any groups nor among the two treatments group. Nevertheless, some potential can be seen in TC and LDL levels, whereby the hypocholesterolemic effect could be seen better in hyperlipidemia patients. One of the reasons that cannot be ignored is the authors felt that low dose of NS seeds (1 g/day) maybe insufficient to see any significant changes in lipid profile of healthy patients (Pelegrin et al., 2019).

A systematic review and metaanalysis of randomized placebo-controlled trials have been done to evaluate the effects of NS on plasma lipid in humans. Authors examined 17 RCT studies which corresponds to plasma lipid concentrations in humans. Metaanalysis of the RCTs shown supplementation with NS able to significantly reduce plasma concentration of TC, LDL-C, and TG. The result was found greater with seed oil, whereas seed powder was discovered to be correlated with a substantial increase in HDL-C levels (Sahebkar et al., 2016).

Another study of systematic review and metaanalysis of randomized controlled trial has been done to evaluate the effects of NS in the treatment of nonalcoholic fatty liver disease. Study reported that supplementation of NS in nonalcoholic fatty liver disorder had no effect on lowering TC, TG, and LDL but showed a substantial increase in HDL values (Tang et al., 2021).

13.4 Antiatherogenic effects of *Nigella sativa*

Antiatherogenic potential of NS has been evaluated on hypercholesterolemia-induced rabbits. This study endeavored on the impact of NS dust and NS oil in comparison to simvastatin. A total of 25 adult New Zealand male white rabbits were split up into five groupings ($n = 5$): normal group (NC), positive control group (hypercholestrolemic, no treatment), fed with 1000 mg/kg NS powder, fed with 500 mg/kg NS oil, and fed with 10 mg/kg simvastatin for 8 weeks. Supplementation of either NS oil or NS powder was proven to reduce significantly ($P < .05$) TC and LDLC levels, meanwhile increased ($P < .05$) HDL levels after 8 weeks when equated to positive control group. Plaque development and intima:media ratios were evaluated and significantly inhibited ($P < .05$) in NS supplementation groups (oil and powder). This study portrays hypocholesterolemic, antiatherogenic, and cardioprotective properties of NS (Al-Naqeep et al., 2011).

Another analysis was done to gauge the activity of NS as a prevention on atherosclerosis progression. A total of 15 adult Dutch Polish male rabbits induced with hypercholesterolemia were used. The rabbits were distributed into three groups ($N = 5$) which were normal diet, hypercholesterolemic diet (1% cholesterol), and hypercholesterolemic + NS (1% cholesterol diet supplemented with 5% NS powder) for 8 weeks. Results shown supplementation of NS powder reduced significantly ($P < .05$) on TC, LDL, and reduced aortic fatty acid streak formation in hypercholesterolemic induced rats when compared to control group (1% cholesterol diet only group). HDLs were seen some positive results regardless not significant ($P > .05$) (Asgary et al., 2013).

13.5 Antihyperglycemic effects of *Nigella sativa*

While there are many studies that suggested the antihyperglycemic properties of NS, there are also studies that reported conflicting result and suggested that NS do not have antihyperglycemic properties. Normal rats treated with petroleum ether extract of NS for 4 weeks did not show reduction in plasma glucose as the plasma glucose remains stable and did not show significant difference with control group (Le et al., 2004). Same conclusion was arrived by Rahman et al. and Awadi et al. (1988), which reported that streptozotocin (STZ)-induced rats treated with NS extract for one weeks show no improvement in Oral Glucose Tolerance Test. Despite reported that diabetic rats treated daily with 500 mg/kg of NS oil for 30 days did not show any significant reduction in serum glucose concentration (191.17 ± 1.13 mg/dL) when compared with untreated diabetic rats (247.31 ± 11.74 mg/dL), however, it also reported that diabetic rats treated daily with a combination of lipoic acid, NS oil, and carnitine for 30 days significantly showed reduction of serum glucose concentration (164.61 ± 1.5 mg/dL) when compared with untreated diabetic rats (247.31 ± 11.74 mg/dL) (Salama et al., 2011).

Recent decades show contradiction result from previously reported study as most of the studies reported that NS do possess antihyperglycemic properties and has the potential to be utilized as therapeutic herbal and dietary remedy in controlling diabetes mellitus (DM), which is the critical cause of hyperglycemia.

NS ethanol extracts have been found to inhibit α-glucosidase activity, which is the important enzyme that catalyze hydrolysis of starch to simple sugar which in turn increase blood glucose level. In vitro study revealed that at 800 ppm, NS ethanol extract inhibits 62% of α-glucosidase activity with IC_{50} value at 610 ppm. Combination of NS and *Cinnamomum cassia* ethanol extract enhances the inhibition properties as the value of IC_{50} (260 ppm) is much lower compared to NS ethanol extract alone. In vivo study revealed that after 28 days, type 1 diabetic rats treated with combination of this extract, plasma glucose level (400 mg/dL) decreases significantly when compared with untreated diabetic rats (600 mg/dL), but combination of both extract (200 mg/kg) with metformin (25 mg/kg) enhanced the antihyperglycemic activity as plasma glucose levels were much lower compared with diabetic rats treated with extracts alone and untreated diabetic rats with the presence of improvements in pancreatic morphological changes. This suggested that there is synergistic effect between both extract and metformin, which means that the combination of extracts has the potential to be an adjunct treatment with the present antidiabetic treatment (Kaur et al., 2018).

Normal albino rats treated intraperitoneally with 50 mg/kg NS seed caused significant decline in blood glucose level after seven days (3.66 ± 0.44 mmol/L) when compared with untreated normal albino rats (5.8 ± 0.44 mmol/L). Increase in the dosage of NS seed to 100, 200, and 300 mg/kg caused significant reduction of serum glucose level after 14 days of treatment. Hypoglycemic effect of NS seed is vanished in this study when the treatment was increased to 400 and 500 mg/kg for 14 days. This concludes that NS seeds show antihyperglycemic effect in optimum dosage (Hawsawi et al., 2001).

Many in vivo studies regarding antihyperglycemic activity of NS seed were on diabetic rats due to main causes of hyperglycemia is type 1 and type 2 DM. Administration of NS methanol extract for 30 days had been effectively lowered diabetic rats blood glucose level. Diabetic rats treated with 500 mg/kg of NS methanol extract have significantly reduced blood glucose level after 30 and 60 min of glucose (2.5 g/kg) ingestion in contrast to untreated diabetic rats. Furthermore, it also suppressed the rise of blood glucose level due to sucrose load effectively at 30 and 60 min with increases of insulin secretion in isolated rat islets that treated with 25−200 μg/mL of NS in presence of 30 and 11 mM glucose with highest insulin level (9.97 ± 0.45) was when the islet pancreas treated with 200 μg/mL of NS methanol extract in presence of 11 mM glucose (Hannan et al., 2019).

NS seeds are also well-known for its richness in bioactive compound and its oil. NS oil extract has been observed to have antihyperglycemic activity along with its most popular bioactive compound, thymoquinone. Only after 10 days of intraperitoneally treatment with 2 mg/kg of NS aqueous extract, there is significant decline in serum glucose concentration (469 ± 42 mg/dL) in STZ-diabetic rats compared with STZ-diabetic untreated rats (655 ± 5 mg/dL), and after 30 days of treatment, serum glucose concentration of treated STZ-diabetic rats with NS aqueous extract has depleted to the level that it is not significant compared to the basal level of normal rats. Although STZ-diabetic rats treated with 0.2 mg/kg NS oil did not show great depletion in serum glucose concentration after 10 days of treatment, serum insulin level begin to increase following 20 days of treatment. Thymoquinone which is the main component of NS volatile oil quinone had been attributed with many medicinal properties, and it has been proved that serum glucose concentration in STZ-diabetic rats treated with 3 mg/kg of thymoquinone has decreased significantly only after 10 days of treatment with the level of serum glucose concentration decrease greatly after 30 days of treatment that it is not significant compared with control normal rat groups. Besides that, serum insulin level restores to normal when STZ-diabetic rats treated with thymoquinone for 30 days. This proved that one of the main bioactive compounds in NS that contribute to the antihyperglycemic properties is thymoquinone by increasing the serum insulin level (Abdelmeguid et al., 2010).

Besides thymoquinone, six indazole-type alkaloids have been identified from defatted and aqueous extract of NS. All indazole-type alkaloid had been said to increase glucose consumption in hepG2 cells with 17-O-(β-D-glucopyranosyl)-4-O-methyl nigellidine exhibited stronger capability to coordinate glucose consumption compared to metformin through 5′-adenosine monophosphate (AMP)-activated protein kinase (AMPK) pathway (Yuan et al., 2014).

Randomized trial of type 2 diabetes patients reported that after daily intake of 3 g of NS oil soft gel capsules for 12 weeks decreased their fasting blood sugar, HbA1c, LDL-c, and TG levels without any compromise to their TC, HDL-c, insulin secretion, and insulin resistance. Supplementation of NS also had been proved to reduce blood glucose level significantly in patients with diabetes (Heshmati et al., 2015). Supplementation of NS for 3 months with dosage 2 gm/day clearly cause significant reduction of FBG, glycosylated hemoglobin, HbA_{1C}, and blood glucose level 2 h postprandially (2 hPG) accompany with reduction in insulin resistance and increased β-cell function. But increasing the dosage of supplementation to 3 gm/day did not cause any significance difference compared to patients supplement with 2 gm/day of NS (Bamosa et al., 2010). Highly significant reduction of FBG, postprandial blood glucose, and HbA_{1C} in type 2 diabetic patients after 6 months daily intake of NS tea (5 gm/day) supplementation when compared with before supplementation of NS tea (El-Sharmy et al., 2011). This implied that NS might act as cotherapy drugs in order to improve type 2 diabetic patient conditions to avoid any effect such as hyperglycemia which in prolonged duration will lead to serious damage to multiple organs such as nervous system, kidney, and liver.

13.6 Mechanism of action of *Nigella sativa* as antihypercholesterolemic and antihyperglycemic agent

Arylesterase is one of the paraoxonase 1 (PON1) enzymes. PON1 arylesterase is known to protect HDL and LDL from lipid peroxidation. Patient with cardiovascular disease such as high blood pressure and atherosclerosis has been reported to possess low level of arylesterase (Yildiz et al., 2008). Arylesterase has been known as a prognostic biomarker in the development of cardiovascular diseases. From previous studies reported in this chapter, NS have shown potentials as an antihypercholesterolemic agent in animal models as well as humans. A few studies reported the mechanisms of action of NS is by increasing arylesterase in hypercholesterolemic animal models (Ahmad et al. 2013a, 2013b). On the other hand, HMG-CoA reductase is a rate-limiting enzyme of cholesterol biosynthesis (Zhong et al., 2021). Ahmad et al. (2013) reported supplementation of thymoquinone (one of the key bioactive compounds in NS) reduced HMG-CoA reductase serum levels in hyperlipidemia-induced Wistar rats (Ahmad et al., 2013b).

Hyperglycemia is a condition when there is a high blood glucose level in body which is characterized by blood glucose level of more than 125 mg/dL while fasting (FBG) or more than 180 mg/dL 2 h postprandial, which is what diabetic patients were termed (Ellen et al., 2019). Major risk factors for a person to develop hyperglycemia are family history of type 2 diabetes, weight greater than 120% of appropriate body weight, hyperlipidemia or hypertension, and presence of polycystic ovarian syndrome. Glucose homeostasis which maintains normal glucose level in body is the balance of insulin and glucagon hormone and is of critical importance to human health. It was maintained within hepatic glucose secretion and peripheral glucose uptake and utilization. Many factors cause hyperglycemia such as lowered insulin secretion, decline in glucose utilization, and excessive glucose production. There are also other secondary factors that contribute to hyperglycemia such as pancreas damage (chronic pancreatitis, hemochromatosis, pancreatic cancer, and cystic fibrosis), endocrine disorder such as Cushing syndrome, acromegaly, and pheochromocytoma, which cause peripheral insulin resistance. Prolonged hyperglycemia or left untreated which is usually caused in diabetic person will lead to various microvascular and macrovascular complications such as neuropathy, retinopathy, and nephropathy (Brealy et al., 2009). Therefore, many studies had been done extensively in order to find any suitable treatment to prevent hyperglycemia in body.

One of the extracts that has been studied extensively is NS. It has been proposed that NS along with its bioactive component such as thymoquinone, alkaloids, saponin, and many more has various mechanisms to suppress hyperglycemia or to restore blood glucose level in diabetic patients which include activation of AMPK pathway, stimulation of insulin production, enhancement of insulin sensitivity, and inhibition of intestinal amylase.

AMPK also known as master regulator of metabolism had been studied extensively for its association with type 2 diabetes and hyperglycemia. Most of the physiological effects due to peripheral activation of AMPK proved to be beneficial to type 2 diabetes patients. Physiological effect that happens due to activation of AMPK such as in skeletal muscle and cardiac muscle where it will stimulate glucose uptake, fatty acid oxidation, glucose transporter type (GLUT) 4 translocation, mitochondrial biogenesis, glycolysis, and inhibit glycogen and protein synthesis, hepatic gluconeogenesis, and secretion of insulin from pancreatic β-cells while stimulates the hypothalamus to increase food intake (Coughlan et al., 2014) . Diabetic *Meriones shawi* rats treated with NS seed ethanol extract for 48 mg/kg/day for 4 weeks had been reported to restore plasma glucose concentration and glucose tolerance, increase insulinemia and acetyl-CoA carboxylase phosphorylation in liver and skeletal muscle, which is one of the main target downstream AMPK pathway activations. This leads to inhibition of conversion of acetyl-CoA to malonyl-CoA, which will allow long chains of fatty acid to enter mitochondria for oxidation and subsequently increase fatty acid oxidation in liver that will lead to reduction in fatty acid in liver and increased hepatic insulin sensitivity. Besides that, activation of AMPK also caused decrease in hepatic gluconeogenesis. NS seed ethanol extract also caused activation of AMPK in skeletal muscle that will lead to increase in GLUT4 synthesis and translocation. All this action by activation of AMPK will in turn decrease glycemia (Benhaddou-Andaloussi et al., 2011). Other than AMPK activation, NS seed extract also increases insulin secretion and β-cell proliferation in pancreas that led to decreased glycemia. NS have been successfully increased serum insulin level after diabetic rats treated with four NS seed ethanol extracts for 4 weeks. The same mechanism has also been reported by study done by Hanan et al., (2019) where insulin secretion was increased when diabetic rat islet treated with NS methanol extract range from 25 to 200 μg/mL accompanied with 3 and 11 mM glucose in a dose-dependent manner (Hanan et al., 2019).

Hyperglycemia due to type 1 diabetes is classified by insufficiency of insulin which may be caused by damage to pancreatic β-cells caused by factors such as inflammatory or autoimmune disease. Total damage to β-cells in type 1 diabetic led to hyperglycemia as there is little to no insulin secreted to reduce blood glucose concentration. Protective effect, regeneration, and partial regeneration of NS extract had been reported in various in vivo study. Histopathological finding of STZ-induced rats treated with hydroalcoholic NS extracts (5−20 mg/kg) for 32 days suggests that NS extracts protected most of the Islet of Langerhans compared with untreated STZ-induced rats that have small and shrunken islet with the islet cells mostly necrotized and degenerated (Alimohammadi et al., 2016). Same conclusion was arrived when STZ-rats treated with NS seed volatile oil (0.2 mL/kg) for 4 weeks as immunohistochemical staining revealed that STZ-rats treated with NS seed oil have increased of insulin-immunoreactive β-cells and granules immunopositivity compared with untreated STZ-rats. Besides that, protective effect of NS seed volatile oil can be seen when electron microscopic findings revealed that there is fair increase in lowered secretory vesicles with granules and small damage accompany with loss of cristae in β-cells mitochondria in contrast to untreated STZ-rats that have severe destruction of β-cells in Langerhans islet (Kanter et al., 2009). This illuminates effectiveness of NS seed volatile oil in the prevention of β-cells damage in STZ-rats. Protective effect of NS seed toward β-cells will lead to increase in serum insulin concentration which in turn will cause decrease in glycemia as can be seen when STZ-rats treated with NS extract and NS seed oils manage to lower serum glucose level to same level in normal rats and restored serum insulin concentration to normal range same as normal rats. Main bioactive component of NS seed volatile oil, thymoquinone has long been said to contribute to NS seed many medicinal properties. As can be seen, it has been proved that thymoquinone has protective effect toward β-cells, increase serum insulin concentration, lower blood glucose level, and most importantly it has been reported to have reduced oxidative stress in diabetes rat models.

Thymoquinone, main component of NS seed volatile oil had been associated with broad therapeutic spectrum. Thymoquinone has been proved to have antihyperglycemic properties as treatment of STZ-rats for 30 days with 3 mg/kg of thymoquinone not only lower serum glucose concentration but also completely returned serum glucose and insulin level to normal. Thymoquinone is able to normalize serum glucose and insulin concentration due to its ability to normalized levels of acetyl-CoA, malonyl-CoA, fatty acid synthase, and fatty acid binding protein that increase due to hyperglycemia (Abdelmeguid et al., 2010; Gray et al., 2016). Besides lowering blood glucose levels, thymoquinone had been found to have antioxidant properties as it had been reported that thymoquinone weakened oxidative stress including decrease in nitric oxide and lipid peroxidase in STZ-rats treated with 50 mg/kg of thymoquinone for 4 weeks (Faisal et al., 2021). Besides thymoquinone, others bioactive components of NS seeds had been identified for antihyperglycemic and antidiabetic properties such as indazoles-type alkaloids that exhibit antihyperglycemic properties via AMPK pathway activation (Yuan et al., 2014).

References

Abd Ghafar, S.A., Ismail, M., Saiful Yazan, L., Fakurazi, S., Ismail, N., Chan, K.W., Md Tahir, P., 2013. Cytotoxic activity of kenaf seed oils from supercritical carbon dioxide fluid extraction towards human colorectal cancer (HT29) cell lines. Evidence-based Complementary and Alternative Medicine 2013, 1–8.

Abdelmeguid, N.E., Fakhoury, R., Kamal, S.M., Al Wafai, R.J., 2010. Effects of Nigella sativa and thymoquinone on biochemical and subcellular changes in pancreatic β-cells of streptozotocin-induced diabetic rats. Journal of Diabetes 2 (4), 256–266.

Adeloye, D., Abaa, D.Q., Owolabi, E.O., Ale, B.M., Mpazanje, R.G., Dewan, M.T., Omoyele, C., Ezeigwe, N., Alemu, W., Harhay, M.O., Auta, A., Adewole, I.F., 2020. Prevalence of hypercholesterolemia in Nigeria: a systematic review and meta-analysis. Public Health 178, 167–178.

Aguilar-Ballester, M., Hurtado-Genovés, G., Taberner-Cortés, A., Herrero-Cervera, A., Martínez-Hervás, S., González-Navarro, H., 2021. Therapies for the treatment of cardiovascular disease associated with type 2 diabetes and dyslipidemia. International Journal of Molecular Sciences 22 (2), 660.

Ahmad, A., Husain, A., Mujeeb, M., Khan, S.A., Najmi, A.K., Siddique, N.A., Anwar, F., 2013. A review on therapeutic potential of Nigella sativa: a miracle herb. Asian Pacific Journal of Tropical Biomedicine 3 (5), 337–352.

Ahmad, S., Beg, Z.H., 2013a. Elucidation of mechanisms of actions of thymoquinone-enriched methanolic and volatile oil extracts from Nigella sativa against cardiovascular risk parameters in experimental hyperlipidemia. Lipids in Health and Disease 12 (1), 1–12.

Ahmad, S., Beg, Z.H., 2013b. Hypolipidemic and antioxidant activities of thymoquinone and limonene in atherogenic suspension fed rats. Food Chemistry 138 (2–3), 1116–1124.

Al-Awadi, F.M., Gumaa, K.A., 1988. Studies on the activity of individual plants of an antidiabetic plant mixture. Journal of Ethnopharmacology 22 (3), 315.

Al-Bukhari, M.I., 1976. In: Al-Bukhari, S. (Ed.), The Collection of Authentic Sayings of Prophet Muhammad (Peace Be upon Him), Division 71 on Medicine, second ed. Hilal Yayinlari, Ankara.

Alimohammadi, S., Hobbenaghi, R., Javanbakht, J., Kheradmand, D., Mortezaee, R., Tavakoli, M., Akbari, H., 2016. Protective and antidiabetic effects of extract from Nigella sativa on blood glucose concentrations against streptozotocin (STZ)-induced diabetic in rats: an experimental study with histopathological evaluation. Diagnostic Pathology 11 (1), 1-1.

Al-Naqeep, G., Al-Zubairi, A.S., Ismail, M., Amom, Z.H., Esa, N.M., 2011. Antiatherogenic potential of Nigella sativa seeds and oil in diet-induced hypercholesterolemia in rabbits. Evidence-based Complementary and Alternative Medicine 2011, 213628.

Asgary, S., Ghannadi, A., Dashti, G., Helalat, A., Sahebkar, A., Najafi, S., 2013. Nigella sativa L. improves lipid profile and prevents atherosclerosis: evidence from an experimental study on hypercholesterolemic rabbits. Journal of Functional Foods 5 (1), 228–234.

Bamosa, A.O., Kaatabi, H., Lebdaa, F.M., Elq, A.M., Al-Sultanb, A., 2010. Effect of Nigella sativa seeds on the glycemic control of patients with type 2 diabetes mellitus. Indian Journal of Physiology & Pharmacology 54 (4), 344–354.

Benhaddou-Andaloussi, A., Martineau, L., Vuong, T., Meddah, B., Madiraju, P., Settaf, A., Haddad, P.S., 2011. The in vivo antidiabetic activity of Nigella sativa is mediated through activation of the AMPK pathway and increased muscle Glut4 content. Evidence-based Complementary and Alternative Medicine 2011, 538671.

Brealey, D., Singer, M., 2009. Hyperglycemia in critical illness: a review. Journal of Diabetes Science and Technology 3 (6), 1250–1260.

Coughlan, K.A., Valentine, R.J., Ruderman, N.B., Saha, A.K., 2014. AMPK activation: a therapeutic target for type 2 diabetes? Diabetes, Metabolic Syndrome and Obesity: Targets and Therapy 7, 241.

El-Shamy, K.A., Mosa, M.M.A., El-Nabarawy, S.K., El-Qattan, M., 2011. Effect of Nigella sativa tea in type 2-diabetic patients as regards glucose homeostasis, liver and kidney functions. Journal of Applied Science and Research 7 (12), 2524–2534.

Ellen Davis, R.N., Natalie Mandolfo, M.S.N., July 2019. Hyperglycemia and cancer: a state-of-the-science review. In: Oncology Nursing Forum, vol 46, No. 4. Oncology Nursing Society, pp. 459–472.

Faisal Lutfi, M., Abdel-Moneim, A.M.H., Alsharidah, A.S., Mobark, M.A., Abdellatif, A.A., Saleem, I.Y., Alsharidah, M., 2021. Thymoquinone lowers blood glucose and reduces oxidative stress in a rat model of diabetes. Molecules 26 (8), 2348.

Gray, J.P., Zayasbazan Burgos, D., Yuan, T., Seeram, N., Rebar, R., Follmer, R., Heart, E.A., 2016. Thymoquinone, a bioactive component of Nigella sativa, normalizes insulin secretion from pancreatic β-cells under glucose overload via regulation of malonyl-CoA. American Journal of Physiology-Endocrinology and Metabolism 310 (6), E394–E404.

Hannan, J.M.A., Ansari, P., Haque, A., Sanju, A., Huzaifa, A., Rahman, A., Azam, S., 2019. Nigella sativa stimulates insulin secretion from isolated rat islets and inhibits the digestion and absorption of (CH2O) n in the gut. Bioscience Reports 39 (8). BSR20190723.

Hawsawi, Z.A., Ali, B.A., Bamosa, A.O., 2001. Effect of Nigella sativa (black seed) and thymoquinone on blood glucose in albino rats. Annals of Saudi Medicine 21 (3–4), 242–244.

Hedayatnia, M., Asadi, Z., Zare-Feyzabadi, R., Yaghooti-Khorasani, M., Ghazizadeh, H., Ghaffarian-Zirak, R., et al., 2020. Dyslipidemia and cardiovascular disease risk among the MASHAD study population. Lipids in Health and Disease 19 (1), 1–11.

Heshmati, J., Namazi, N., Memarzadeh, M.R., Taghizadeh, M., Kolahdooz, F., 2015. Nigella sativa oil affects glucose metabolism and lipid concentrations in patients with type 2 diabetes: a randomized, double-blind, placebo-controlled trial. Food Research International 70, 87–93.

Ibrahim, R.M., Hamdan, N.S., Mahmud, R., Imam, M.U., Saini, S.M., Abd Rashid, S.N., et al., 2014. A randomised controlled trial on hypolipidemic effects of Nigella Sativa seeds powder in menopausal women. Journal of Translational Medicine 12 (1), 1–7.

Iqbal, M.J., Butt, M.S., Qayyum, M.M.N., Suleria, H.A.R., 2017. Anti-hypercholesterolemic and anti-hyperglycaemic effects of conventional and supercritical extracts of black cumin (Nigella sativa). Asian Pacific Journal of Tropical Biomedicine 7 (11), 1014–1022.

Kanter, M., Akpolat, M., Aktas, C., 2009. Protective effects of the volatile oil of Nigella sativa seeds on β-cell damage in streptozotocin-induced diabetic rats: a light and electron microscopic study. Journal of Molecular Histology 40 (5–6), 379–385.

Kaur, G., Invally, M., Khan, M.K., Jadhav, P., 2018. A nutraceutical combination of Cinnamomum cassia & Nigella sativa for Type 1 diabetes mellitus. Journal of Ayurveda and Integrative Medicine 9 (1), 27–37.

Le, P.M., Benhaddou-Andaloussi, A., Elimadi, A., Settaf, A., Cherrah, Y., Haddad, P.S., 2004. The petroleum ether extract of Nigella sativa exerts lipid-lowering and insulin-sensitizing actions in the rat. Journal of Ethnopharmacology 94 (2–3), 251–259.

Mohamad, S., Ibrahim, N.H., Yusof, H., 2014. Blood pressure and lipid lowering effects of Nigella sativa seeds and honey mixture. Journal of Nursing and Health Science 3 (1), 89–96.

Nelson, R.H., 2013. Hyperlipidemia as a risk factor for cardiovascular disease. Primary Care: Clinics in Office Practice 40 (1), 195–211.

Pelegrin, S., Galtier, F., Chalançon, A., Gagnol, J.P., Barbanel, A.M., Pélissier, Y., et al., 2019. Effects of Nigella sativa seeds (black cumin) on insulin secretion and lipid profile: a pilot study in healthy volunteers. British Journal of Clinical Pharmacology 85 (7), 1607–1611.

Ramadan, M.F. (Ed.), 2020. Black Cumin (Nigella Sativa) Seeds: Chemistry, Technology, Functionality, and Applications. Springer Nature.

Sabzghabaee, A.M., Dianatkhah, M., Sarrafzadegan, N., Asgary, S., Ghannadi, A., 2012. Clinical evaluation of Nigella sativa seeds for the treatment of hyperlipidemia: a randomized, placebo controlled clinical trial. Medical Archives 66 (3), 198–200.

Sahebkar, A., Beccuti, G., Simental-Mendía, L.E., Nobili, V., Bo, S., 2016. Nigella sativa (black seed) effects on plasma lipid concentrations in humans: a systematic review and meta-analysis of randomized placebo-controlled trials. Pharmacological Research 106, 37–50.

Salama, R.H.M., 2011. Hypoglycemic effect of lipoic acid, carnitine and Nigella sativa in diabetic rat model. International Journal of Health Sciences 5 (2), 126.

Shoaei-Hagh, P., Kamelan Kafi, F., Najafi, S., Zamanzadeh, M., Heidari Bakavoli, A., Ramezani, J., et al., 2021. A randomized, double-blind, placebo-controlled, clinical trial to evaluate the benefits of Nigella sativa seeds oil in reducing cardiovascular risks in hypertensive patients. Phytotherapy Research 2021, 1–13.

Sultan, M.T., Aslam, F., Rasool, B., Imran, M., Ahmad, A.N., Tariq, H.B., et al., 2020. Nigella sativa ameliorates oxidative stress induced adverse effects in rodent modeling studies: indices of serum chemistry and hematology. Food Science and Technology 41, 404–411.

Sultan, M.T., Butt, M.S., Ahmad, R.S., Batool, R., Naz, A., Suleria, H., 2011. Supplementation of powdered black cumin (Nigella sativa) seeds reduces the risk of hypercholesterolemia. Functional Foods in Health and Disease 1 (12), 516–524.

Tang, G., Zhang, L., Tao, J., Wei, Z., 2021. Effect of Nigella sativa in the treatment of nonalcoholic fatty liver disease: a systematic review and meta-analysis of randomized controlled trials. Phytotherapy Research 2021, 1–11.

Tavakoli-Rouzbehani, O.M., Abbasnezhad, M., Kheirouri, S., Alizadeh, M., 2021. Effects of Nigella sativa oil supplementation on selected metabolic parameters and anthropometric indices in patients with coronary artery disease: a randomized, double-blind, placebo-controlled clinical trial. Phytotherapy Research 2021, 1–12.

Ulasli, M., Gurses, S.A., Bayraktar, R., Yumrutas, O., Oztuzcu, S., Igci, M., Igci, Y.Z., Cakmak, E.A., Arslan, A., 2014. The effects of Nigella sativa (Ns), Anthemis hyalina (Ah) and Citrus sinensis (Cs) extracts on the replication of coronavirus and the expression of TRP genes family. Molecular Biology Reports 41 (3), 1703–1711.

World Health Organization, 2021. Mean Cholesterol - Global Health Observatory (GHO) Data available from: https://www.who.int/data/gho/indicator-metadata-registry/imr-details/2384.

Yildiz, A., Gur, M., Demirbag, R., Yilmaz, R., Akyol, S., Aslan, M., Erel, O., 2008. Paraoxonase and arylesterase activities in untreated dipper and non-dipper hypertensive patients. Clinical Biochemistry 41 (10–11), 779–784.
Yuan, T., Nahar, P., Sharma, M., Liu, K., Slitt, A., Aisa, H.A., Seeram, N.P., 2014. Indazole-type alkaloids from Nigella sativa seeds exhibit antihyperglycemic effects via AMPK activation in vitro. Journal of Natural Products 77 (10), 2316–2320.
Zhong, S., Li, L., Liang, N., Zhang, L., Xu, X., Chen, S., Yin, H., 2021. Acetaldehyde Dehydrogenase 2 regulates HMG-CoA reductase stability and cholesterol synthesis in the liver. Redox Biology 41, 101919.

CHAPTER FOURTEEN

Anticancer activity of black cumin seed bioactive compounds

Suzy Munir Salama
Indigenous Knowledge and Heritage Center, Ghibaish College of Science and Technology, Ghibaish, Sudan

Abbreviations

Bax Bcl-2-associated X protein
BCl-2 B-cell lymphoma 2
BCUSs black cumin seeds
COL carvacrol
DNA deoxy ribonucleic acid
DU145 human prostate cancer cell line
HepG-2 human liver cancer cell line
HL-60 human leukemia cell line
LON longifolene
Nf-κB nuclear factor kappa-light-chain-enhancer of activated B cells
PC-3 human prostate cancer cell line
SCLC human small cell lung cancer
THY thymol
TQU thymoquinone
WEHI-3 mouse leukemia cell line.

14.1 Introduction

In the recent days, cancer is considered one of the crucial problems of health that causes high mortality worldwide. Recent publications revealed significant increase in the global prevalence and mortality due to cancer diseases by the year 2020. Ferlay et al. (2021) reported that an average of 19.3 million cancer cases were reported with special reference to breast cancer, lung cancer, and prostate cancer. Authors added that lung, gastric, and hepatic cancers recorded higher figures of deaths (Ferlay et al., 2021). Metastasis of cancer illness is the main reason of high mortality due to the activity of sick cells which can send chemicals or cross the boundaries of blood vessels to pass to different body parts spreading the disease (Mehraj et al., 2022). High toxicity and low potentiality of the chemotherapeutic medicines in treating various kinds of cancers could be attributed to the alteration of

Biochemistry, Nutrition, and Therapeutics of Black Cumin Seed
ISBN: 978-0-323-90788-0
https://doi.org/10.1016/B978-0-323-90788-0.00005-6

pathways of cell signaling resulting in serious adverse symptoms and failure of chemotherapy (Majdalawieh et al., 2017). Noteworthy, more than 55% of the drugs used in treating cancer ailments are plant-based products for their anticancer efficacy of the biotic compounds extracted from medicinal plant species (Mehraj et al., 2022).

Scientists directed their studies on the natural remedies as alternative and safe products compared to synthetic compounds for treating human disorders (Adib-Hajbaghery and Rafiee, 2018). Black cumin seeds (BCUS) are seeds belong to the annual herb *Nigella sativa* that grows widely in the countries at the borders of Mediterranean Sea and Asian countries such as India and Pakistan. *N. sativa* seeds are used by Arab and Asian natives for decades as food spice rich in nutrients added to different food recipes viz bread, pickles, and yoghurt (Khoddami et al., 2011) for maintaining health and treating various diseases including high blood pressure, kidney disorders, heart diseases, and diabetes (Gali-Muhtasib et al., 2006; Khan and Afzal, 2016). BCUSs are mentioned in Quran by Prophet Mohammed as the seeds of healing power and in Holly Bible as Gith of Pliny (Gali-Muhtasib et al., 2006; Chopra, 1956; Nadkarni, 1976). These black seeds are traditionally named in Arab countries as *Habbat-albaraka* and *Alhabba-alsooda*. Based on studies, researchers referred the high nutritive constituents of BCUS to their contents of plant-based oil, minerals, fiber, protein, and vitamins. Additionally, many biologically active compounds have been identified from BCUS and proved for their biological activities against many disorders (Mehraj et al., 2022). This chapter exposes the recent reports on the anticancer activities of biotic compounds isolated from BCUS.

14.2 Discussion

14.2.1 Anticancer phytochemical compounds of BCUS

BCUSs were found to contain many active constituents including approximately fixed oil 38%, volatile oil 1.1%, alkaloids 2.1%, sterols 30%, and tocopherols 18.5 mg/100 g seeds (Mehraj et al., 2022; Sultan et al., 2012). The major anticancer compounds reported by the studies of researchers constitute thymoquinone (TQU), alpha-hederin (α-HD), thymol (THY), longifolene (LON), and carvacrol (COL) and illustrated in Fig. 14.1.

14.2.2 Anticancer of BCUS bioactive compounds

14.2.2.1 Thymoquinone

TQU is an essential oil that constitutes 50% of the volatile oils of BCUS and was proved for its bioactivity as anticancer. Yi et al. (2008) and Ahmad et al.

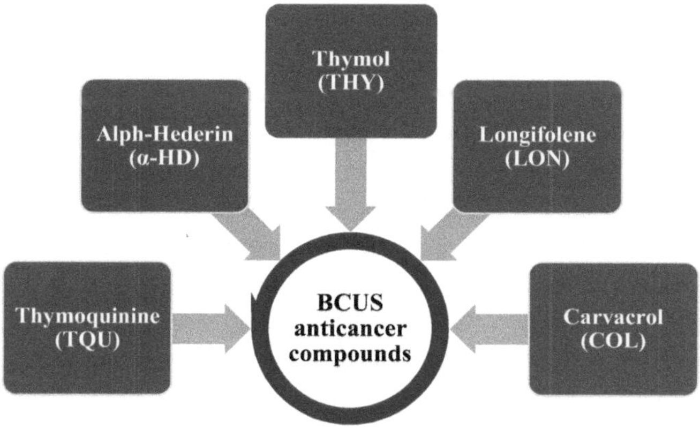

Figure 14.1 Anticancer compounds isolated from black cumin seeds (BCUSs).

(2021) published that TQU can be considered a promising drug in cancer therapy through lowering the proliferation activity of endothelial cells that line blood vessels and inhibiting angiogenesis of experimentally induced prostate cancer in experimental animals and cell culture as well (Ahmad et al., 2021; Yi et al., 2008). In liver cancer cell line HepG-2, TQU showed elevation in the hepatic level of glutathione transferase enzyme that increases the antioxidative power of liver cells and consequently protects the cells against carcinogenic agents (Nagi and Almakki, 2009). In the cancer of pancreas, TQU reported control on proliferation of sick cells via its significant reducing potential to mucin 4 expression in cancer cells without showing effect on healthy cells (Torres et al., 2010). Recent studies explained the potency of TQU in treating cancer cells in terms of its prooxidant activity in the diseased cells increasing the production of reactive oxygen species and consequently apoptosis of tumorous cells (Mehraj et al., 2022). Peng et al. (2013) concluded that TQU could inhibit DNA nuclear factor-kappa in the human osteosarcoma cell line exhibiting anticancer, antiangiogenic, and apoptotic activities (Peng et al., 2013). In vitro study on mouse leukemic cell line WEHI-3 coupled with in vivo experiment on the anticancer activity of TQU on blood cancer induced in mice has shown considerable antileukemic activity through the overexpression of the apoptosis regulator protein (Bax gene) and down-expression of the anti-apoptotic protein (BCl-2 gene) (Ali Salim et al., 2014). TQU induced caspase-8 apoptotic pathway in the mitochondria of human leukemia (HL-60) cell line leading to their death (El-Mahdy et al., 2005) and melanoma 518A2 cell lines through the induction of mitochondrial apoptosis and DNA damage of cells (Effenberger et al., 2010). Jafri et al. (2010) stated that treating lung cancer cell line of human small cell lung cancer with TQU or a composite of TQU with the effective drug cisplatin has recorded high anticancer activity. Authors added that the combined agents TQU + cisplatin have significantly reduced the lung tumor size in mice (Jafri et al., 2010). These outcomes were confirmed by Alaufi et al., (2017) who published that the combined mixture of TQU and cisplatin could effectively induce cytotoxicity and stimulate apoptosis in human oral squamous cancerous cell line without affecting normal epithelium in vitro study (Alaufi et al., 2017). A study conducted on experimental induction of colon cancer in rats reported that the treatment of sick rats with TQU has showed remarkable suppression in the tumor progression through down-expression of vascular endothelium growth factor (Asfour et al., 2013). Ke et al. (2015) reported that the treatment of many kinds of human hepatoma cell lines with prepared concentration of TQU could exert significant inhibition of cell growth and stimulation of cellular apoptosis depending on TQU dose compared to normal liver cell. Additionally, TQU reported obvious inhibition of hepatocellular carcinoma in mice induced by human liver cancer cells (Ke et al., 2015). Recent docking study on TQU extracted from BCUS published that TQU could down-regulate P38 which was reported for its crucial role in hepatocellular carcinoma disease (Tabassum and Ahmad, 2020). According to the potency of TQU in animal and cell culture experiments, it is considered as a potent chemopreventive candidate in the treatment of cancer (Gomathinayagam et al., 2020).

14.2.2.2 Alpha-hederin

In vitro and dose-dependent treatment of different types of cell lines, murine leukemia, human mammary adenocarcinoma, colon cancer, human leukemia, human endothelium, mouse fibrosarcoma, mouse's lung cancer and liver cancer cell culture with the steroidal coumarin α-HD at high concentration 50 mg/mL and double this dose have led to the death of all treated cancer cells via the stimulation of caspase-3 and mitochondrial apoptotic pathway showing anticancer efficacy in vitro (Huat and Swamy, 2003). Cheng et al. (2014) exposed that α-HD has cytotoxic activity on breast cancer cell culture, and its anticancer mechanism is attributed to its stimulation to caspase-3 and caspase-9 as a result cancer cells stop growth and start apoptotic death (Cheng et al., 2014).

14.2.2.3 Thymol

Satooka and Kubo (2012) reported that the monoterpene phenol THY suggested cytotoxic activity against melanoma cell line and recorded 50% inhibition concentration of about 60 μg/mL. Mechanistically, THY elevated the grade of reactive oxygen subjects in cancerous cells leading to their death (Satooka and Kubo, 2012). As per Islam et al. (2019) documents from literature have stated that THY exhibit anticancer activity through its antioxidant and apoptotic efficacies in *in vitro* cancer cell culture and studies on mice and rats (Islam et al., 2019). The mechanistic efficacy of THY as anticancer is regarded to its antiproliferative activity (Agarwal et al., 2020).

14.2.2.4 Longifolene

Cytotoxic effect of the essential oil LON isolated from BCUS was first reported by Bourgou et al. (2010) on lung and colon adenocarcinoma cell culture recording 50% inhibition concentration of approximately 60 and 80 μM, respectively (Bourgou et al., 2010).

14.2.2.5 Carvacrol

The monoterpenoid COL was studied for its antitumor potential in vitro. Results of the experiment revealed that COL induced programmed cell death of two cell culture models of prostate cancer DU145 and PC-3 via augmentation in the generation of reactive oxygen agents and mitochondrial apoptosis of cancerous cells depending on the dose of the administered COL. In the same study, COL altered the overactivity of Notch pathway in the prostate cancer cell line tested (Khan et al., 2019) and revealed promising chemotherapeutic candidate in treating prostate cancer. Bourgou et al. (2010) published that COL exhibits cytotoxic potential on human lung and colon adenocarcinoma recording 50% inhibition concentration of approximately 45 μg per mL in both cancer cells (Bourgou et al., 2010).

14.3 Conclusion

In conclusion, many bioactive compounds of BCUS were proved for their therapeutic efficacy against cancer diseases in experimental animal models and cell culture screenings. The anticancer active ingredients studied and reported by the studies were TQU, α-HD, THY, LON, and COL. TQU was proved by many studies as the most potent anticancer active constituent of BCUS. Further studies are required on BCUS to elucidate more active ingredients, which need screening for their bioactivity, especially cancer fighters.

References

Adib-Hajbaghery, M., Rafiee, S., 2018. Medicinal plants use by elderly people in Kashan, Iran. Nursing and Midwifery Studies 7 (2), 67–73. http://eprints.kaums.ac.ir/id/eprint/3146.

Agarwal, S., Tripathi, S., Mishra, N., 2020. Pharmacological potential of thymol. Innovations in Food Technology. Springer, pp. 489–500. https://doi.org/10.1007/978-981-15-6121-4_34.

Ahmad, M.F., Ahmad, F.A., Ashraf, S.A., Saad, H.H., Wahab, S., Khan, M.I., et al., 2021. An updated knowledge of Black seed (*Nigella sativa* Linn.): review of phytochemical constituents and pharmacological properties. Journal of Herbal Medicine 25. https://doi.org/10.1016/j.hermed.2020.100404.

Alaufi, O.M., Noorwali, A., Zahran, F., Al-Abd, A.M., Al-Attas, S., 2017. Cytotoxicity of thymoquinone alone or in combination with cisplatin (CDDP) against oral squamous cell carcinoma *in vitro*. Scientific Reports 7 (1), 1–12. https://doi.org/10.1038/s41598-017-13357-5.

Ali Salim, L.Z., Othman, R., Abdulla, M.A., Al-Jashamy, K., Mohd Ali, H., Hassandarvish, P., et al., 2014. Thymoquinone inhibits murine leukemia WEHI-3 cells *in vivo* and *in vitro*. PLoS One 9 (12), e115340. https://doi.org/10.1371/journal.pone.0115340.

Asfour, W., Almadi, S., Haffar, L., 2013. Thymoquinone suppresses cellular proliferation, inhibits VEGF production and obstructs tumor progression and invasion in the rat model of DMH-induced colon carcinogenesis. Pharmacology & Pharmacy 4. https://doi.org/10.4236/pp.2013.41002.

Bourgou, S., Pichette, A., Marzouk, B., Legault, J., 2010. Bioactivities of black cumin essential oil and its main terpenes from Tunisia. South African Journal of Botany 76 (2), 210–216. https://doi.org/10.1016/j.sajb.2009.10.009.

Cheng, L., Xia, T.-S., Wang, Y.-F., Zhou, W., Liang, X.-Q., Xue, J.-Q., et al., 2014. The anticancer effect and mechanism of α-hederin on breast cancer cells. International Journal of Oncology 45 (2), 757–763. https://doi.org/10.3892/ijo.2014.2449.

Chopra, R.N., 1956. Glossary of Indian Medicinal Plants. CSIR, New Delhi, p. 175. https://doi.org/10.1086/402350.

Effenberger, K., Breyer, S., Schobert, R., 2010. Terpene conjugates of the *Nigella sativa* seed-oil constituent thymoquinone with enhanced efficacy in cancer cells. Chemistry and Biodiversity 7 (1), 129–139. https://doi.org/10.1002/cbdv.200900328.

El-Mahdy, M.A., Zhu, Q., Wang, Q.E., Wani, G., Wani, A.A., 2005. Thymoquinone induces apoptosis through activation of caspase-8 and mitochondrial events in p53-null myeloblastic leukemia HL-60 cells. International Journal of Cancer 117 (3), 409–417. https://doi.org/10.1002/ijc.21205.

Ferlay, J., Colombet, M., Soerjomataram, I., Parkin, D.M., Piñeros, M., Znaor, A., et al., 2021. Cancer statistics for the year 2020: an overview. International Journal of Cancer 149 (4), 778–789. https://doi.org/10.1002/ijc.33588.

Gali-Muhtasib, H., El-Najjar, N., Schneider-Stock, R., 2006. The medicinal potential of black seed (*Nigella sativa*) and its components. Advances in Phytomedicine 2, 133–153. https://doi.org/10.1016/S1572-557X(05)02008-8.

Gomathinayagam, R., Ha, J.H., Jayaraman, M., Song, Y.S., Isidoro, C., Dhanasekaran, D.N., 2020. Chemopreventive and anticancer effects of thymoquinone: cellular and molecular targets. Journal of Cancer Prevention 25 (3), 136–151. https://doi.org/10.15430/JCP.2020.25.3.136.

Huat, B.T.K., Swamy, S., 2003. Intracellular glutathione depletion and reactive oxygen species generation are important in α-hederin-induced apoptosis of P388 cells. Molecular and Cellular Biochemistry 245 (1), 127–139. https://doi.org/10.1023/A:1022807207948.

Islam, M.T., Khalipha, A.B., Bagchi, R., Mondal, M., Smrity, S.Z., Uddin, S.J., et al., 2019. Anticancer activity of Thymol: a literature-based review and docking study with Emphasis on its anticancer mechanisms. International Union of Biochemistry and Molecular Biology Life 71 (1), 9–19. https://doi.org/10.1002/iub.1935.

Jafri, S.H., Glass, J., Shi, R., Zhang, S., Prince, M., Kleiner-Hancock, H., 2010. Thymoquinone and cisplatin as a therapeutic combination in lung cancer: *In vitro* and *in vivo*. Journal of Experimental & Clinical Cancer Research 29 (1), 1–11. http://www.jeccr.com/content/29/1/87.

Ke, X., Zhao, Y., Lu, X., Wang, Z., Liu, Y., Ren, M., et al., 2015. TQ inhibits hepatocellular carcinoma growth *in vitro* and *in vivo via* repression of Notch signaling. Oncotarget 6 (32), 32610–32621. https://doi.org/10.18632/oncotarget.5362.

Khan, M.A., Afzal, M., 2016. Chemical composition of *Nigella sativa* Linn: part 2 recent advances. Inflammopharmacology 24 (2), 67–79. https://doi.org/10.1007/s10787-016-0262-7.

Khan, F., Singh, V.K., Saeed, M., Kausar, M.A., Ansari, I.A., 2019. Carvacrol induced program cell death and cell cycle arrest in androgen-independent human prostate cancer cells *via* inhibition of notch signaling. Anti-Cancer Agents in Medicinal Chemistry 19 (13), 1588–1608. https://doi.org/10.2174/1871520619666190731152942.

Khoddami, A., Ghazali, H.M., Yassoralipour, A., Ramakrishnan, Y., Ganjloo, A., 2011. Physicochemical characteristics of nigella seed (*Nigella sativa* L.) oil as affected by different extraction methods. Journal of the American Oil Chemists' Society 88 (4), 533–540. https://doi.org/10.1007/s11746.

Majdalawieh, A.F., Fayyad, M.W., Nasrallah, G.K., 2017. Anti-cancer properties and mechanisms of action of thymoquinone, the major active ingredient of *Nigella sativa*. Critical Reviews in Food Science and Nutrition 57 (18), 3911–3928. https://doi.org/10.1080/10408398.2016.1277971.

Mehraj, T., Elkanayati, R.M., Farooq, I., Mir, T.M., 2022. A review of *Nigella sativa* and its active principles as anticancer agents. Black Seeds (*Nigella Sativa*) 91–118. https://doi.org/10.1002/ijc.33588.

Nadkarni, K., 1976. Crocus sativus, *Nigella sativa*. Indian Materia Medica, pp. 386–411.

Nagi, M.N., Almakki, H.A., 2009. Thymoquinone supplementation induces quinone reductase and glutathione transferase in mice liver: possible role in protection against chemical carcinogenesis and toxicity. Phytotherapy Research: An International Journal Devoted to Pharmacological and Toxicological Evaluation of Natural Product Derivatives 23 (9), 1295–1298. https://doi.org/10.1002/ptr.2766.

Peng, L., Liu, A., Shen, Y., Xu, H.-Z., Yang, S.-Z., Ying, X.-Z., et al., 2013. Antitumor and anti-angiogenesis effects of thymoquinone on osteosarcoma through the NF-κB pathway. Oncology Reports 29 (2), 571–578. https://doi.org/10.3892/or.2012.2165.

Satooka, H., Kubo, I., 2012. Effects of thymol on B16-F10 melanoma cells. Journal of Agricultural and Food Chemistry 60 (10), 2746–2752. https://doi.org/10.1021/jf204525b.

Sultan, M.T., Butt, M.S., Ahmad, R.S., Pasha, I., Ahmad, A.N., Qayyum, M.M.N., 2012. Supplementation of *Nigella sativa* fixed and essential oil mediates potassium bromate induced oxidative stress and multiple organ toxicity. Pakistan Journal of Pharmaceutical Sciences 25 (1), 175–181. https://doi.org/10.1186/s12906-015-0853-7.

Tabassum, H., Ahmad, I.Z., 2020. Molecular docking and dynamics simulation analysis of thymoquinone and thymol compounds from *Nigella sativa* L. that inhibits P38 protein: probable remedies for hepatocellular carcinoma. Medicinal Chemistry 16 (3), 350–357. https://doi.org/10.2174/1573406415666190416165732.

Torres, M.P., Ponnusamy, M.P., Chakraborty, S., Smith, L.M., Das, S., Arafat, H.A., et al., 2010. Effects of thymoquinone in the expression of mucin 4 in pancreatic cancer cells: implications for the development of novel cancer therapies. Molecular Cancer Therapeutics 9 (5), 1419–1431. https://doi.org/10.1158/1535-7163.MCT-10-0075.

Yi, T., Cho, S.-G., Yi, Z., Pang, X., Rodriguez, M., Wang, Y., et al., 2008. Thymoquinone inhibits tumor angiogenesis and tumor growth through suppressing AKT and extracellular signal-regulated kinase signaling pathways. Molecular Cancer Therapeutics 7 (7), 1789–1796. https://doi.org/10.1158/1535-7163.MCT-08-0124.

CHAPTER FIFTEEN

Antioxidant activity and phenolic content of black cumin seeds

Norsharina Ismail[1], Siti Aisyah Abd Ghafar[2], Md Zuki Abu Bakar[1,3]
[1]Natural Medicines and Products Research Laboratory, Institute of Bioscience, Universiti Putra Malaysia, Serdang, Selangor, Malaysia
[2]Department of Basic Sciences, Faculty of Dentistry, Universiti Sains Islam Malaysia, Kuala Lumpur, Selangor, Malaysia
[3]Department of Veterinary Pre-Clinical Sciences, Faculty of Veterinary Medicine, Universiti Putra Malaysia, Serdang, Selangor, Malaysia

Abbreviations

ABTS 2,20-azino-bis-3-ethylbenzothiazoline-6-sulfonic acid
BHA butylated hydroxy anisole
BHT butylated hydroxy toluene
COVID-19 coronavirus 2019
CP cold press
DPPH 2, 2-diphenyl-1-picrylhydrazyl
FRAP ferric reducing antioxidant power
GAE gallic acid equivalents
HPLC high performance liquid chromatography
HPLC−PDA high-performance liquid chromatography with photodiode array detector
IC_{50} half-maximal inhibitory concentration
LC-ESI-QTOF/MS liquid chromatography coupled with electrospray ionization-quadrupole time-of-flight and mass spectrometry
PRFs phenolic-rich fractions
QE quercetin equivalents
ROS reactive oxygen species
SFE supercritical fluid extraction
TE trolox equivalents
TFC total flavonoid content
TPC total phenolic content

15.1 Introduction

Plant-based foods have a high energy density, and they are consumed all over the world. They comprise several bioactive substances that possess health-promoting properties in addition to their nutritional value. Polyphenols, phytosterols, biogenic amines, and carotenoids, for instance, are among the compounds that have been found to improve human health in incidents

Biochemistry, Nutrition, and Therapeutics of Black Cumin Seed
ISBN: 978-0-323-90788-0
https://doi.org/10.1016/B978-0-323-90788-0.00006-8

of cancer, cardiovascular disease, and diabetes, including in those with gastrointestinal, immunological, and neurological problems (Samtiya et al., 2021). The most frequent plant-derived natural antioxidants include polyphenols (phenolic acids, flavonoids, anthocyanins, stilbenes, and lignans), carotenoids such as carotenes and xanthophylls, as well as vitamins C and E. Polyphenols and carotenoids, in particular, are natural antioxidants with antiaging, antiinflammatory, antiviral, antimicrobial, and anticancer properties (Li et al., 2016; Zhang et al., 2016; Xu et al., 2017).

On the one hand, natural antioxidants have been exploited as active packaging and coatings to enhance food quality and shelf life. Synthetic antioxidants, for instance, butylated hydroxy toluene (BHT), butylated hydroxy anisole (BHA), and tert-butyl hydroquinone are detrimental to human health. Hence, natural antioxidants are being explored largely for food packaging. Several studies have shown that naturally derived antioxidant-based active packaging ingredients can deliver active substances to food at controlled rates, reducing lipid oxidation while maintaining food quality and nutritional value. Plant extracts in aqueous and alcoholic solvents, herbs and spices essential oils, and polyphenolics derived from various waste biosources are all good sources of natural antioxidants (Barbosa-Pereira et al., 2013; De Abreu et al., 2011).

15.1.1 Black cumin seeds

The Ranunculaceae family includes *Nigella sativa* L., widely known as black cumin seeds, and is commonly used as a spice in Mediterranean cuisine. It is considered as a medicinal spice due to the inclusion of terpenoids, flavonoids, tannins, alkaloids, and polyphenols, which make it a great natural antioxidant. On the scientific research of pharmacognostical properties, chemical arrangement, and pharmacological activity of black cumin seeds, as reviewed by Areefa et al. (2020), the authors summarized that numerous studies validate its significance in folk medicine as an antihypertensive, anti-inflammatory, immune stimulant, analgesic, antioxidant, antimicrobial, antifungal, and as an anticancer agent.

A recent research focused on the pharmacological activities of black cumin seed and its active metabolites, which could be a good starting point for developing a new Coronavirus 2019 (COVID-19) prevention and treatment strategy. It is on the basis that the essential oil and black cumin seed extracts have significant therapeutic results toward immune disorder, autophagy dysfunction, oxidative stress, ischemia, and inflammation in a variety of COVID-19 comorbidities, including diabetes, cardiovascular disorders, Kawasaki-like diseases, and a variety of bacterial and viral contagions (Islam et al., 2021).

15.2 Antioxidant and oxidative stress

Free radicals generate oxidative stress, which is the main risk factor for illnesses such as heart disease, diabetes, hypertension, and cancer (Burits and Bucar, 2000). The production of several free radicals, including superoxide and hydroxyl radicals, leads to the formation of reactive oxygen species (ROS), which is correlated with oxidative stress. Free radicals might be generated during metabolic processes and then eliminated by the intrinsic antioxidant system. Toxins may disrupt the balance of prooxidants and antioxidants, leading to an increase in oxidation free radicals (Shankar and Mehendale, 2014).

Despite a substantial literature devoted to the accumulation and origins of ROS, recent antioxidant-based remedies lack specificity for defective tissues, cells, and organelles, and hence might fail to achieve an efficient concentration at the site of pathological oxidative stress. Furthermore, antioxidants focus on massive quantities of reactive oxygen intermediates and are incapable in influencing specialized intermediates in the oxidative reaction, resulting in a number of unfocused treatment options. Food high in antioxidants, on the other hand, may have an important function in the immune system, the creation of cellular energy, and the scavenging of ROS. The wide range of processes in which antioxidant molecules are engaged implies that antioxidants may play a preventive function in the development of age-related illnesses. As a result, an antioxidant can be an effective method for both health span and longevity extension. Despite the fact that antioxidants are not medications, they hold great promise and could be used to battle age-related disorders in the future. The potential for antioxidants to replace conventional medicines in the treatment of age-related disorders could be significant, and long-term clinical trials are necessary to assess this (Tan et al., 2018).

Antioxidants are categorized as primary or secondary antioxidants based on their reactive mechanism for decreasing ROS levels in the body. Primary antioxidants directly stabilize radicals through processes such as hydrogen transfer or single electron transfer, such as employing 2, 2-diphenyl-1-picrylhydrazyl (DPPH), 2,20-azino-bis-3-ethylbenzothiazoline-6-sulfonic acid (ABTS), and galvinoxyl radicals. Whereas secondary antioxidant activity corresponds to the lowering of ROS levels in the body, such copper reduction uses the Fenton reaction to avoid the production of radicals (Galano et al., 2016; Couttolenc et al., 2020).

15.3 Phenolic compounds

Polyphenols are plant-derived secondary metabolites that make up the biggest group of phytochemicals (Li et al., 2014). Phenolic compounds are a class of small molecules with at least one phenol unit in their composition. They have been categorized into subgroups established from their chemical structures, such as phenolic acids, flavonoids, tannins, coumarins, lignans, quinones, stilbenes, and curcuminoids (Gan et al., 2018). Among these, the most important dietary phenolic components are phenolic acids, flavonoids, and tannins. They drew interest because of their wide range of bioactivities, particularly for antioxidative properties, which might be explained by phenolic compounds' redox capacity. Polyphenols display their antioxidant activity in two ways: by acting as radical scavengers to counteract cellular damage caused by ROS and by preventing ROS from being generated in the first place (Costea et al., 2019). Classification and characterization of natural polyphenols isolated from various food items is a requirement for investigators due to safety matters, standards, regulations, and authorization of synthetic antioxidants (Zhang et al., 2018).

15.4 Antioxidant properties and phenolic constituents of black cumin seeds

The phenolic components were eluted in part due to the solvents used and the extraction duration. The solubility of phenolic compounds can vary greatly due to the complexity of their physicochemical structures (Muzolf-Panek and Stuper-Szablewska 2021). Notably, polar solvents such as methanol, ethanol, and water, as well as other solvents such as chloroform, dichloromethane, ethyl acetate, and hexane, were the most regularly used for the extraction of phenolic compounds from black cumin seeds (Table 15.1). Nevertheless, ethanol and ethanol—water solvents are preferable for food-based applications as they minimize the risk of hazardous solvent residue contamination. Since combinations of water and ethanol affect the polarity of the solvent and hence influence extraction efficiency, mixtures of solvents are commonly used to enhance extraction yields. Generally, employing 50% aqueous ethanol to extract phenolic components from herbs

Table 15.1 Extraction methods, antioxidant activity, and phenolic content of black cumin seeds.

Extraction methods	Extracts	Antioxidant tests	Phenolic content/ phenolic compounds/ others	References
Soxhlet extraction	Aqueous, methanolic, ethanolic, chloroform, and dichloromethane crude extracts	DPPH and FRAP	Total polyphenols contents	Dalli et al. (2021)
Solvent extraction	Water, 50% aqueous Ethanol and pure ethanol	DPPH, ABTS, and FRAP	TFC, TPC, protocatechuic acid, *t*-cinnamic acid, *p*-hydroxybenzoic acid, kaempferol, catechin, and luteolin	Muzolf-Panek & Stuper-Szablewska (2021)
Incubator shaker	Ethanol extract	DPPH, ABTS, and FRAP	Cinnamic acid, protocatechuic acid, *p*-hydroxybenzoic acid, and chlorogenic acid	Feng et al. (2020)
Solvent extraction	Water—methanol extract	DPPH	Total phenolics	Soleimanifar et al. (2019)
Solvent extraction	Water—methanol extract	DPPH, ABTS, FRAP, and β-carotene bleaching assay	TPC, TFC, thymoquinone, diosgenin, and phenolic acids, comprising chlorogenic acid, caffeic acid, and kaempferol	Hameed et al. (2019)

(Continued)

Table 15.1 Extraction methods, antioxidant activity, and phenolic content of black cumin seeds.—cont'd

Extraction methods	Extracts	Antioxidant tests	Phenolic content/ phenolic compounds/ others	References
SFE and CP	*Nigella sativa* oil	DPPH and FRAP	TPC, caryophyllene, thymoquinone, 1,4-cyclohexadiene, longifolene, carvacrol	Mohammed et al. (2016)
Extracted with n-hexane and further fractionated	Black cumin seed total oil, neutral lipid fraction, and polar lipid fraction	DPPH, ABTS, FRAP, and β-carotene bleaching test	TPC and TFC	Mosbah et al. (2015)
Solvent extraction	70% ethanolic extracts	DPPH, ABTS, and FRAP	TPC, TFC, quercetin, and kaempferol	Toma et al. (2015)
Sonication	Aqueous methanolic extract	DPPH	TPC, TFC, and thymoquinone	Uras et al. (2014)
Shaking water bath	70% methanolic extract	DPPH, FRAP, β-carotene, and linoleic acid system	TPC	Şen et al. (2010)
Solvent extraction	Methanol extract	DPPH, FRAP, superoxide anion radical scavenging assay, phosphomoly bdenum method, and linoleic acid peroxidation assay	TPC	El-Beshbishy et al. (2009)

Soxhlet extraction	Ethyl acetate, hexane, and water fractions	DPPH, β-carotene –linoleate bleaching, and inhibition of corn oil oxidation	Hydroxybenzoic, syringic, and *p*-coumaric acids	Mariod et al. (2009)
Soxhlet extraction	Methanol extract	DPPH, superoxide anion radical scavenging	Vanillic acid	Bourgou et al. (2008)
Soxhlet extractor using n-hexane	Crude oils	Oxidative stability (peroxide and p-anisidine values)	Total phenolics	Ramadan (2004)

and spices was shown to be highly effective (Muzolf-Panek and Stuper-Szablewska, 2021). At high temperatures, solubility and diffusion increase. Excessively high temperatures might cause solvent evaporation, resulting in destruction of thermolabile components and undesirable impurity extracts (Zhang, Lin et al., 2018).

On the other hand, the oil extraction was obtained conventionally using a cold press (CP) extraction method that involves neither heat nor solvents. This approach received better consumer acceptance as it is more natural and safer (Kiralan et al., 2014). Nonetheless, this technique gave minimal yields, and the remaining meal included 10%–12% oil, making it unsuitable for food processing (Soto et al., 2007; Zúñiga et al., 2003). Supercritical fluid extraction (SFE) is an ecologically approachable technology for extracting plant oils that has several advantages over standard techniques employed in the oil industry (Akanda et al., 2012). SFE is the recommended way for extracting antioxidant chemicals from black cumin oil as it produces a higher concentration of thymoquinone (Solati et al., 2014). The supercritical fluid-extracted black cumin oil had a significant quantity of thymoquinone and total phenolic components. In comparison to CP, the SFE method might be considered the best method for extracting black cumin oil since it improves the yield quality (Mohammed et al., 2016).

A few in vitro tests were utilized to investigate the antioxidant activity of black cumin seeds, including the radical scavenging activity test, DPPH, ABTS, and ferric reducing antioxidant power (FRAP) assays (Table 15.1). The most frequently used methods to measure radical scavenging abilities are the DPPH and ABTS tests. Nevertheless, as these radicals are nonphysiological radicals, they are not biologically relevant (Shalaby and Shanab, 2013). Both assays work on the same principles, relying on single electron transfer pathways via stable DPPH• and ABTS•+ free radicals, respectively, with the possibility of hydrogen atom donation. The spectrophotometer could easily measure the reduction of these free radicals in the presence of antioxidants in the visible region.

In comparison with DPPH•, ABTS•+ cation radicals have the advantage of being soluble in both aqueous and organic solvents, allowing testing of both nonpolar and polar samples. These cation radicals can be employed at various pH levels, which is highly valuable for investigating the efficacy of different plant components. DPPH• radicals, on the other hand, are acidic pH sensitive and only dissolve in organic solvents. The decolorization of the DPPH• radical solution due to the acidic pH renders determining the antioxidant's radical scavenging efficacy impossible. Furthermore, when evaluating the properties of samples containing anthocyanins, the results of the DPPH test may be underestimated due to color interference between these flavonoids and the DPPH solution (Shalaby and Shanab, 2013). Additionally, the FRAP assay, which measures the reduction of ferric iron to ferrous iron, is usually used to investigate the samples' metal chelating capabilities (Benzie and Strain, 1999).

The drawbacks of this associated research include the use of a variety of extraction methods for spices and herbs (Assefa et al., 2017), making comparisons of phenolic substances and antioxidant activity between different studies of these plant extracts difficult. Furthermore, the antioxidant activity of extracted compounds can be distinguished by a number of mechanisms, including radical scavenging capacity or metal chelating capabilities. Even if the method for measuring antioxidant activity is identical, the particular method may differ greatly depending on the modifications made (Muzolf-Panek and Stuper-Szablewska, 2021).

The Folin–Ciocalteu method was applied to measure the total phenolic content (TPC) of black spices, which was conveyed as gallic acid equivalents (GAEs)/g of sample. Meanwhile, a spectrophotometric approach based on the formation of a flavonoid–aluminum complex was employed to establish the total flavonoid content (TFC). Total polyphenols were highest in the aqueous extract of black cumin seeds (51.17 ± 1.95 μg GAE/mg dry weight). The DPPH antiradical activity revealed that the aqueous extract had the lowest value of all the extracts analyzed, together with a half-maximal inhibitory concentration (IC_{50}) of 1489.34 ± 28.64 μg/mL, whereas chloroform and dichloromethane extracts had no activity. Meanwhile, the flavonoids in the ethanolic extract were highly concentrated (43.06 ± 1.25 μg quercetin equivalents (QE)/mg dry weight). With an IC_{50} of 1026.437 ± 3.61 μg/mL in the FRAP test, the methanolic extract has the smallest value (Table 15.1) (Dalli et al., 2021).

Among the three black spices, black pepper had the highest overall polyphenol content, followed by black cardamom and black cumin (Table 15.1) (Feng et al., 2020). Twenty distinct spices, extending from 12.03 to 22.88 mg GAE/g, have earlier been found to have significant variances in TPC values, which were significantly greater than the present black spices (Feng et al., 2020). It was generally recognized that the geographical environment and harvest season had an impact on the polyphenol content of spices (Škrovánková et al., 2017). In the black species, TFC values ranged from 0.41 ± 0.01 to 3.97 ± 0.01 mg QE/g. Of the three black spices, black pepper (3.97 ± 0.01 mg QE/g) possesses the highest flavonoid compounds, shadowed by black cardamom (0.73 ± 0.01 mg QE/g) and black cumin (0.41 ± 0.01 mg QE/g). The TFC of the black cumin sample in this investigation was more than double that of raw black cumin seeds from India, which could be due to a varietal difference or a different solvent extraction ratio (Liang et al., 2018).

Long-time extraction of 24 h with 50% aqueous ethanol produced the highest total phenolics and antioxidant activities in 13 herbs and spices used for food flavoring and preservation. The extracts used in this study were classified into four groups based on their antioxidant activity. Cloves were categorized as having very high antioxidant activity. All spices and bay leaves exhibited high antioxidant capacity. Basil, oregano, thyme, and rosemary were under moderate antioxidant activity, whereas nutmeg, black seed, cardamom, caraway, garlic, and onion showed the lowest antioxidant activity. Chemometric studies confirmed this distinction. The impact of phenolic components on the antioxidant activity of extracts was demonstrated using principal component analysis. Finally, the principal components were used to execute a noticeable discrimination of samples according to the general discrimination analysis, revealing a great variety of plant materials based on phenolic profiles and antioxidant activities (Muzolf-Panek and Stuper-Szablewska, 2021).

The DPPH, FRAP, and ABTS tests were manipulated to investigate the antioxidant capability of three black spices. Black pepper has the greatest DPPH value, followed by black cardamom and black cumin. When it came to ABTS, three black spices outperformed the DPPH and FRAP assays in terms of antioxidant capacity. The antioxidant capabilities of black pepper were substantially higher than those of black cumin. These three black spices had stronger ABTS radical scavenging activity than Korean black pepper, cumin, and cardamom; the difference might be described by various sample preparation procedures (Assefa et al., 2017). The black cardamom had the highest FRAP capacity. Three black spices had FRAP values that were within the range of Serbian black spices (Nikolic et al., 2019).

The TPC of black cumin seed extract produced using water—methanol solvent was determined to be 955.77 mg/kg. The IC_{50} of the black cumin seed extract was 104.76 mg/mL in a DPPH test, while the IC_{50} of BHT was 8.06 mg/mL. At a concentration of 100 mg/mL, the black cumin seed extract prevented the development of oxidation primary products in unrefined soybean oil in the incubation assay. The increase in black cumin seed oil's relative oxidative stability was concluded due to the presence of phenolic components. Black cumin seed extract can also be compared with BHT in terms of antioxidant properties at greater concentrations, allowing it to be used in edible oils (Soleimanifar et al., 2019).

According to the findings, some treatments with spice combinations had more antioxidant activity than those with a single spice. TPC, TFC, DPPH, β-carotene bleaching assay, FRAP, and ABTS assay values for methanolic extract of black cumin seeds were 292.5 ± 9.14 mg GAE/100 g, 188.8 ± 5.69 mg QE/100 g, IC_{50} μg/mL of 46.3 ± 0.32, 76.96 ± 0.81%, 5.53 ± 0.08 mg/Trolox equivalent g, and 67.18 ± 0.82 μmol Trolox equivalent/g, respectively. Finally, the extraction of spices was dependent on the solvent used, and they demonstrated outstanding antioxidant power not just individually but also in combination, suggesting that they can be used to combat oxidative stress-related diseases (Hameed et al., 2019).

The black cumin seeds oil was extracted using SFE and CP techniques, and the physicochemical parameters, antioxidant activity, and thermal behavior were examined. The IC_{50} of DPPH antioxidant activity of SFE and cold-pressed oils were 1.58 and 2.30 mg/mL, respectively. On the other hand, the FRAP activity was 538.67 and 329.00 mmol/100 mL for SFE oil and CP oil, respectively. SFE oil showed 160.51 mg/100 mL, while CP oil exhibited 94.40 mg/100 mL of TPC, expressed as GAE. This study found that black cumin oil extracted by SFE might yield a high level of natural antioxidants (Mohammed et al., 2016).

With the utilization of n-hexane, the total oil from black cumin seeds was extracted and divided into two lipid fractions: neutral and polar. The total oil had the greatest TPC and TFC levels. In all antioxidant experiments, total oil exhibited the highest antioxidant capacity, with IC_{50} of DPPH (83.023 ± 4.58 μg/mL), ABTS (80.09 ± 10.51 μg/mL), reducing power assay (85.93 ± 20.01 μg/mL), and β-carotene bleaching assay (60.62%) (Mosbah et al., 2015).

TPC and TFC levels, as well as primary antioxidant activity, were determined in aqueous methanolic extracts of different parts of black cumin. As a result, the TPC value of the seed extract was significantly greater than that of the aerial part extract. The TFC ranged between 106.10 ± 0.10 and 118.06 ± 0.02 mg rutin equivalent/g extract. In a DPPH radical scavenging assay, the extracts' inhibitory was reported to be 95.09%–90.71% (Uras, Silahtaroȷ;lu, Ilçim and Kökdil, 2014).

The black cumin samples from Konya had the most potent radical scavenging activity when compared with different origins from Isparta, Corum, Burdur, Afyon, and Samsun. Black cumin samples of Konya origin at 1 mg/

mL exhibited 94.59% in the β-carotene bleaching method and showed 44.44% in the DPPH assay at 0.5 mg/mL. The TPC of black cumin extracts and their antioxidant activity were found to be positively correlated. Their antioxidant activity was either comparable to or greater than that of positive BHA and BHT controls. These encouraging findings suggest that black cumin seeds could be used as a health-promoting element in dietary supplements and nutraceuticals (Şen et al., 2010).

The greater TPC was elicited by curcumin, followed by black cumin and grape seed extracts. Each sample showed the same pattern when tested for DPPH and superoxide scavenging activities, as well as the phosphomolybdenum method. However, antioxidant activity using the ferric thiocyanate-linoleic system showed an increased order of grape seed extract > curcumin > black cumin extract. There was a strong linear relationship between TPC and the above parameters in the investigated samples, while the DPPH test was slightly lower than in the other analyses. The high TPC of these samples may be related to their reducing power and antioxidative potential and thus could be useful as food preservatives and health supplements (El-Beshbishy et al., 2009).

The presence of antioxidant-active components in phenolic-rich fractions (PRFs) of black cumin seedcake was determined. According to the overall phenolic content of the PRFs, the ethyl acetate fraction has the highest phenolic content (78.8 mg/g), followed by the other fractions. The PRFs showed promise as natural antioxidants and could be employed to enhance corn oil's oxidative stability. Some of the black cumin PRFs were more effective than BHA at preventing the production of corn oil's primary and secondary oxidation products. Because these PRFs function as free-radical terminators and chelating agents, the use of these natural antioxidants to stabilize edible oils is a possibility (Mariod et al., 2009).

The essential oil and methanol extracts produced from Tunisian black cumin seeds exhibit substantial antioxidant activity by scavenging free radicals. These activities could be linked to the essential oil's high terpenic hydrocarbon content and the methanol extract's high phenolic acid and flavonoid content. These data show that black cumin seeds from Tunisia could be a good source of natural antioxidants (Bourgou et al., 2008).

The oxidative stability of black cumin, coriander (*Coriandrum sativum* L.), and niger (*Guizotia abyssinica* Cass.) crude and stripped seed oils was studied and compared. Crude oils are more resistant to oxidation than their refined counterparts. The polar portions are entirely separated from the oil during stripping. Coriander, black cumin, and niger seeds were ranked in order of oxidative stability. The proportion of polar lipids in crude oil was shown to be related to the oil's oxidative stability. As a result, the carryover of crude oil's polar lipids may be a major role in their improved oxidative stability (Ramadan, 2004).

15.5 Effect of processing methods on the antioxidant activity and phenolic content of black cumin seeds

Microwave radiation may increase the amount of bioactive phenolics in black cumin seeds, which have physiological action and antioxidant potency. Microwave radiation's effect on black cumin seed antioxidant activity as a purpose of pretreatment time and extract concentration was examined. As a result, microwave pretreatment lowered TFC considerably as pretreatment duration increased. In all extracts, antioxidant activity improved as pretreatment time and extract concentration increased. The most effective extract was an aqueous methanolic extract from radiated seeds. The outcomes of several antioxidative assays were equivalent to those obtained from Fourier-transform infrared spectroscopy of extracts, suggesting the presence of multiple functional groups. The antioxidant efficacy of the sample extracts was graded as follows: 70% methanol in water >70% ethanol in water > methanol > ethanol. Treatment time and extract concentration were discovered to be important components in influencing the product's overall quality. The findings of this study are significant in terms of employing black cumin seed to generate high-antioxidant food products (Hossen and Ali, 2019).

The application of biogein and humic acid encouraged vegetative development, with data indicating a significant increase in plant height, number of branches per plant, and seed yield per plant as compared to untreated plants (control). Using these fertilizers also boosted oil content, yield extract, total phenolic, and total flavonoid levels. In terms of antioxidant activity, organic and biofertilizer treatments have been proven to increase antioxidant activity. The application of biogein and humic acid resulted in an increase, decrease, appearance, and disappearance of various chemicals when compared to the control sample. The preceding findings show that biogein and humic acid have a stimulating influence on growth and bioactive substances, which is evident in black cumin yield productivity (Abou El-Leel et al., 2019).

The antioxidant activities of black cumin seeds were studied using superheated steam heating instead of ordinary hot air heating. When the time of heating was extended from 0 to 30 min for both treatments, the TPC, TFC, and DPPH tests of black cumin seed increased considerably. At practically all heating durations, the TPC, TFC, and DPPH activities of the black cumin seed heated under superheated steam were considerably better than those of conventional hot air heating. Based on the findings, it can be assumed that superheated steam can be utilized as a substitute for roasting black cumin seeds while still maintaining increased antioxidant characteristics (Liang et al., 2018).

In pot experiments, the reactions of black cumin seeds to high zinc concentrations were examined. There was also a considerable increase in TPC and TFC, including increased antioxidant activity. Overall, these data imply that black cumin seeds might be employed in phytoremediation, and that Zn abundance could be a useful solution to enhance the nutritional quality of this essential species (Marichali et al., 2016).

Extraction yield, TPC, and DPPH activity were measured in control and irradiation (2, 4, 8, 10, 12, and 16 kGy) samples extracted in acetone, methanol, and water. The findings demonstrated that the extraction yields for all of the test solvents increased with higher radiation dose. Finally, gamma irradiation increased free radical scavenging activity while improving extraction yield and TPC. The antioxidant activity and TPC of black cumin seed after irradiation were also influenced by the type of solvent used for extraction (Khattak et al., 2008).

15.6 Phenolic compounds in black cumin seeds

Because polyphenols have such a vast structural diversity, reliable identification and quantification of these phenolic compounds have proven to be challenging. One of the most recent approaches to identifying and characterizing polyphenols is liquid chromatography paired with electrospray ionization-quadrupole time-of-flight and mass spectrometry (LC-ESI-QTOF/MS). The polyphenolic components of black cumin that have a specific phenolic composition, principally flavonoids and phenolic acids, were successfully identified using this mass spectrometry analysis (Feng et al., 2020).

The qualitative evaluation of LC-ESI-QTOF/MS using positive and negative ionization modes of phenolic components in black cumin detected four subclasses of phenolic acids (i.e., hydroxycinnamic acids, hydroxybenzoic acids, hydroxyphenylacetic acids, and hydroxyphenylpropanoic acids). The most abundant compounds in black cumin were hydroxycinnamic acids, which include cinnamic acid, chlorogenic acid, and *p*-coumaroyl tartaric acid (Feng et al., 2020). Black cumin has also been found to contain cinnamic acid (Singh et al., 2004). Two caffeic acid derivatives (caffeic acid 3-O-glucuronide and caffeic acid 3-sulfate) and ferulic acid 4-O-glucoside were also detected in hydroxycinnamic acids subclasses of black cumin (Feng et al., 2020). Procatechuic acid, 4-O-methylgallic acid, hippuric acid, *p*-hydroxybenzoic acid, vanillic acid 4-sulfate, and paeoniflorin are all hydroxybenzoic acids found in black cumin (Mnif and Aifa, 2015).

Apart from dihydrochalcones and dihydroflavonols, four subclasses of flavonoids (anthocyanins, flavonols, isoflavonoids, and flavones) were qualitatively detected in black cumin in both positive and negative modes (Feng et al., 2020). In black cumin, flavonols were the most common subclass, with 10 distinct flavonols tentatively identified. Kaempferol 3-*O*-glucosyl-rhamnosyl-galactoside, myricetin 3-*O*-rhamnoside, patuletin 3-*O*-glucosyl-(1−[6)-[apiosyl (1−[2)]-glucoside, kaempferol 3,7-*O*-diglucoside, and kaempferol 3,7,4′-*O*-triglucoside were found (Feng et al., 2020). The black cumin seedcake also contained myricetin (Kadam and Lele, 2017).

Six anthocyanins have been identified, namely malvidin 3,5-*O*-diglucoside, delphinidin 3-*O*-glucosyl-glucoside, peonidin 3-*O*-sambubioside-5-*O*-glucoside, cyanidin 3-*O*-diglucoside-5-*O*-glucoside, cyanidin 3-*O*-(2-*O*-(6-*O*-(E)-caffeoyl-D-glucoside)-D-glucoside)-5-*O*-D-glucoside, and peonidin 3-*O*-rutinoside (Feng et al., 2020). Furthermore, a total of eight isoflavonoids derivatives were detected, including 3′,4′,7-trihydroxyisoflavan, 3′-*O*-methylequol, 2′-hydroxyformononetin, 3′,4′,7-trihydroxyisoflavanone, 6″-*O*-acetyldaidzin, 3′,4′,5,7-tetrahydroxyisoflavanone, quercetin, and kaempferol (Feng et al., 2020). Five compounds have been provisionally identified in flavones subclass, namely cirsilineol, gardenin B, apigenin 6,8-di-C-glucoside, diosmin, and kaempferol-3-glucoside (Feng et al., 2020).

For quantification, high-performance liquid chromatography with a photodiode array detector (HPLC−PDA) might be utilized (Spínola et al., 2015). Based on LC-ESI-QTOF/MS characterization and previously described antioxidant activities, HPLC−PDA was used to quantify eight polyphenols, including four phenolic acids (i.e., protocatechuic acid, p-hydroxybenzoic acid, chlorogenic acid, and caffeic acid) and four flavonoids (i.e., kaempferol-3-glucoside, quercetin, kaempferol, and diosmin) (Feng et al., 2020). The *p*-hydroxybenzoic acid (22.86 ± 0.01 mg/g) was the most abundant phenolic acid in black cumin, followed by protocatechuic acid (1.39 ± 0.01 mg/g) and chlorogenic acid (0.02 ± 0.01 mg/g). Caffeic acid was not identified in this analysis (Feng et al., 2020). The content of *p*-hydroxybenzoic acid in black cumin observed by Feng et al. (2020) was greater than the Iranian black cumin sample (0.188 ± 0.21 mg/100 g) detected by Mariod et al. (2009). The quantity of the discovered protocatechuic acid in black cumin (Feng et al., 2020) was substantially greater than that in black cumin (0.13 mg/g) described by Ani et al. (2006). Meanwhile, kaempferol (9.81 ± 0.07 mg/g) was observed to be the most abundant flavonoid in black cumin, followed by quercetin (1.83 ± 0.02 mg/g), kaempferol-3-glucoside (0.39 ± 0.01 mg/g), and diosmin (0.21 ± 0.01 mg/g) (Feng et al., 2020). The existence of ferulic acid, kaempferol, and quercitrin in the ethanolic extract was confirmed by Toma et al. (2015).

15.7 Conclusion

The antioxidant capacity of black cumin was found to be highly associated with its polyphenolic composition. Furthermore, studies on the isolation and quantification of individual phenolic compounds are needed to clarify their various antioxidant mechanisms, as well as the presence of probable synergism, if any, among the compounds in black cumin, as well as the properties of these phenolics on antioxidant status in animal models to appraise their possible benefits. These findings will help researchers better understand phenolic compounds in the future, as well as assist the broad usage of black cumin in the food, nutrition, and pharmaceutical industries.

References

Abou El-Leel, O.F., Maraei, R.W., Aly, A.A.E.-H., 2019. Studying the response of *Nigella sativa* plants to different fertilizers. Analele Universitatii Din Oradea, Fascicula Biologie 26 (1), 14–20.

Akanda, M.J.H., Sarker, M.Z.I., Ferdosh, S., Manap, M.Y.A., Rahman, N.N.N.A., Kadir, M.O.A., 2012. Applications of supercritical fluid extraction (SFE) of palm oil and oil from natural sources. Molecules 17 (2), 1764–1794. https://doi.org/10.3390/molecules17021764.

Ani, V., Varadaraj, M.C., Naidu, K.A., 2006. Antioxidant and antibacterial activities of polyphenolic compounds from bitter cumin (*Cuminum nigrum L.*). European Food Research and Technology 224, 109–115. https://doi.org/10.1007/s00217-006-0295-z.

Areefa, A., Mohd, A., Shah, C.S., 2020. A review on Nigella sativa (Kalonji) seeds: a universal healer. Cell Medicine 10 (2), 11. https://doi.org/10.5667/CellMed.2020.0011.

Assefa, A.D., Saini, R.K., Keum, Y.S., 2017. Fatty acids, tocopherols, phenolic and antioxidant properties of six citrus fruit species: a comparative study. Journal of Food Measurement and Characterization 11 (4), 1665–1675. https://doi.org/10.1007/s11694-017-9546-x.

Barbosa-Pereira, L., Cruz, J.M., Sendón, R., Rodríguez Bernaldo de Quirós, A., Ares, A., Castro-López, M., Paseiro-Losada, P., 2013. Development of antioxidant active films containing tocopherols to extend the shelf life of fish. Food Control 31 (1), 236–243. https://doi.org/10.1016/j.foodcont.2012.09.036.

Benzie, I.F.F., Strain, J.J., 1999. Ferric reducing/antioxidant power assay: direct measure of total antioxidant activity of biological fluids and modified version for simultaneous measurement of total antioxidant power and ascorbic acid concentration. Methods in Enzymology 299, 15–27. https://doi.org/10.1016/S0076-6879(99)99005-5.

Bourgou, S., Ksouri, R., Skandrani, I., Chekir-Ghedira, L., Marzouk, B., 2008. Antioxidant and antimutagenic activities of the essential oil and methanol extract from Tunisian *Nigella sativa* (Ranunculaceae). Italian Journal of Food Science 2, 191–202.

Burits, M., Bucar, F., 2000. Antioxidant activity of *Nigella sativa* essential oil. Phytotherapy Research 14 (5), 323–328. https://doi.org/10.1002/1099-1573(200008)14:5<323::AID-PTR621>3.0.CO;2-Q.

Costea, T., Nagy, P., Ganea, C., Szöllősi, J., Mocanu, M.M., 2019. Molecular mechanisms and bioavailability of polyphenols in prostate cancer. International Journal of Molecular Sciences 20 (5). https://doi.org/10.3390/ijms20051062.

Couttolenc, A., Díaz-Porras, Á., Espinoza, C., Medina, M.E., Trigos, Á., 2020. On the primary and secondary antioxidant activity from hydroxy-methylcoumarins: experimental and theoretical studies. Journal of Physical Organic Chemistry 33 (1), 1–10. https://doi.org/10.1002/poc.4025.

Dalli, M., Azizi, S., Kandsi, F., Gseyra, N., 2021. Evaluation of the in vitro antioxidant activity of different extracts of *Nigella sativa* L. seeds, and the quantification of their bioactive compounds. Materials Today Proceedings 45 (2021), 7259–7263. https://doi.org/10.1016/j.matpr.2020.12.743.

De Abreu, D.A.P., Losada, P.P., Maroto, J., Cruz, J.M., 2011. Lipid damage during frozen storage of Atlantic halibut (Hippoglossus hippoglossus) in active packaging film containing antioxidants. Food Chemistry 126 (1), 315–320. https://doi.org/10.1016/j.foodchem.2010.10.048.

El-Beshbishy, H.A., Mohamadin, A.M., Abdel-Naim, A.B., 2009. In vitro evaluation of the antioxidant activities of grape seed (*Vitis vinifera*) extract, blackseed (*Nigella sativa*) extract and curcumin. Journal of Taibah University Medical Sciences 4 (1), 23–35. https://doi.org/10.1016/S1658-3612(09)70078-2.

Feng, Y., Dunshea, F.R., Suleria, H.A.R., 2020. LC-ESI-QTOF/MS characterization of bioactive compounds from black spices and their potential antioxidant activities. Journal of Food Science & Technology 57 (12), 4671–4687. https://doi.org/10.1007/s13197-020-04504-4.

Galano, A., Mazzone, G., Alvarez-Diduk, R., Marino, T., Alvarez-Idaboy, J.R., Russo, N., 2016. Food antioxidants: chemical insights at the molecular level. Annual Review of Food Science and Technology 7 (December), 335–352. https://doi.org/10.1146/annurev-food-041715-033206.

Gan, R.Y., Chan, C.L., Yang, Q.Q., Li, H.B., Zhang, D., Ge, Y.Y., Corke, H., 2018. Bioactive compounds and beneficial functions of sprouted grains. In: Sprouted Grains: Nutritional Value, Production, and Applications. https://doi.org/10.1016/B978-0-12-811525-1.00009-9.

Hameed, S., Imran, A., Nisa, M.U., Arshad, M.S., Saeed, F., Arshad, M.U., Asif Khan, M., 2019. Characterization of extracted phenolics from black cumin (*Nigella sativa* Linn), coriander seed (*Coriandrum sativum* L.), and fenugreek seed (*Trigonella foenum-graecum*).

International Journal of Food Properties 22 (1), 714–726. https://doi.org/10.1080/10942912.2019.1599390.

Hossen, J., Ali, A., 2019. Effect of microwave radiation on the antioxidant activity of black cumin seed. Acta Scientiarum Polonorum, Technologia Alimentaria 18 (3), 257–268. https://doi.org/10.17306/J.AFS.2019.0671.

Islam, M.N., Hossain, K.S., Sarker, P.P., Ferdous, J., Hannan, M.A., Rahman, M.M., Uddin, M.J., 2021. Revisiting pharmacological potentials of *Nigella sativa* seed: a promising option for COVID-19 prevention and cure. Phytotherapy Research 35 (3), 1329–1344. https://doi.org/10.1002/ptr.6895.

Kadam, D., Lele, S.S., 2017. Extraction , characterization and bioactive properties of *Nigella sativa* seedcake. Journal of Food Science & Technology 54 (12), 3936–3947. https://doi.org/10.1007/s13197-017-2853-8.

Khattak, K.F., Simpson, T.J., Ihasnullah, 2008. Effect of gamma irradiation on the extraction yield, total phenolic content and free radical-scavenging activity of *Nigella sativa* seed. Food Chemistry 110 (4), 967–972. https://doi.org/10.1016/j.foodchem.2008.03.003.

Kiralan, M., Özkan, G., Bayrak, A., Ramadan, M.F., 2014. Physicochemical properties and stability of black cumin (*Nigella sativa*) seed oil as affected by different extraction methods. Industrial Crops and Products 57, 52–58. https://doi.org/10.1016/j.indcrop.2014.03.026.

Li, A.N., Li, S., Zhang, Y.J., Xu, X.R., Chen, Y.M., Li, H.B., 2014. Resources and biological activities of natural polyphenols. Nutrients 6 (12), 6020–6047. https://doi.org/10.3390/nu6126020.

Li, Y., Zhang, J.J., Xu, D.P., Zhou, T., Zhou, Y., Li, S., Li, H.B., 2016. Bioactivities and health benefits of wild fruits. International Journal of Molecular Sciences 17 (8). https://doi.org/10.3390/ijms17081258.

Liang, L.C., Zzaman, W., Yang, T.A., Easa, A.M., 2018. Effect of conventional and superheated steam roasting on the total phenolic content, total flavonoid content and DPPH radical scavenging activities of black cumin seeds. Pertanika Journal of Tropical Agricultural Science 41 (2), 663–676.

Marichali, A., Dallali, S., Ouerghemmi, S., Sebei, H., Casabianca, H., Hosni, K., 2016. Responses of *Nigella sativa* L. to zinc excess: focus on germination, growth, yield and yield components, lipid and terpene metabolism, and total phenolics and antioxidant activities. Journal of Agricultural and Food Chemistry 64 (8), 1664–1675. https://doi.org/10.1021/acs.jafc.6b00274.

Mariod, A.A., Ibrahim, R.M., Ismail, M., Ismail, N., 2009. Antioxidant activity and phenolic content of phenolic rich fractions obtained from black cumin (*Nigella sativa*) seedcake. Food Chemistry 116, 306–312. https://doi.org/10.1016/j.foodchem.2009.02.051.

Mnif, S., Aifa, S., 2015. Cumin *(Cuminum cyminum L.)* from traditional uses to potential biomedical applications. Chemistry and Biodiversity 12, 733–742.

Mohammed, N.K., Abd Manap, M.Y., Tan, C.P., Muhialdin, B.J., Alhelli, A.M., Hussin, A.S.M., 2016. The effects of different extraction methods on antioxidant properties, chemical composition, and thermal behavior of black seed (Nigella sativa l.) oil. Evidence-based Complementary and Alternative Medicine 2016. https://doi.org/10.1155/2016/6273817.

Mosbah, A., Sobhi, W., Benboubetra, M., 2015. Antioxidant and anti-hemolitic activities of Algerian *Nigella sativa* total oil and its neutral lipid fraction. International Journal of Pharmacy and Pharmaceutical Sciences 7 (11), 50–54.

Muzolf-Panek, M., Stuper-Szablewska, K., 2021. Comprehensive study on the antioxidant capacity and phenolic profiles of black seed and other spices and herbs: effect of solvent and time of extraction. Journal of Food Measurement and Characterization. https://doi.org/10.1007/s11694-021-01028-z (0123456789).

Nikolic, J.S., Mitic, V.D., Stankov Jovanovic, V.P., Dimitrijevic, M.V., Stojanovic, G.S., 2019. Chemometric characterization of twenty three culinary herbs and spices according to antioxidant activity. Journal of Food Measurement and Characterization 13 (3), 2167–2176. https://doi.org/10.1007/s11694-019-00137-0.

Ramadan, M.F., 2004. Oxidative stability of black cumin *(Nigella sativa L.),* coriander *(Coriandrum sativum L.)* and Niger *(Guizotia abyssinica Cass.)* crude seed oils upon stripping. European Journal of Lipid Science and Technology 106, 35–43. https://doi.org/10.1002/ejlt.200300895.

Samtiya, M., Aluko, R.E., Dhewa, T., Moreno-Rojas, J.M., 2021. Potential health benefits of plant food-derived bioactive components: an overview. Foods 10 (4), 1–25. https://doi.org/10.3390/foods10040839.

Şen, N., Kar, Y., Tekeli, Y., 2010. Antioxidant activities of black cumin (*Nigella sativa* L.) seeds cultivating in different regions of Turkey. Journal of Food Biochemistry 34 (Suppl. 1), 105–119. https://doi.org/10.1111/j.1745-4514.2009.00309.x.

Shalaby, E.A., Shanab, S.M.M., 2013. Comparison of DPPH and ABTS assays for determining antioxidant potential of water and methanol extracts of *Spirulina platensis.* Indian Journal of Marine Sciences 42 (5), 556–564.

Shankar, K., Mehendale, H.M., 2014. Oxidative stress. Encyclopedia of Toxicology, , third ed.vol 3. https://doi.org/10.1016/B978-0-12-386454-3.00345-6

Singh, U.P., Singh, D.P., Maurya, S., Maheshwari, R., Singh, M., Dubey, R.S., Singh, R.B., 2004. Investigation on the phenolics of some spices having pharmacotherapeutic properties. Journal of Herbal Pharmacotherapy 4 (4), 27–42.

Škrovánková, S., Mlček, J., Orsavová, J., Juríková, T., Dřímalová, P., 2017. Polyphenols content and antioxidant activity of paprika and pepper spices. Potravinarstvo Slovak Journal of Food Sciences 11 (1), 52–57. https://doi.org/10.5219/695.

Solati, Z., Baharin, B.S., Bagheri, H., 2014. Antioxidant property, thymoquinone content and chemical characteristics of different extracts from *Nigella sativa* L. seeds. JAOCS, Journal of the American Oil Chemists' Society 91 (2), 295–300. https://doi.org/10.1007/s11746-013-2362-5.

Soleimanifar, M., Niazmand, R., Jafari, S.M., 2019. Evaluation of oxidative stability, fatty acid profile, and antioxidant properties of black cumin seed oil and extract. Journal of Food Measurement and Characterization 13 (1), 383–389. https://doi.org/10.1007/s11694-018-9953-7.

Soto, C., Chamy, R., Zúñiga, M.E., 2007. Enzymatic hydrolysis and pressing conditions effect on borage oil extraction by cold pressing. Food Chemistry 102 (3), 834–840. https://doi.org/10.1016/j.foodchem.2006.06.014.

Spínola, V., Pinto, J., Castilho, P.C., 2015. Identification and quantification of phenolic compounds of selected fruits from Madeira Island by HPLC-DAD – ESI-MS^n and screening for their antioxidant activity. Food Chemistry 173, 14–30. https://doi.org/10.1016/j.foodchem.2014.09.163.

Tan, B.L., Norhaizan, M.E., Liew, W.P.P., Rahman, H.S., 2018. Antioxidant and oxidative stress: a mutual interplay in age-related diseases. Frontiers in Pharmacology 9 (OCT), 1–28. https://doi.org/10.3389/fphar.2018.01162.

Toma, C., Olah, N., Vlase, L., Mocan, A., Medicine, D., Street, R., Street, V.B., 2015. Comparative studies on polyphenolic composition, antioxidant and diuretic effects of *Nigella sativa* L. (black cumin) and *Nigella damascena* L. (Lady-in-a-Mist) seeds. Molecules 2015 (20), 9560–9574. https://doi.org/10.3390/molecules20069560.

Uras, S.S., Silahtaroɭlu, S., Ilçim, A., Kökdil, G., 2014. Fatty acid, tocopherol, mineral composition, total phenolic, flavonoid, thymoquinone content, and antioxidant activity of *Nigella sativa* L. Ankara Universitesi Eczacilik Fakultesi Dergisi 39 (3), 173–186.

Xu, D.P., Li, Y., Meng, X., Zhou, T., Zhou, Y., Zheng, J., Li, H.B., 2017. Natural antioxidants in foods and medicinal plants: extraction, assessment and resources. International Journal of Molecular Sciences 18 (1), 20–31. https://doi.org/10.3390/ijms18010096.

Zhang, J.J., Li, Y., Zhou, T., Xu, D.P., Zhang, P., Li, S., Li, H.B., 2016. Bioactivities and health benefits of mushrooms mainly from China. Molecules 21 (7), 1–16. https://doi.org/10.3390/molecules21070938.

Zhang, Q.W., Lin, L.G., Ye, W.C., 2018. Techniques for extraction and isolation of natural products: a comprehensive review. Chinese Medicine 13 (1), 1–26. https://doi.org/10.1186/s13020-018-0177-x.

Zhang, Y.Y., Zhang, F., Thakur, K., Ci, A.T., Wang, H., Zhang, J.G., Wei, Z.J., 2018. Effect of natural polyphenol on the oxidative stability of pecan oil. Food and Chemical Toxicology, vol 119. https://doi.org/10.1016/j.fct.2017.10.001.

Zúñiga, M.E., Soto, C., Mora, A., Chamy, R., Lema, J.M., 2003. Enzymic pre-treatment of *Guevina avellana mol* oil extraction by pressing. Process Biochemistry 39 (1), 51–57. https://doi.org/10.1016/S0032-9592(02)00286-8.

CHAPTER SIXTEEN

Antibacterial, antimicrobial, and antimalarial effect of black seed oil

Rohazila Mohamad Hanafiah[1], Nur Fatimah Zaharah Salehuddin[1], Siti Aisyah Abd Ghafar[1], Mohd Asyraf Kassim[2]

[1]Department of Basic Sciences, Faculty of Dentistry Universiti Sains Islam Malaysia, Kuala Lumpur, Selangor, Malaysia

[2]School of Industrial Technology, Universiti Sains Malaysia, Gelugor, Pulau Pinang, Malaysia

Abbreviations

ATCC: American Type Culture Collection
BBMV: broad bean mosaic virus
BSO: black seed oil
FTIR: fourier-transform infrared spectroscopy
HCV: hepatitis C virus
HIV: human immunodeficiency virus
IFN-γ: interferon gamma
MCMV: murine cytomegalovirus
MRSA: methicillin-resistant *Staphylococcus aureus*
MTT: 3-(4,5-dimethylthiazol-2-yl)-2,5-diphenyltetrazolium bromide
NDV: newcastle disease virus
PBMC: peripheral blood mononuclear cells
PPR: petits ruminants
TLC: thin-layer chromatography
TQ: thymoquinone
UV: ultraviolet
ZYMV: zucchini yellow mosaic virus

16.1 Introduction

Nigella sativa is a perennial plant, and it's under the Ranunculaceae family. It could be found in Eastern Europe, Middle East, and Western Asia. The length of this plant is about 2–3.5 mm and the width is 1–2 mm (Mohammed et al., 2019). The plant has tiny black seeds which are flat, trigonous, and sharp in appearance (Fararh et al., 2004). The characteristic of black seeds is similar to sesame seeds, and it's the most remarkable component of the herb in terms of their beneficial health effects.

Biochemistry, Nutrition, and Therapeutics of Black Cumin Seed
ISBN: 978-0-323-90788-0
https://doi.org/10.1016/B978-0-323-90788-0.00012-3

The extraction of black seed oil (BSO) from *N. sativa* contains many components, for example, carbohydrates, proteins, glucose, rhamnose, xylose, arabinose, and vitamins (El-Naggar et al., 2010). It also contains crude fiber, minerals (calcium, iron, and potassium), fatty acids, aliphatic alcohols, terpenoids, unsaturated hydroxy ketones, and alkaloids (Sewara et al., 2019). Besides that, several compounds such as thymoquinone (TQ), dithymoquinone, thymohydroquinone, thymol, carvacrol, nigellimine-N-oxide, nigellicine, nigellidine, and alpha-hederin were found in BSO extracts. Several researchers found fixed oil, proteins, alkaloids, saponins, and essential in BSO extracts (Burtis and Bucar, 2000). All the compounds are used for numerous nutritional and pharmaceutical functions. BSO can be combined in tea, coffee, and dough. It also can be mixed with honey or drizzled for salads. Additionally, the BSO is extensively utilized as a spice, carminative, condiment for food industrial, and scented agent for cosmetics business.

The varieties of chemical compounds in BSO are useful for therapeutic application such as antiviral, antibacterial, antifungal, and anticancer activities (Pereira et al., 2004). Report from researchers stated that BSO extracts possess antimicroorganism activity against pathogen bacteria. The BSO is usually used in Asian countries for several medicinal properties. Other than that, *N. sativa* plants were used traditionally to treat a number of illnesses such as asthma, hypertension, diabetes, inflammation, cough, bronchitis, headache, eczema, fever, dizziness, and influenza in Indian, Arabian countries, and Europe. They also used *N. sativa* as carminative, diuretic, lactagogue, and vermifuge (Ali and Blunden, 2003).

16.2 Antibacterial activities of black seed oil

There are a lot of review articles that mention about antibacterial activity of the BSO extracts against pathogenic bacteria. They found that BSO bioactive compound which is TQ has antibacterial activities toward *Bacillus subtilis*, with an average inhibition zone of 15.74 mm (Mohammed et al., 2019). They used disc diffusion assay to measure the antibacterial activity. *B. subtilis* can cause bacteremia, endocarditis, and pneumonia on immunocompromised patient.

Moreover, the BSO extracts from methanol were reported to have antibacterial activities against some pathogenic bacteria such as *Streptococcus pyogenes, Pseudomonas aeruginosa, Klebsiella pneumoniae,* and *Proteus vulgaris.* Methanol extracts of BSO displayed substantial antibacterial activity at the concentration of 50 mg/mL ($P \leq .01$) against *S. pyogenes* with a greater inhibition zone of 19 mm, while a 15 mm zone of inhibition was observed in *P. aeruginosa, K. pneumonia,* and *P. vulgaris*. The study also mentions that both aqueous and methanol extracts of black seed demonstrated a larger inhibition on gram-positive bacteria (*S. pyogenes*) compared with gram-negative bacteria (*P. aeruginosa, K. pneumoniae,* and *P. vulgaris*) (Nor Aishah et al., 2013). BSO extracts also reported to have antibacterial activities against *Staphylococcus aureus* and *Staphylococcus epidermidis*. The zone inhibitions of BSO extracts against both bacteria were 23 mm (*S. epidermidis*) and 46 mm (*S. aureus*), respectively (Salman et al., 2008). Both bacteria are gram-positive, and they can cause skin infection on immunocompromised patient. Study from Turkey reported that BSO extraction possessed antibacterial activities against *Aeromonas hydrophila*. They used three concentrations of BSO extract (0.5%, 1.0%, and 2.0%), and all the concentrations showed antibacterial activities against those bacteria (Arici et al., 2005).

Some researchers also used animal study to determine antibacterial activities of BSO extracts against pathogenic bacteria. Study from Hosseinzadeh et al. (2007) investigates the antibacterial activity of BSO extracts on male mice affected intraperitoneally with *S. aureus* or *Escherichia coli*. In this study, they found the inhibitory activity of the methanol to extract at a dose of 2.14 g/kg in mice infested with *S. aureus* was 87.5%. The doses of 1.2 and 2.14 g/kg in mice infested with *E. coli* were 100% contrasted with mice, which consumed saline (the negative control). Other than that, the BSO extracts also revealed substantial antibacterial activity toward methicillin-resistant *Staphylococcus aureus* (MRSA) isolated from diabetic patients (Emeka et al., 2015). The cytotoxic test shown that BSO extracts were not toxic to proliferation of gingival fibroblast (Ugur et al., 2016). The extractions of BSO from ethanol and *n*-hexane were reported that possessed antibacterial activities toward gram-positive and negative bacteria such as *Bacillus cereus, B. subtilis, E. coli, S. epidermidis, K. pneumonia,* and *Salmonella typhimurium* (Khan and Kou, 2016). Other than that, BSO extracts displayed an extensive spectrum of activity toward gram-positive and negative bacteria, including *B. cereus* (ATCC 14579), *Listeria monocytogenes* (ATCC 19115), *Enterococcus faecalis* (ATCC 29212), *Micrococcus luteus* (NCIMB 8166), *S. aureus* (ATCC 25923), *S. epidermidis* (CIP 106510), *P. aeruginosa* (ATCC 27853), *E. coli* (ATCC 35218), *Salmonella enterica, Serovar typhimurium* (ATCC 14028), *Vibrio alginolyticus* (ATCC 33787), and *Vibrio parahaemolyticus* (ATCC 17802). The bioactive compound from BSO (TQ) was reported capable to inhibit bacterial biofilm development (Chaieb et al., 2011).

TQ acquired from seeds of *N. sativa* uncover antibacterial activities and antibiofilm activities toward various strains of gram-positive and gram-negative bacteria, comprising *Bacillus* spp, *Listeria* spp, *Enterococcus* spp, *Micrococcus* spp, *Staphylococcus* spp, *Pseudomonas* spp, *Escherichia* spp, *Salmonella* spp, and *V. parahaemolyticus*. The BSO methanol extract also shown a greater inhibition zone on gram-positive (*S. pyogenes*) contrasted with gram-negative bacteria (*P. aeruginosa, K. pneumoniae*, and *P. vulgaris*). Besides that, TQ revealed bactericidal activities toward gram-positive bacteria with minimum inhibitory concentrations (MICs) ranging from 8 to 32 μg/mL and demonstrated the biofilm activities against *S. aureus* and *S. epidermidis*, respectively (Abdallah, 2017). Furthermore, the BSO extracts showed antibacterial activities toward *Helicobacter pylori*. This study contributed a scientific foundation for the evaluation of possible uses of this priceless seed for the treatment of *H. pylori*-induced gastric ulcers (Salem et al., 2010).

BSO extracts have antibacterial activity against *L. monocytogenes* with significantly better inhibition zone contrasted to gentamicin ($P < .01$). The averages zones of inhibition created by BSO extracts and gentamicin were 31.50 ± 1.0 and 14.80 ± 0.50 mm, respectively (Nair et al., 2005). Other than that, all tested strains of MRSA were susceptible to BSO extracts at a concentration of four mg/discs. The MIC of the BSO extracts against MRSA was 0.2–0.5 mg/mL (Hannan et al., 2008).

16.3 Antimicroorganism activities of black seed oil

Microbial resistance to conventional antibiotics has increased the use of medicinal plants that have significant outcomes for a variety of pathogenic infections. This is because herbal medicines are less harmful and the active compounds can be used with other compounds (Emeka et al., 2015). *N. sativa* extracts have shown to be antimicrobial toward *Streptococcus faecalis, S. aureus, E. coli, P. aeruginosa,* and *Candida albicans* (Islam et al., 2019). From previous research, an average inhibition zone of 15.74 mm showed solid bacterial inhibitory impacts were reflected toward *B. subtilis* (Mohammed et al., 2019). Moreover, BSO also has antimicrobial activity toward MRSA (Mohammed et al., 2019).

For ages, BSO has been used to encourage health, particularly in the Middle East and Southeast Asia (Abd-Awn et al., 2012). Study from Mohammed et al. (2019) investigates antimicrobial activity by performed antimicrobial susceptibility test against *P. aeruginosa* (ATCC 9027), *E. coli* (ATCC 8739), *S. aureus* (ATCC 6538P), *S. enterica* (NCTC 6017*), B. subtilis* (ATCC 6633*),* methicillin-resistant *Staphylococcus aureus* (MRSA), and *B. cereus* (Mohammed et al., 2019).

Fourier-transform infrared spectroscopy (FTIR), thin-layer chromatography, and ultraviolet-visible spectroscopy techniques were performed to determine functional groups, substituents, and the existence of conjugated double bonds in BSO. TQ is one of the main active components from extracted BSO conducted by FTIR (Emeka et al., 2015; Islam et al., 2019; Mohammed et al., 2019). TQ is a major bioactive constituent of BSO that has been extensively employed in the treatment of diseases including rheumatism, bronchitis, asthma, parasitic infections, and numerous others (Emeka et al., 2015; Mohammed et al., 2019). Moreover, the essential oils from black seeds were discovered to have significant amounts of TQ and other compounds, such as thymol and dithymoquinone. All these compounds have been implicated to inflammation prevention, antioxidant, antimicrobial, anticarcinogenic, and antiulcer activities (Singh et al., 2014).

Scientifically, nearly all human diseases can be treated by BSO. This is because BSO extract showed highest inhibitory zone for gram-positive bacteria of *B. subtilis* (Mohammed et al., 2019). In contrast, antibacterial activity toward gram-negative bacteria was not identified (Mohammed et al., 2019). Moreover, BSO extract has also been found to have antibacterial activity against MRSA (Mohammed et al., 2019). Research have provided evidence for antiviral activity of BSO, for example, the addition of black seed exhibited the highest antimicrobial effects toward *Lactococcus lactis*, *P. aeruginosa*, *L. monocytogenes*, and *S. aureus* with 22.30 mm, 9.60 mm, 9.50 mm, and 9.0 mm inhibition zone, respectively (Abdel-Shafi, 2013).

Many studies have found that some plant photochemical compounds, such as phenols, flavonoids, alkaloids, terpenoids, saponins, tannins, and anthraquinones have antimicrobial properties. These compounds can kill bacteria or fungal cells by inhibiting the growth, affecting cellular membrane permeability, interfering with some metabolic processes, and modulating signal transduction. TQ, p-cymene, longifolene, and thymohydroquinone were found to be essential for BSO (Singh et al., 2014). Extracted thymohydroquinone from BSO was showed to have significant effect against gram-positive bacteria (Abd-Awn et al., 2012). *N. sativa* was well-known for its use as a food preservative and as a medicinal herb (Hafez, 2008). Thus, in order to avoid spoilage, it was recommended to use BSO as an antimicrobial agent in food processing. This is because 2.0% concentration of BSO can inhibit the growth of 24 pathogenic, spoilage, and lactic acid bacteria (Abdallah, 2017).

On the other study of antimicrobial activity of BSO, *N. sativa* seed oil has antimicrobial effect toward MRSA isolated from 34 patients with diabetic abrasions in Southeast Nigeria (Emeka et al., 2015). The antimicrobial activity was investigated using disc diffusion assay. In this study, they found that eight of the 19 isolates were susceptible to varying concentrations of BSO, whereas 11 were resistant (Emeka et al., 2015). In contrast, the ethanol extract of black seeds was shown to be particularly efficient against *S. aureus* in other finding (Emeka et al., 2015).

No cytotoxicity against human peripheral blood mononuclear cells (PBMCs) of BSO is showed by using 3-(4,5-dimethylthiazol-2-yl)-2,5-diphenyltetrazolium bromide (MTT) assay in vitro (Mohammed et al., 2019). In this experiment, they discovered that essential oils made from plant parts have varying impacts on human and animal which also known as in vitro and in vivo. Furthermore, no adverse effect was observed on PBMCs growth at the dosages and intervention times frames (Mohammed et al., 2019). The cytotoxic activity of BSO on PBMCs was evaluated after 24, 48, and 72 h of incubation by using MTT assay at different concentrations of 25, 50, and 100 μg/mL (Mohammed et al., 2019). Moreover, *N. sativa* had a low cytotoxic activity at low concentration of 100 μg/mL, and the high cytotoxicity effect was measured at 400 μg/mL. Thus, the cytotoxicity produced is 95% (Dorra et al., 2019).

It is predicted that essential oils and their constituents can interact with the bacterial membrane and cause interference via lipophilic products (Mohammed et al., 2019). Essential oils allow the bacterial cell membrane to disintegrate and release the cell substance, resulting in the death of bacterial cell (Mohammed et al., 2019). According to Iraqi studies, they discovered inhibitory zones for BSO extract, which increased as the concentration of extract increased. The efficacy of antimicroorganisms is considered by the mixture of fatty acids, volatile oils, and trace elements (Abd-Awn et al., 2012).

Based on Iraqi studies, they reported that *N. sativa* seeds contain 0.5%—1.6% of yellowish volatile oil, 35.6%—41.6% of fixed oil, 22.7% of proteins, amino acids, minerals, and vitamins. The purpose of this study was to compare the impact of different concentrations of BSO extract on the sensitivity of mutans streptococci and the adhesion of oral pathogens to the tooth surface to chlorohexidine gluconate. Moreover, they also found that the diameter of inhibition zone of different concentrations of BSO extract was increased as the extract concentration increased against mutans streptococci in oral pathogens (Abd-Awn et al., 2012).

According to a study conducted by Islam et al., different BSO shown varying degrees of antifungal activity. On the other hand, the essential oil was extremely effective. In addition, several components of BSO have moderate inhibitory effects against pathogenic yeast, such as *C. albicans*, dermatophytes, nondermatophytes, and some aflatoxin growing fungi. Interestingly, the main active ingredient of BSO that possess antioxidant, antiinflammatory, antiviral, antimicrobial, immunomodulatory, and anticoagulant activity is TQ. Isolated TQ from *N. sativa* shown potent antifungal effect against *Aspergillus niger, Fusarium solani,* and *Scopulariopsis brevicaulis*. Moreover, the antifungal activity of two new defensins peptides, named Ns-D1 and Ns-D2 was found to be potent and diverse (Abdallah, 2017).

BSO was the most effective oil at preventing spore germination, and the hyphae seemed to be fragmented (Hafez, 2008). From other research, the extracted BSO was investigated for antifungal activity against different pathogenic fungus, including *Aspergillus flavus, A. niger, Fusarium moniliforme, Fusarium graminearum,* and *Penicillium viridicatum*. Inverted petri plate and poison food media techniques were used for this research by maintaining the cultures in Czapek agar medium, and percentage of mycelial inhibition was calculated to evaluate fungi toxicity (Singh et al., 2014). As for the results for *F. moniliforme* and *P. viridicatum*, volatile oil inhibited the growth zone by more than 90% (Singh et al., 2014).

To control fungal infections, a variety of alternative techniques have been developed, including mineral salts, essentials oils, plant extracts, and biological agents. Study from the Department of Botany (Plant Pathology Branch) in Hungary investigated the antifungal activities of BSO toward powdery mildews of *Podosphaera xanthii* (cucumber) and *Blumeria graminis* f.sp. *hordei* (barley). From the research, cucumber powdery mildew disease severity was lessened from 52% to 7.7%, 18.6%, and 20% on leaves treated with BSO, while for barley powdery mildew disease, severity was decreased from 63.4% to 9.4%, 16%, and 16.4%, respectively (Hafez, 2008).

There are a lot of findings that prove BSO as an antiviral agent against numerous viruses. For example, *N. sativa* against murine cytomegalovirus (MCMV) infections, avian influenza (H9N2), peste des petits ruminants (PPR) virus, broad bean mosaic virus, human immunodeficiency virus (HIV), hepatitis C virus (HCV), zucchini yellow mosaic virus (ZYMV), and Newcastle disease virus (NDV) (Islam et al., 2019). BSO displayed potent antimicrobial action against phytopathogenic fungi, such as *Pythium vexans, Rhizoctonia solani, L. monocytogenes,* and *Colletotrichum capsici*. Moreover, BSO can control *Erysiphe cichoracearum* which caused squash powdery mildew disease. Except for *Botrytis fabae, N. sativa* oil was effective against all the tested fungi (Hafez, 2008).

Antimicrobial properties of BSO are efficient toward a wide range of bacteria, fungi, and viruses. From previous research, it was confirmed the potency of BSO as an anti-HIV agent and other viruses as well. Study from Abdallah et al. in 2017 investigates the inhibition viral titers in the spleen and liver in mice with MCMV was successful. Scientifically, black seed significantly reduced the pathogenicity of influenza virus (H9N2) in turkey fowls, and this plant product also can increase immunological response (Abdallah, 2017). Besides that, there are researchers who conducted clinical study on patients in Egypt infected by HCV. After treatment period with *N. sativa* for three months with dose of 450 mg three times daily, patients showed varying degrees of recovery and viral load reduction, as well as improvements in oxidative stress, clinical condition, and glycemic control in diabetic patients (Abdallah, 2017).

Secondary metabolites, such as alkaloids, flavonoids, tannins, and terpenoids found in medicinal plants, make these plants a rich source of antimicrobial agents. There is also research that provided evidence for antiviral activity of BSO by plaque reduction method, for example, the consequences of *N. sativa* extracts as inhibitor of ZYMV infectivity in vitro and in vivo. From these studies, *N. sativa* were inhibited the development of ZYMV symptoms on squash plants by 85% and 80%, respectively. In the early stages of infection, BSO functions directly by interacting with virus particles and preventing the nucleic acid virus from release (Islam et al., 2019).

Study from Egypt in 2019 investigated about the treated influenza A by BSO. The objective of the research was to investigate the in vitro antiviral activity of the medicinal herbs of *N. sativa* against influenza virus (H5N1) as well as their antioxidant impact spectrophotometrically (Dorra et al., 2019). Besides that, BSO acts as an antiinfluenza virus activity for existing respiratory problem that has been isolated from various birds of several countries, known as avian influenza virus (H9N2). BSO may be used as a feed additive to decrease pathogen load in poultry (Islam et al., 2019).

In addition, BSO can increase the number of macrophages, enhance the function and production of interferon gamma when they act against cytomegalovirus (Islam et al., 2019). Black seed has an immunomodulatory and therapeutic activity against PPR virus in infected goats. Thus, BSO will inhibit the development of clinical symptoms and reduce clinical signs and histopathological disorders (Islam et al., 2019). Ethanolic extract of BSO can decrease viral load and mortality in embryonated chicken eggs for NDV (Islam et al., 2019).

16.4 Antimalarial activities of black seed oil

Malaria is one of the major public health problems, which is a serious mosquito-borne disease caused by a parasite. This disease is generally caused by several types of malaria parasites including *Plasmodium falciparum*, *Plasmodium malariae*, *Plasmodium ovale*, *Plasmodium vivax*, and *Plasmodium knowlesi*. The most common plasmodium parasite caused malaria is *P. falciparum*, which sometimes could be potentially life-threating and may cause serious illness such as kidney failure and coma. Other types of malaria parasites generally cause mild illness, but it can remain dormant in the patient body for months after infection.

Development of Plasmodium sp. in human body involved a series of stages. The malaria parasite enters the human bloodstream in sporozoites form that injected by female mosquito during taking blood from human. The sporozoites migrate to the liver and multiply to form merozoites. Later, merozoite formed will release into blood stream and infect red blood cells. The merozoites in the red blood cells initiate erythrocytic phase and asexual multiplication cycle and evolve into trophozoites. In the red blood cell, the parasite also differentiates to form gametocytes. This form is nonpathogenic and more transmissible parasite form and take about 8 to days for maturation. The gametocytes then become microgametes (male) and macrogametes (female) during the process of gametogenesis. In the mosquito midguts, the gametes form zygote and subsequently develop into motile ookinete. This ookinete later divided into 1000 of sporozoites and can last for 8–15 days. During the maturation period, the sporozoites will migrate to saliva gland and can cause infection to human being.

The effectiveness of malaria treatment is totally depending on the potential of drug to eliminate the parasite and cause mild or no toxicity effect to the host. The drug introduced must have the ability to inhibit the growth of parasite and can operate on the targeted biochemical metabolisms of the parasite and host. To date, mode of the reaction of many drugs introduced is not well understood (Mojab et al., 2012).

Current antimalarial drug and therapeutic are relatively insensitive and imprecise (White 1989; Raphemot et al., 2016). Development of effective antimalarial drug that is cheaper and can be available for people especially in developing country is necessary to ensure the sustainable public health. Antimalarial drug from various types of traditional and herbal plants has been identified as an alternative approach to prevent and cure malaria. According to Willcox and Bodeker (2004) in the assessment conducted on the application of 1200 herbal plants to cure malaria indicated that some of the herbal plants may be safe and effective for the malaria treatment. Several studies reported on the potential of antimalarial drug using herbal and plant extract such as *Zanthoxylum tsihanimposa*, *Helianthus annuus*, *Garuga floribunda*, *Alectryon serratus*, *Lepisanthes rubiginosum*, etc. (Widyawaruyanti et al., 2014; Ekasari et al., 2019; Randrianariavelojosia et al., 2003; Dell'Agli et al., 2012).

To date, current investigation on the potential of antimalarial drug from *N. sativa* or black seed has gained great attention globally. This traditional plant source has been used for 1000 years to treat various disease. Early study has proved that the *N. sativa* exhibit antiparasite activity (Agarwal et al., 1979). Abdulelah et al. (2007) in their study on the antimalarial drug by *N. sativa* extract using different solvents including ethanol and chloroform indicated that both oral treatment and intraperitoneal treatment of seed extract showed a significant suppression of *Plasmodium berghei* activity in mice. Similar study by Okeola et al. (2011) in mice infested with *Plasmodium yoelii nigeriensis* also reported that methanolic extract of *N. sativa* seeds restored the activity of red blood cells antioxidant activities and exhibited effective in clearing plasmodium parasite compared to chloroquine. A study on the potential of BSO extract on antiparasitic activity conducted via oral administration of methanol and aqueous extract showed that a significant suppression and decrease on the parasitemia was observed for mice treated with the black seed extract (Ashcroff et al., 2018). Overall, according to this analysis, it clearly indicated that *N. sativa* seed extract has strong antioxidant and antimalaria activity, which can be potentially used as a future alternative approach to treat malaria infection.

References

Abd-Awn, B.H., Al-Dhaher, Z.A., Al-Dafaai, R.R., 2012. The effect of black seed oil extracts on mutans streptococci in comparison to chlorhexidine gluconate (in vitro). Journal of Baghdad College of Dentistry 24 (4), 1–5.

Abdel-Shafi, S., 2013. Preliminary studies on antibacterial and antiviral activities of five medicinal plants. Journal of Plant Pathology Microbiology 4, 190–194.

Abdallah, E.M., 2017. Black Seed (*Nigella sativa*) as antimicrobial drug: a mini-review. Novel Approches in Drug Designing and Develop 3 (2), 1–5.

Abdulelah, H.A.A., Zainal-Abidin, B.A.H., 2007. In vivo anti-malarial tests of nigella sativa (black seed) different extracts. American Journal of Pharmacology and Toxicology 2 (2 SE-Research Article). https://doi.org/10.3844/ajptsp.2007.46.50.

Agarwal, R., Kharyam, M.D., Shrivastava, R., 1979. Antimicrobial and anthelmintic activity of the essential oil fof Nigella sativa. Indian Journal Experimental Biology 17 (11), 1264–1265.

Ali, B.H., Blunden, G., 2003. Pharmacological and toxicological properties of *Nigella sativa*. Phytotherapy Research 17 (4), 299–305.
Arici, M., Sagdic, O., Gecgel, U., 2005. Antibacterial effect of Turkish black cumin (*Nigella sativa L.*) oils. Grasas y Aceites 56 (10), 259–262.
Ashcroft, O., Salaudeen, F., Mohammed, K., Spencer, T., Garba, M., Nataala, U., Umaru, M., Iduh, M., 2018. Anti-malarial effect of *Nigella sativa* seeds (black seed) extract on mice infected with *Plasmodium bergei*. European Journal Of Pharmaceutical And Medical Research 5, 131–137.
Burtis, M., Bucar, F., 2000. Antioxidant activity of *Nigella sativa* essential oil. Phytotherapy Research 14, 323–328.
Chaieb, K., Kouidhi, B., Jrah, H., Mahdouani, K., Bakhrouf, A., 2011. Antibacterial activity of Thymoquinone, an active principle of *Nigella sativa* and its potency to prevent bacterial biofilm formation. BMC Complement Alternative Medicine 11, 29–33.
Dell'Agli, M., Sanna, C., Rubiolo, P., Bassilico, N., Colombo, E., Scaltrito, M.M., Ndiath, M.O., Maccarone, L., Taramelli, D., Bicchi, C., Ballero, M., Bosisio, E., 2012. Anti-plasmodial and insecticidal activities of the essential oils aromatic plants growing in the Mediterranean area. Malaria Journal 11 (219), 1–10.
Dorra, N., El-Berrawy, M., Sallam, S., Mahmoud, R., 2019. Evaluation of antiviral and antioxidant activity of selected herbal extracts. Journal of High Institute of Public Health 49 (1), 36–40.
Ekasari, W., Pratiwi, D.W., Amanda, Z., Suciati, Widyawaruyanti, A., Arwati, H., 2019. Various parts of Helianthus annuus plants as new sources of antimalarial drugs. Evidence-Based Complementary and Alternative Medicine 2019, 1–7.
El-Naggar, T., Gómez-Serranillos, M.M., Palomino, O.P., Arce, C., Carretero, M.E., 2010. *Nigella sativa L.* seed extract modulates the neurotransmitter amino acids release in cultured neurons *in vitro*. Journal of Biomedicine and Biotechnology 2010, 8–12.
Emeka, L.B., Emeka, P.M., Khan, T.M., 2015. Antimicrobial activity of *Nigella sativa L.* seed oil against multi-drug resistant *Staphylococcus aureus* isolated from diabetic wounds. Pakistan Journal of Pharmaceutical Science 28 (6), 1985–1990.
Fararh, K.M., Atoji, Y., Shimizu, Y., Shiina, T., Nikami, H., Takewaki, T., 2004. Mechanisms of the hypoglycaemic and immunopotentiating effects of *Nigella sativa L.* oil in streptozotocin-induced diabetic hamsters. Research in Veterinary Science 77 (2), 123–129.
Hafez, Y.M., 2008. Effectiveness of the antifungal black seed oil against powdery mildews of cucumber (*Podosphaera xanthii*) and barley (*Blumeria graminis* f. sp. *hordei*). Acta Biologica Szegediensis 52 (1), 17–25.
Hannan, A., Saleem, S., Chaudhary, S., Hussain, M., 2008. Anti-bacterial activity of *Nigella sativa* against clinical isolates of methicillin resistant *Staphylococcus aureus*. Journal of Ayub Medical College, Abbottabad: JAMC 20, 72–74.
Hosseinzadeh, H., Fazly, B., Bibi, S., Haghi, M., 2007. Antibacterial activity of total extracts and essential oil of *nigella sativa L.* Seeds in mice. Pharmacology Online 2, 429–435.
Islam, M.T., Khan, M.R., Mishra, S.K., 2019. An updated literature-based review: phytochemistry, pharmacology and therapeutic promises of *Nigella sativa* L. Oriental Pharmacy and Experimental Medicine 19 (2), 115–129.
Khan, A.R., Kou, K., 2016. Wide spectrum antibacterial activity of *Nigella sativa L.* seeds. IOSR Journal of Pharmacy 6 (7), 12–16.
Mohammed, S.J., Amin, H.H., Aziz, S.B., Sha, A.M., Hassan, S., Abdul Aziz, J.M., Rahman, H.S., 2019. Structural characterization, antimicrobial activity, and *in vitro* cytotoxicity effect of black seed oil. Evidence-Based Complementary and Alternative Medicine 2019, 1–8.
Mojab, F., 2012. Antimalaria natural products: a review. Aviecenna Journal of Phytomedicine 2 (2), 52–62.

Nair, M., Vasudevan, P., Venkitanarayanan, K., 2005. Antibacterial effect of black seed oil on *Listeria monocytogenes*. Food Control 16, 395–398.

Nor' Aishah, H., Nawahwi, M.Z., Malek, A., 2013. Antimicrobial activity of *Nigella sativa* seed extract (Aktiviti Antimikrob Ekstrak *Nigella sativa*). Sains Malaysiana 42, 143–147.

Okeola, V.O., Adaramoye, O.A., Nnehi, C.M., Falade, C.O., Farombi, E.O., Ademowo, O.G., 2011. Antimalarial and antioxidant activities of methanolic extract of Nigella sativa seeds (black cumin) in mice infected with Plasmodium yoelli nigeriensis. Parasitology Research 108 (6), 1507–1512.

Pereira, P., Huerta, B., Borge, C., Astorga, R., Romero, R., Perea, A., 2004. Antimicrobial activity of five essential oils against origin strains of the Enterobacteriaceae family. Acta Pathologica, Microbiologica et Immunologica Scandinavica 113 (1), 1–6.

Randrianariavelojosia, M., Rasidimanana, V.T., Rabarison, H., Cheplogoi, P.K., Ratsimbason, M., Mulholland, D.A., Mauclere, P., 2003. Plant traditionally prescribed to treat tazo (malaria) in the eastern region of Madagascar. Malaria Journal 2 (25), 1–9.

Raphemot, R., Posfai, D., Derbyshire, E.R., 2016. Current therapies and future possibilities for drug development against liver-stage malaria. Journal of Clinical Investigation 126 (6), 2013–2020.

Salem, E.M., Yar, T., Bamosa, A.O., 2010. Comparative study of *Nigella Sativa* and triple therapy in eradication of *Helicobacter Pylori* in patients with non-ulcer dyspepsia. Saudi Journal of Gastroenterology 16 (3), 207–214.

Salman, M.T., Khan, R., Shukla, I., 2008. Antimicrobial activity of *Nigella sativa Linn.* seed oil against multi-drug resistant bacteria from clinical isolates. Natural Product Radiance 7, 1–8.

Sewara, J., Mohammed, H.H., Amin, S.B., Aziz, A., Heshu, S.R., 2019. Structural characterization, antimicrobial activity, and *in vitro* cytotoxicity effect of black seed oil. Evidence-Based Complementary and Alternative Medicine 2019 9, 1–9.

Singh, S., Das, S.S., Singh, G., Schuff, C., de Lampasona, M.P., Catalan, C.A., 2014. Composition, *in vitro* antioxidant and antimicrobial activities of essential oil and oleoresins obtained from black cumin seeds (*Nigella sativa L.*). BioMedicine Research International 2 (25), 1–9.

Ugur, A.R., Dagi, H.T., Ozturk, B., Tekin, G., Findik, D., 2016. Assessment of *In vitro* antibacterial activity and cytotoxicity effect of *Nigella sativa* oil. Pharmacognosy Magazine 12 (S4), S471–S474.

White, N.J., Krishna, S., 1989. Treatment of malaria: some considerations and limitations of the current methods of assessment. Transactions of the Royal Society of Tropical Medicine and Hygiene 83 (6), 767–777.

Widyawaruyanti, A., Devi, A.P., Fatria, N., Tumewu, L., Tantular, I.S., Hafid, A.F., 2014. In vitro anti-malarial activity screening of several Indonesia plant using HRP2 Assay. International Journal of Pharmacy and Pharmaceutical Sciences 6 (6), 125–128.

Willcox, M.L., Bodeker, G., 2004. Traditional herbal medicine for malaria. Bio Medicine of Journal 329 (7475), 1156–1159.

Further reading

El-Shafi, S.A., 2013. Preliminary studies on antibacterial and antiviral activities of five medicinal plants. Journal of Plant Pathology Microbiology 4 (190), 2–5.

Hassan, M.Q., Akhtar, M., Ahmed, S., Ahmad, A., Najmi, A.K., 2017. *Nigella sativa* protects against isoproterenol induced myocardial infarction by alleviating oxidative stress, biochemical alterations and histological damage. Asian Pacific Journal of Tropical Biomedicine 7 (4), 294–299.

Udu, R., Oyweri, J., Gathirwa, J., 2021. Anti malaria effect of nigella sativa seed extract on mice infected with plasmodium bergei ANKA in a mouse model. Journal of Pathogen 2021, 1–10.

CHAPTER SEVENTEEN

Neuroprotective effects of black cumin seed and seed oil

Hakan Ak[1], İsmail GÜLŞEN[2], İhsan CANBEK[3]
[1]Kırşehir Ahi Evran University, School of Medicine, Department of Neurosurgery, Kırşehir, Turkey
[2]Lokman Hekim Hospital, Clinic of Neurosurgery, Van, Turkey
[3]Afyonkarahisar Health Sciences University, Department of Neurosurgery, Afyonkarahisar, Turkey

17.1 Introduction

The black cumin (BC) (*Nigella sativa* [NS]) belongs to the Ranunculaceae family, which is grown worldwide (Ermumcu & Şanlıer, 2017). BC has been extensively applied for above 2000 years as a traditional medicine for many diseases (Razavi and Hosseinzadeh, 2014). Considering that it has been used for many purposes for a long time, it is of great importance to know the content of NS seed in detail (Kabir et al., 2019). It is mainly composed of fat, protein, and carbohydrates. It also contains moisture and ash in lesser amounts. The amount of oil varies between approximately 22% and 40% (Atta, 2003). BC seeds and their oils have many nutritionally essential components such as polyunsaturated fatty acids, phytosterols, besides many biological activities, for example, antioxidants, antidiabetic, antitumor, and other active constituents (Butt and Sultan, 2010). The biological activity of BC is mostly based on p-cymene (7%—15%), thymoquinone (TQ) (30%—48%), and carvacrol (6%—12%). These three compounds form about 48%—70% of the biological activity. Monounsaturated and saturated fatty acids constitute the remaining part (Cheikh-Rouhou et al., 2007). This chapter discusses the neuroprotective effects of BC seed, and seed oil will be discussed under specific titles in detail.

17.2 Effects of thymoquinone on neuroinflammation

TQ (2-Isopropyl-5-methyl-1,4-benzoquinone) is the most important element of the BC with several effects on different systems and organs including antiinflammatory, antioxidant, immunomodulatory, and neuroprotective effects.

Biochemistry, Nutrition, and Therapeutics of Black Cumin Seed
ISBN: 978-0-323-90788-0
https://doi.org/10.1016/B978-0-323-90788-0.00002-0

Neuroinflammation is the inflammatory response of the central nervous system's (CNS's) which is composed of the spinal cord and brain. Chemokines, reactive oxygen species (ROS), cytokines, and secondary messengers are produced through such neuroinflammation process. There are biochemical, immune, psychological, and physiological consequences in the neuroinflammatory responses. Inflammation causes edema, damage of tissue, immune cells recruitment, and subsequently death of cell (DiSabato et al., 2016).

Neuroinflammation is accepted as a main contributing factor in the pathogenesis of severeal neurodegenerative diseases. Microglia activation is accepted to mainly cause to initiate and progress the neuroinflammation (Farkhondeh et al., 2018). TQ inhibits the neuroinflammation but underlying mechanism for this effect is not exactly known (Velagapudi, El-Bakoush et al., 2017). Literature related to this subject consists of mostly experimental studies.

Presence of a strong relationship between inflammation and oxidative stress is known, and an unbalanced redox state has contribution to the pathogenesis of inflammatory processes. Inflammation leads to an increased level of ROS causing cellular oxidative damage. Moreover, ROS produced by microglia has important effects on adjacent neurons. Adenosine monophosphate-activated protein kinase (AMPK) regulates inflammatory responses as a known sensor of energy balance. Sirtuin 1 (SIRT1) is related to the cellular processes such as neuroinflammation. SIRT1 activation is a significant factor in achieving antiinflammatory activity. The impacts of cellular ROS, AMPK, and SIRT1 in the antineuroinflammatory activity of TQ in BV2 microglia were investigated in a study. The LKB1 (liver kinase B1) and phosphor-AMPK proteins levels increased with TQ. Moreover, the NAD+ levels increased, the cytoplasmic levels decreased, and the SIRT1 protein's nuclear accumulation increased with through TQ. Suppressing the SIRT1 and AMPK expressions through RNAi or pharmacological antagonists caused to abolish the TQ's antiinflammatory activity. TQ-induced increase in SIRT1 was reversed by the AMPK's pharmacological antagonism. It was shown that gp91phox and p40phox protein was inhibited to reduce cellular ROS generation with TQ (Velagapudi, El-Bakoush et al., 2017).

The TQ's inhibitory effects of lipopolysaccharide (LPS)-induced neuroinflammation in BV2 microglia were achieved by interfering with NF-κB signaling (Velagapudi, Kumar et al., 2017). NF-κB binds to DNA and the NF-κB activation is related with inflammation in microglia as a transcription factor. Activated NF-kB induces proinflammatory cytokines such as cyclooxygenase-2 (COX-2), microsomal prostaglandin E (PGE2) synthase-1, and inducible nitric oxide synthase (iNOS) (Zhang and Jiang, 2015). TQ increased the nuclear localization, Nrf2 transcriptional activity of, DNA binding, and increased protein levels of heme oxygenase-1 (HO-1) and NAD(P)H quinone dehydrogenase 1 (NQO1) to activate the NF-E2-related factor 2/antioxidant responsive element (Nrf2/ARE) signaling. The release of interleukin-6 (IL-6), IL-1b, and tumor necrosis factor-alpha (TNF-α) was inhibited by the TQ treatment (5, 2.5, and 10 mM), reducing the messenger ribonucleic acid release and levels of IL-6, TNF-a, PGE2, and IL-1b in the microglia cells. NF-kB-dependent neuroinflammation was inhibited by TQ treatment by reducing kB inhibitor phosphorylation, iNOS protein levels, and NF-kB binding to the DNA. TQ increased the nuclear factor (erythroid-derived 2)-like 2 (NFE2L2 or Nrf2) protein nuclear accumulation, ARE transcriptional activity, and Nrf2 binding to the ARE consensus binding site (Velagapudi, Kumar et al., 2017).

The effects of TQ on oral sodium nitrite (SN)-induced brain injury were studied by measuring cytokines, fibrotic elements, oxidative stress, and apoptotic markers in tissue of brain. SN was found to cause excessive neuronal shrinkage, degeneration, and vacuolation in hippocampal sections, but most of these effects were found to be improved by TQ (50 mg/kg). In the evaluation of oxidative stress markers, TQ dose-dependently decreased MDA levels and significantly increased glutathione (GSH) activities. TQ restored the decreased cytochrome c oxidase in brain tissue dose- dependently. TQ inhibited higher C-reactive protein (CRP) levels. The increased levels of transforming growth factor-beta, a fibrosis factor, induced by SN were dose-dependently inhibited by TQ. TQ decreased the caspase-3 brain tissue activity, which is an important apoptosis marker. The cultivated brain cells' survival rate increased by concomitantly adding 500 μg of TQ and SN (Hamdan et al., 2019).

The specific oxidative stress markers levels, profiling 84 key genes associated with oxidative stress and the chosen antioxidant enzymes' activities through RT2 polymerase chain reaction (PCR) array were assessed to investigate the TQ's antioxidant effects (12.5 μ) on H_2O_2-activated BV-2 microglia or LPS/IFNγ. The nitric oxide (NO) and superoxide's cellular production in the H_2O_2-and LPS/IFNγ-activated microglia was significantly reduced by TQ. The lipid hydroperoxides levels were significantly reduced by TQ, significantly increasing the antioxidant GSH levels in the

H_2O_2-activated microglia and LPS/IFNγ-activated BV-2 cells. The H_2O_2-activated and LPS/INFγ-activated microglia which was treated with TQ had lower SOD and CAT activities. Based on the chosen 84 gene's RT2 PCR array profiling associated with the oxidative stress, TQ treatment in the LPS/IFNγ-activated microglia enhances the down- or up-regulation of specific genes associated with defense of the cells' natural antioxidant against activation of LPS/IFNγ, down-regulates specific prooxidant genes, and up-regulates specific antioxidant genes (Cobourne-Duval et al., 2018).

The quantitative proteomic analysis was done using Q-Exactive/Orbitrap Proteomic MS/LC-MS (liquid chromatography-mass spectrometry) on the IFNγ/LPS-activated BV-2 microglial cells treated with TQ (12.5 μ) to globally assess protein expressions changes and the TQ's ability to suppress the inflammatory response using ELISArray for the inflammatory cytokines. Moreover, the effects of TQ on the NFκB signaling targets' gene expression were assessed with the RT2 PCR array (Cobourne-Duval et al., 2018). This study showed significant increase of the expression of biliverdin reductase A, glutaredoxin-3, mitochondrial Lon protease, and 3-mercaptopyruvate sulfurtransferase as the antioxidant neuroprotective proteins through TQ treatment. Moreover, the expression of inflammatory cytokines, IL-10, IL-4, IL-6, IL-2, and IL-17a significantly reduced TQ in the activated microglia. TQ treatment is down-regulated many NFκB signaling targets genes, such as complement factor B (CFB), IL6, chemokine (C−C) motif ligand 5 (CCL5), and chemokine (C−C motif) ligand 3 (CXCL3).

Although the effectiveness of TQ in LPS-activated macrophages has been demonstrated in different studies, there is a lack of studies which evaluated the TQ's effects on immunological response of glial/microglia cell. For this purpose, a study was performed to investigate the TQ's propensity to inhibit BV2 microglial cells activated by LPS in relation to a given control, TQ reduced IL-6, granulocyte colony-stimulating factor, selective iNOS inhibitor N6-(1-iminoethyl)-L-lysine, dihydrochloride L-NIL, IL-1b, IL-12p40/70, monocyte chemotactic protein 5 (MCP-5)/chemokine (C−C motif) ligand 12 (CCL12), CCL2/MCP-1, and IFN-induced protein 10 (IP-10)/C-X-C motif chemokine 10 (Cxcl 10) in BV-2 murine microglial cells stimulated by LPS. These findings showed a general antiinflammatory impact of TQ on activated microglia attenuating the NO, cytokines, and iNOS (Taka et al., 2015).

17.3 Effects of thymoquinone on migraine pain

Migraine is a common neurological disorder. Although an irresistible headache is a cardinal symptom, several complaints such as nausea, vomiting, photophobia, and phonophobia may accompany the headache. The migraine's pathophysiology is not stated clearly; however, meningeal neurogenic inflammation was accepted as a main responsible factor (Koyuncu Irmak et al., 2019).

Meningeal mast cells (MMCs) and calcitonin gene-related peptide (CGRP) caused to activate the inflammatory processes sensitizing the trigeminal nociceptors. MMCs are proinflammatory immune cells that have a close relationship with nerves and blood vessels in the meninges (Levy et al., 2019; Kilinc et al., 2018). Although TQ has multiple antiinflammatory effects on neuroinflammation, whether the presence of any effect on CGRP release and MMCs is not known (El Gazzar, 2007).

An experimental study investigated the TQ's effects on the release of CGRP in strategic structures related to migraine and on the MMCs in GTN-induced ex vivo meningeal preparations and in vivo migraine model. CGRP levels stimulated by GTN in plasma and structures related to migraine, including brainstem and trigeminal ganglion were ameliorated by TQ. Despite the TQ's failure to inhibit the CGRP release from trigeminal meningeal afferents induced by GTN in the ex vivo meningeal preparations, MMCs degranulation was inhibited by it, preventing the increasing number of MMCs in the in vivo migraine model induced by GTN. The main conclusion achieved from this study was that release of CGRP in brainstem and trigeminal ganglion neurons is modulated and MMCs are stabilized by the TQ. It was stated that TQ may be a potent agent in preventing meningeal neurogenic inflammation and ultimately migraine (Kilinc et al., 2018).

17.4 Effects of thymoquinone on depression

Depressive disorders are serious mood disorders that last at least two weeks. They are associated with significantly reduced quality of life and role, comorbidity, and even sometimes mortality (Cuijpers et al., 2019). Neuroinflammation is involved in depression, and inflammatory indices including IL-6, TNF-a, and CRP may be affected during pathogenesis (Howren et al., 2009; Dowlati et al., 2010).

BC in animal models may be effective in the depressive behavior. With this aim, NS and TQ's hydroalcoholic extract effects were studied in depression-like behavior induced by LPS in rats through forced swimming test and open-field apparatus. NS and TQ were separately administrated to decrease times of immobility and coadministration of both compounds induced the lowest immobility times. The treated animals' crossing number was improved by NS and TQ in forced swimming test. In conclusion, TQ and NS improved memory and learning impairments in rats to protect against depression-like behavior induced by LPS (Hosseini et al., 2012). Cisplatin (CP) is a powerful antineoplastic agent treating various human malignancies. However, CP has neurotoxic, nephrotoxic, and ototoxic potential at repeated doses. Prophylactic treatment with TQ diminished the CP-dependent neurotoxicity as well as impairment in learning and memory (Kandeil et al., 2020).

The anxiolytic role of TQ on arsenic-induced hippocampal toxicity was investigated in an experimental study conducted on rats. Arsenic is a metalloid, which is mostly common, and the human exposure's main route is drinking polluted water. Arsenic exposure leads to behavioral disturbances and concomitant psychiatric disorders. TQ supplementation (5 and 2.5 mg/kg/day, p.o,) significantly improved the anxiogenic changes associated with arsenic. This improvement was majorly due to its anxiolytic and antioxidant potential. Positive changes of the cytokine levels and hippocampal oxidative stress with lower DNA fragmentation were expressed by the TQ pretreated rats. It was concluded that TQ may manage depressive and anxiety outcomes of arsenic intoxication as a powerful therapeutic agent (Firdaus et al., 2018).

17.5 Effects of thymoquinone on epilepsy

Epilepsy is characterized by the epileptic seizures which is defined as the temporary manifestation of symptoms or signs caused by the brain's non-normal synchronized neuronal activity (Fisher et al., 2005). It may present as a variable combination of somatosensory, motor, autonomic, behavioral, or specific sensory disorders. The seizure can usually last a few seconds or a few minutes. It can rarely last more than 20 min uninterrupted. This condition is called status epilepticus (SE) (Cho and Kim, 2019). It has been reported that oxidative stress is due to excessive free radical release as a main mechanism for the progression and initiation of epilepsy (Shin et al., 2011).

In the penicillin-induced epilepsy model, TQ (50, 10, and 100 mg/kg) showed prolonging of the first spike-wave latency. Also, the epileptiform activity amplitude and the spike-wave frequency were reduced by TQ (Beyazcicek et al., 2016). TQ (80 and 40 mg/kg, ip) shortened the myoclonic seizures duration seizure in mice model induced by pentylenetetrazole (PTZ) and increased the duration of the seizures onset by increasing the gamma-aminobutyric acid (GABA)ergic tone through opioid receptor (Hosseinzadeh and Parvardeh, 2004). TQ increased brain total antioxidant capacity and decreased ROS/reactive nitrogen species formation in rats treated with prilocaine. The NFκB-p65, AQP4, and NFκB-p50 expression was reduced by TQ in the brain tissue (Akgül et al., 2021).

In an SE rat model, TQ (10 mg/kg) has reduced the brain damage through the antioxidant pathway (Shao et al., 2017). This study also investigated the TQ's antioxidant effect at HO-1, SOD, and Nrf2 levels in the cortex and hippocampus. SE latency was found to be prolonged, and the TQ group was found to have lower seizure severity. In the lithium-prilocaine SE model, the same findings were demonstrated and the TNF-α and COX-2's protein levels were significantly down-regulated by the TQ pretreatment in the brain, as mediated by the NF-κB signaling pathway (Shao et al., 2016).

Temporal lobe epilepsy as a more common variant of epilepsy characterized by reactive astrogliosis, higher oxidative stress, and neuronal loss. In this type of epilepsy, Dariani et al. investigated the impacts of TQ pretreatment (10 mg/kg) in rats' intrahippocampal kainate model. In this model, the impacts were lipid peroxidation, seizure activity, hippocampal neuronal loss particularly in CA3, cornu ammonis-1 (CA1), the hilar regions of the hippocampus, mossy fiber sprouting, and astrogliosis (Dariani et al., 2013).

The TQ (40 mg/kg, orally) and vitamin C's (250 mg/kg ip) individual and combined effects in generalized seizures induced by pentlyenetetrazole (PTZ) were studied in an experimental study (Ullah et al., 2015). Epileptiform and polyspike discharge was significantly reduced by vitamin C or TQ supplementation. Epileptic seizures reduced the GABA B1 receptor expression, did not change the PKA expression, reduced calmodulin/calcium-dependent protein kinase II (CaMKII), and inhibit the phosphorylation of cAMP response element-binding protein (CREB) in hippocampus and cortex. All of these effects reversed with the supplementation of TQ and vitamin C (Ullah et al., 2015).

The TQ's effects in intractable pediatric seizures were investigated in the pilot study of Akhondian et al. In this study, 1 mg/kg oral TQ were given to children with a history of refractory epilepsy. It was determined that TQ decreased the seizure frequency (Akhondian et al., 2011). A characteristic feature of this study is not being an experimental study.

A bioanalytical UHPLC-based method was used to quantify the TQ-loaded PLGA nanoparticles to cure epilepsy. Optimized THQ-PLGA-NPs are nanosized, promising higher therapeutic efficacy (Ahmad et al., 2020).

17.6 Effects of thymoquinone on Parkinson's disease

Parkinson's disease is clinically characterized by tremor at rest, rigidity, postural imbalance, and bradykinesia as a neurodegenerative disease related to age (Cacabelos, 2017). In this disease, proteinous spherical bodies known as Lewy bodies are formed. The disease is divided into six neuropathological disease stages. In stage 1–2, known as the presymptomatic stage, inclusion bodies are limited to the pontine tegmentum/medulla oblongata and anterior olfactory nucleus/olfactory bulb. Once the disease progresses, the substantia nigra and the other nucleus of the forebrain and midbrain start to be affected in stage 3–4. At stage 5–6, the neocortex is included (Braak et al., 2004; Sveinbjornsdottir, 2016).

Despite acceptance of the dopaminergic nigrostriatal neurons degeneration in Parkinson's disease with Lewy bodies as the main neuropathological change of motor impairment, cholinergic, GABA-ergic, glutamatergic, noradrenergic, adrenergic, and tryptaminergic nerve cells may experience the same damage in their skeletons (Braak and Braak, 2000). Additionally, inflammation and oxidative stress response lead to PD pathogenesis (Lee Mosley et al., 2006).

There are various studies showing the effects of TQ on Parkinson's disease. The TQ's neuroprotective impacts have been demonstrated in a hemiparkinsonian rat model induced by 6-hydroxydopamine. Pretreatment with TQ (10 and 5 mg/kg) significantly improved turning behavior, decreased MDA levels, and prevented loss of substantia nigra pars compacta (SNc) neurons. It has been concluded that TQ provides neuroprotection against 6-OHDA neurotoxicity, partially because of the effect of reducing lipid peroxidation, and therefore may benefit against neurodegenerative diseases including PD (Sedaghat et al., 2014).

TQ pretreatment (15 and 7.5 mg/kg) leads to improvement in motor defects in rotenone-induced PD. TQ also caused changes in Parkin, Drp-1, dopamine, and TH levels (Ebrahimi et al., 2017).

The TQ's neuroprotective impacts on dopaminergic neurons were also investigated in primary mesencephalic cell cultures (Radad et al., 2015). This study concluded that TQ increased lysosomal degradation clearing the injured mitochondria and inhibiting apoptotic cell death mediated by mitochondria.

In a similar study, the efficacy of TQ in αSN-induced synaptic toxicity in human pluripotent stem cell (hiPSC) and rat hippocampal neurons was investigated (Alhebshi et al., 2014). The TQ was found to protect the cultured hippocampal neurons, reduce synaptophysin levels, inhibited synaptic activity, and in addition, administrating the TQ effectively protected these cells against the mutated P123H βSN in hippocampal neurons and synaptic vesicle recycling's αSN-induced inhibition in hippocampal and hiPSC-derived neurons. In another study, TQ prevented dopaminergic neurodegeneration by reducing oxidative stress through the antioxidant response elements/nuclear factor erythroid 2-related factor 2 (ARE/Nrf2) pathway (Dong et al., 2021).

17.7 Effects of thymoquinone on Alzheimer's disease

Alzheimer's disease (AD) presents with cognitive symptoms specific to dementia with an insidious and progressive course. The AD pathogenesis is still not completely clear, and seven different hypotheses have been proposed that may be related to each other. These hypotheses may be listed as amyloid, glutamatergic, oxidative, cholinergic, immunologic, and genetic hypotheses (Monczor, 2005). Various studies, mostly experimental, have been conducted on the effects of TQ on AD.

In an experimental study evaluating the therapeutic effects of TQ on AD, the authors investigate the protective effect of TQ on Alzheimer's amyloid-β (Aβ) peptide caused neurotoxicity in primary neurons of rats. Aβ1-42-induced neurotoxicity was effectively attenuated through TQ treatment improving the cell viability. The generation of ROS and potential depolarization of mitochondrial membrane were inhibited by TQ. Moreover, TQ inhibited aggregation of Aβ1-42 in vitro, restored inhibition of synaptic vesicle recycling, and somewhat reversed the spontaneous firing activity loss. TQ has neuroprotection against Aβ1-42 in rat's cortical and hippocampal neurons (Alhebshi et al., 2013).

The TQ's neuroprotective impacts on morphological damages, Aβ1-40-induced neurotoxicity, DNA condensation, ROS production, and caspase −8, −3, and −9 activations were investigated in primary cerebellar granule neurons (CGNs). TQ pretreatment increased cell survival after subsequent exposure to Ab1−40. TQ significantly decreased the lactate dehydrogenase level as a stable cytoplasmic enzyme and is an indicator of cell death. The extensive neurite network and intact cell bodies during the morphological evaluation were preserved by the TQ. TQ reduced condensed chromatin on Ab1−40-induced DNA condensation in CGNs. An ROS apparent increase was inhibited by the TQ pretreatment, which is a marker of its antioxidant ability. TQ also inhibited caspase−3, −8, and −9 activation. Caspases activation inhibition is known to be a desire for cell surveillance. The inhibitory effects of TQ on caspases −3, −8, and −9 provide the anti-apoptotic activity. In this way, the CGNs were protected by TQ against the cytotoxic effects of Ab1−40 (Ismail et al., 2013).

Another study investigated the TQ's neuroprotective activity against neurotoxicity induced by Aβ in human-induced cholinergic neurons derived from pluripotent stem cell. TQ could decrease the synaptic toxicity induced by Aβ1−42 (Alhibshi et al., 2019). These three studies' results show that TQ provides a neuroprotective effect by breaking the amyloid hypothesis in the AD pathogenesis and therefore shows promise for the future.

The neuroprotective effects of α7 receptor allosteric modulator, TQ, and α7 receptor agonist on the cerebral cortex were studied in AD experimentally induced relative to mesenchymal stem cells (MSCs) activation. It was concluded that α7 nAChR or TQ agonist combined with a positive allosteric modulator (PAM) may contribute to the AD treatment preferred to TQ alone. In exceptional cases, TQ alone or in combination with PAM proved the MSC activation (Abd El Fattah et al., 2016).

AD induced by Aβ was performed in an experimental study to evaluate the healing impacts of TQ on the AD's learning function. Cognitive dysfunction was caused by the Aβ by increasing step-through and initial latency with plaque formation and pyramidal cells degeneration in the hippocampus. This effect was ameliorated through TQ administration by significantly reducing the plaque formation in the hippocampus CA1 region and longer latency time, increasing the number of surviving neurons in the hippocampus. Consequently, TQ has improved the learning function (Poorgholam et al., 2018).

17.8 Effects of thymoquinone on cerebral ischemia

Ischemic brain injury is an important cause of mortality and morbidity. One of the primary factors causing cerebral ischemia pathogenesis is oxidative stress (Chan, 2001; Oskouei et al., 2018).

In the transient forebrain ischemia model, rats were treated with 5 mg/kg oral TQ five days before ischemia and continued throughout reperfusion. TQ reduced neuronal death in the hippocampal CA1 region. TQ treatment significantly decreased highly elevated MDA levels and increased catalase (CAT), superoxide dismutase (SOD), and GSH activities. Inflammation induced by oxidative stress was reduced. TQ prevented up-regulation of iNOS and inhibited formation of peroxynitrite (Al-Majed et al., 2006).

The effects of NS oil (5, 2.5, and 10 mg/kg) and TQ (5, 2.5, and 10 mg/kg) on global cerebral ischemia (GCI) reperfusion injury were studied. These compounds showed protecting effects against ischemia-reperfusion by regulating oxidative stress in the rat hippocampus (Hosseinzadeh et al., 2007). It is logical to say that this study confirms the results of Al-Majed's study.

In a more recent study, the effects of TQ on the brain extramedullary space's properties in temporary middle cerebral artery occlusion in rats were evaluated. TQ inhibited neuron necrosis in the hippocampus CA1 region and increased behavioral performance in the neurological examination. TQ significantly reduced the percentage of hemicerebral infarcts. TQ inhibited the changes in the extracellular space parameters, and thus the brain could be protected from tissue damage caused by ischemic stroke (Fan et al., 2020).

The mitochondria-targeted plastoquinone and TQ's neuroprotective impacts were investigated by creating a middle cerebral artery occlusion. Administrating these low-dose antioxidants in the long term reduced the neurological deficit but did not affect the volume of brain damage (Silachev et al., 2015).

The TQ's effect on small molecule metabolites was investigated using MALDI-MSI in a rat cerebral ischemia model. TQ enhanced the rate of energy-related molecules such as AMP, ADP, creatine, GMP, and antioxidants such as ascorbic acid, taurine, GSH, and other molecules related to metabolism such as glutamine, N-acetyl-L-aspartate, aspartate, glutamate, and sodium ions in the injured areas (Tian et al., 2020).

In another ischemia-reperfusion study investigating the neuroprotective effects of TQ nanoformulation, the cerebral ischemia treatment and neuroprotection are done by the TQ-loaded PLGA-chitosan nanoparticles. The ischemia infarct volume was reduced, and the locomotor activity was enhanced by the TQ. TQ-PLGA NPs treatment significantly increased the CAT's activity and SOD activities (Xiao et al., 2016).

17.9 Impacts of thymoquinone traumatic brain injury

Traumatic brain injury (TBI) is a main cause of mortality and morbidity worldwide. One can divide TBI into secondary injury and primary injury. The treatment of TBI is mostly focused on the chain of events that develop in the secondary stage. After TBI, excitatory amino acids such as aspartate and glutamate increase significantly (Kaur and Sharma, 2018). Free radical production, parenchymal inflammation, lipid peroxidation, increased NO, and intracellular calcium can be encountered causes of neurological damage (Farkhondeh et al., 2018).

The study examining the effect of TQ (5 mg/kg/day) on TBI was performed with the drop weight method on rats. Healing effects of TQ on neural cells after TBI were showed by decreasing the levels of MDA in the neurons' mitochondrial membranes and neuronal nuclei. GSH or SOD peroxidase antioxidant levels were not enhanced by the TQ (Gülşen, Ak, Çölçimen et al., 2016).

17.10 Effects of thymoquinone on multiple sclerosis

Multiple sclerosis (MS) results from the complex relationship between neural cells and the immune system. MS is an inflammatory demyelinating disease of the CNS. Animal models have been essential to elucidate the MS pathogenesis. The most characterized three MS models are experimental allergic/autoimmune encephalomyelitis (EAE), the chronic demyelinating disease which is virally induced, called the toxin-induced demyelination, and Theiler's murine encephalomyelitis virus infection (Procaccini et al., 2015). It is accepted that oxidative stress is the main factor affecting pathogenesis of EAE, which is the most-studied animal model of MS (Gilgun-Sherki et al., 2004).

Symptoms of EAE might be ameliorated by decreasing oxidative stress with TQ (1 mg/kg). TQ improved the paralysis or weakness of hind limb, perivascular inflammation, tail weakness, and low spinal cord GSH levels. This study showed improvement of EAE animals through TQ with modulated oxidative stress (Mohamed et al., 2003).

The NS seeds' effect in EAE-induced Wistar albino rats was studied in histological sections of the cerebellum and medulla. NS increased cerebellum demyelination and led to reduced TGF β1 expression (Noor et al., 2015). In the same model, NS (2.8 g/kg) increased the remyelination in the hippocampus. It was concluded that NS can be applied as a protective or adjuvant treatment even when the treatment started before the appearance of first clinical symptoms (Fahmy et al., 2014).

17.11 Impacts of thymoquinone on the spinal cord injury

The TQ's impacts on spinal cord injury have been demonstrated in different experimental studies.

In the experimental study performed on rats, high-dose and low-dose (30 mg/kg/day and 10 mg/kg/day, respectively) TQ was given before and after spinal cord injury. In this study, only histological evaluation including hemorrhage, spondylosis, and liquefactive necrosis analysis was performed. Although low-dose TQ did not show any benefit, high-dose TQ showed an attenuating effect on the histological findings of spinal cord damage (Üstün et al., 2014).

In an experimental study on spinal cord ischemia-reperfusion injury, TQ (10 mg/kg) was compared with methylprednisolone (30 mg/kg). TQ has an attenuating impact on the ischemia-reperfusion injury of spinal cord according to functional, biochemical, histological, and ultrastructural analyzes. These beneficial effects were achieved by attenuating the oxidative stress, inflammation, and apoptosis (Gökce et al., 2016).

The therapeutic effects of TQ (100, 200, and 400 mg/kg) after spinal cord injury were studied. The effects of TQ at different doses in the neuropathic pain model were investigated by measuring the TNF-α, NO, paraoxonase, total oxidant status, MDA, and interleukin-1β and the mechanical and heat—cold tests. The beneficial effects of TQ in this experimental neuropathic pain were showed, but increasing this TQ dose did not change the effect (Çelik et al., 2014).

In the spinal cord injury's mild contusion model, 30 mg/kg TQs reduced oxidative stress, apoptosis, and inflammatory response through PI3K/Akt and PPAR-γ pathways. TQ increased the Basso, Beattie, and Bresnahan scores in behavioral assessment and reduced the content of water in spinal cord (Chen et al., 2018).

The effectiveness of TQ (10 mg/kg) in reducing peridural fibrosis after laminectomy was demonstrated by an experimental study performed on rats. TQ significantly reduced the vascular volume and new bone volume and significantly decreased the fibrosis volume (Türköz et al., 2021).

17.12 Effects of thymoquinone on peripheral nerve injury

Despite the advances in medicine, the treatment of peripheral nerve damage is still not very satisfactory. After peripheral nerve damage, free oxygen radicals increase and this causes further tissue damage (Yalçin et al., 2013). Various experimental studies were performed to observe the TQ's effects on peripheral nerve damage.

Inflammation contributes to diabetic peripheral neuropathy (DPN) pathogenesis. TQ was shown to alleviate diabetic peripheral neuropathy by modulating inflammation. TQ inhibited proliferation of Schwann cells and protect against apoptosis of Schwann cell, and attenuate morphological changes induced by DPN and demyelination of sciatic nerve. It reduced the expression of IL-1β, COX-2, caspase-3, and IL-6 in cultured isolated Schwann cells or sciatic nerve tissue (Chen et al., 2018).

In peripheral nervous system toxicity induced by acrylamide, TQ (5, 2.5, and 10 mg/kg) showed neuroprotective effects via MAPKinase and apoptosis pathways. TQ concomitant administration (10 and 5 mg/kg) with angiotensin-converting enzyme prevented abnormalities of gait, increasing GSH content while significantly reducing MDA. Moreover, TQ (5 mg/kg) increased the P-ERK/ERK ratio and MBP content, while P-JNK/JNK decreased the apoptotic markers and P-P38/P38 ratios. TQ exerted neuroprotective effects, in part through antiapoptotic mechanisms and antioxidative stress (Tabeshpour et al., 2019).

Sciatic nerve crush injury model was created in our experimental study examining the TQ's effects (5 mg/kg/day) in the acute period in acute peripheral nerve injury (Gülşen, Ak, Kara et al., 2016). Myelin thickness, axon density, and axon diameter were obtained. As a result, TQ did not have any curative effect in the acute period of injury.

Neuroprotective effects of TQ combined with alpha-tocopherol were investigated on lower extremity ischemia-reperfusion muscle and nerve injury (Erkut et al., 2016). The combination of these compounds had a synergistic effect through antiinflammatory, antioxidant, and antiapoptotic mechanisms.

17.13 Neuroprotective effects of thymol

Thymol (2-isopropyl-5-methyl phenol) from the phenolic monoterpenes family has been used for long in traditional medicine for antiinflammatory, antibacterial, antineoplastic, expectorant, antiseptic, and antiviral purposes (Nagoor Meeran et al., 2017). However, promising studies on thymol, including cancer studies, are still ongoing.

17.14 Effects of thymol on Alzheimer's disease

The AD treatment strategy is based on acetylcholinesterase (AchE) inhibitors. The effects of linalool, thymol, carvacrol, TQ, and thymohydroquinone on AchE inhibition were investigated. Thymohydroquinone showed the most powerful AChE inhibiting impact on the concentration range. Thymol has lower AChE inhibitory potential than carvacrol and TQ. However, there is similarity between carvacrol and thymol in terms of structure (Jukic et al., 2007).

Thymol and carvacrol considerably protected PC12 cells against cytotoxicity induced by Aβ25-35 and reversed ROS production induced by Aβ25-35 (Azizi et al., 2020). In addition, carvacrol and thymol were also found to increase PKC (protein kinase C) activity. These results show that thymol and carvacrol show promise in the treatment of AD.

The effects of carvacrol and thymol (0.5, 1, or 2 mg/kg) on the cognitive functions were investigated. Thymol and carvacrol each attenuated the cognitive impairments due to higher cholinergic hypofunction or Aβ levels. In conclusion, antioxidant, anticholinesterase, and antiinflammatory activities might contribute to the beneficial impacts (Azizi et al., 2012). It was also stated that thymol reduced intracellular oxidative stress dependently on dose, exhibits anticholinesterase activity, and increases cell viability of neuronal PC-12 cells (Lee et al., 2015).

17.15 Effects of thymol on anxiety

Anti-anxiety effect of thymol (10, 5, 20 mg/kg i.p.) was studied with light/dark exploration test and elevated plus maze (EPM) behavioral models. 20 mg/kg thymol was found to have a significant antianxiety effect. This effect was thought to be possibly by modulation of GABA, NO-cyclic guanosine 30, 50-monophosphate (NO-cGMP), and 5-hydroxytryptamine (5-HT) pathways (Bhandari and Kabra, 2014).

Antianxiety effect of thymol was showed in a study in which cylinder, open-field, plus maze, hole-board, sodium periorbital-induced hypnosis, and acute toxicity (LD50) were studied (González-Trujano et al., 2017). The antianxiety effect was also demonstrated in an experimental study with an EPM test (Wang et al., 2017).

17.16 Effects of thymol on dementia

The possible beneficial effects of thymol on the aging rat brain were studied (Youdim and Deans, 2000). Antioxidant enzyme activities (GPx and SOD) and the total antioxidant rate of thymol application continued to be significantly higher. Another study of the same authors showed that dietary thymol (42.5 mg/kg body weight/day) provided more polyunsaturated fatty acids levels (Youdim and Deans, 1999).

17.17 Effects of thymol on depression

In an experimental study, the antidepressant effects were tested by giving *Lippia sidoides* essential oil (200 and 100 mg/kg) or corticosterone, or fluoxetine (35 mg/kg) or thymol (50 and 25 mg/kg) [93]. Tail suspension, open field, forced swimming test, sucrose preference, and higher plus maze tests were conducted. Thymol was found to have antidepressant-like effects similar to fluoxetine (Parente et al., 2018). Fluoxetine is an FDA-approved antidepressant that inhibits selective reuptake of serotonin in central neurons, currently in use in the treatment of depression.

Thymol (15 and 30 mg/kg) increased the level of central neurotransmitters in a mouse model and inhibits the expression of proinflammatory cytokines such as IL-6, (IL)-1β, and TNF-α (Deng et al., 2015). Additionally, the activation of nod-like receptor protein 3 (NLRP3) inflammasome and its adapter were inhibited by thymol, subsequently decreasing the caspase-1 expression.

The effects of thymol were studied on the expression of prefrontal cortex TLR 2/4 gene and hippocampus and depressive behavior induced by CUS on a rat model (Amiresmaeili et al., 2018). Thymol showed antidepressive activity without effect on stress-induced TLR-2/4 up-regulation.

17.18 Impacts of thymol on epilepsy

Antioxidant and serum biochemical indices of thymol were examined in kindled rats. Kindling is recognized as a temporal lobe epilepsy's experimental model. Thymol decreased MDA serum levels and hippocampal IL-1β and TNF-α levels, which are increased by epileptic seizures, and increased SOD activity. Thymol showed an anticonvulsant effect with anti-inflammatory and antioxidant mechanisms (Aliabadi et al., 2016).

Sancheti et al. characterized the antiepileptogenic and anticonvulsant potential of thymol (5, 10, 50, 25, and 100 mg/kg) in various experimental models. Maximal electroshock (MES)-induced seizures test, strychnine-induced seizure test, PTZ-induced seizures test, locomotor activity, 4-aminopyridine seizure test, PTZ-induced kindling test, and oxidative stress assessments including MDA and GSH levels in brain tissue were performed. In the MES-induced size model, a reduction in hind limb extension (HLE) duration was seen at all thymol doses. Although 100 mg/kg dose of thymol significantly reduced the duration of HLE, it couldn't provide the full protection obtained by 25 mg/kg phenytoin. In the PTZ-induced seizure test module, 100 mg/kg thymol extended the myoclonic jerk onset, the clonic seizures onset, the death onset, and the HLE onset. Thymol (100 and 50 mg/kg) showed improvement of activity as compared to diazepam in extending the clonic seizures and the myoclonic jerks onset. In the STR-induced 6 seizure test model, 100 mg/kg thymol considerably extended the death onset. In the 4-AP seizure test model, thymol did not protect the animals from 4-AP-induced seizures at any dose. 100 mg/kg of thymol inhibited 80.36% of locomotor activity. 25 mg/kg thymol considerably increased levels of GSH, decreased the score of seizure, and reduced levels of MDA in the PTZ induced kindling model. As a result, considerable anti-epileptogenic and anticonvulsant property of thymol were stated (Sancheti et al., 2014).

The thymol's seizure prevention effect in the partial psychomotor seizure model was evaluated (Mishra and Baker, 2014). Thymol partially inhibited the effects for MES, corneal-kindled, and scMET models.

17.19 Effects of thymol on Parkinson's disease

The effects of thymol (50 mg/kg) against dopaminergic neurodegeneration in a rotenone-induced experimental Parkinson's model were studied

(Javed et al., 2019). Thymol has been shown to protect the endogenous antioxidant defense network and reduce inflammatory mediators, such as cytokines, in this model.

17.20 Other thymol's impacts on nervous system

Sammi et al. studied the cholinergic dysfunction of thymol (Sammi et al., 2017). In their experimental study, although thymol and gamma-terpinene have no activity alone, when thymol (10—100 ppm) is combined with para-cymene or gamma-terpinene, it reduced cholinergic dysfunction via increasing acetylcholine synaptic levels and increasing nicotinic acetylcholine receptor responsiveness.

Thymol has been reported to have an effect on the GABA-A receptor. Compounds with GABAergic activity show neuroprotective effects, mostly through potentiation of GABAA-mediated inhibition. The thymol's pharmacological effect on the natural GABA-A receptor with cortical neuron primary culture was investigated (García et al., 2006). The GABA-induced (5 μM) chloride influx was enhanced by the thymol.

As the analgesic effect of thymol has long been known and is used as a stabilizer in anesthesia applications currently. However, the mechanism underlying these effects is not fully known. It was stated that the thymol's local anesthetic and antinociceptive impacts may be due to the voltage-operated sodium channels blockade (Haeseler et al., 2002). This study concluded that thymol is as potent as lidocaine currently in clinical use as a local anesthetic.

The thymol's effect on spontaneous excitatory transmission was investigated for the spinal substantia gelatinosa (SG) neurons (Xu et al., 2015). The transient receptor potential (TRP) channels which are expressed in heterologous cells are activated by the thymol. However, whether this effect is the same in innate neurons and how thymol affects synaptic transmission is not clear. It has been shown that thymol (1 mM) activated the channels of TRPA1, increasing the L-glutamate's frequent release on SG neurons, generating an external current without activation of TRP in the adult rat spinal cord slices by its antinociceptive potent effect.

The GABA release from GABAergic nerve terminals is inhibited by the GABA-B autoreceptors. The (3H)-GABA release probably was potentiated by the thymol (200 mM) considering its antagonistic impact on GABAb autoreceptors in the rat's neocortical slices (Parker et al., 2014).

In an experimental study evaluating the effect of thymol on sciatic nerve excitability, compound action potential (CAP) was inhibited by thymol dependently on concentration. Moreover, thymol inhibited neuronal excitability (Barbosa et al., 2017).

17.21 Neuroprotective effects of p-cymene

P-cymene (1-Methyl-4-(propane-2-yl) benzene) is a natural compound and an aromatic monocyclic monoterpene, which is found in above 100 plant oils, including BC. It has anti-inflammatory, analgesic, antioxidant, and antitumor effects (Bonjardim et al., 2012; Li et al., 2016; Yvon et al., 2012).

Inflammatory responses are reduced by P-cymene by inhibiting production of the IL-6, TNF-α, and IL-1B, inhibiting NF-KB activities and MAPK (Chen et al., 2014; Zhong et al., 2013).

NF-KB and MAPK activities mainly contribute to the AD pathogenesis and are considered as targets for treatment (Munoz and Ammit, 2010; Tuppo and Arias, 2005; Valerio et al., 2006). The experimental studies found that p-cymene administered at different doses reduces the number of amyloid plaques in a dose-dependent manner and partially disaggregates Aβ1−42 fibrils formed in vitro. In vitro antifibrillar effect of p-cyme was demonstrated (Seifi-Nahavandi et al., 2020).

The dose-dependent antinociceptive efficacy of p-cymene has been demonstrated by several experimental studies. In an experimental study with monoterpenes containing camphene, p-cymene geranyl acetate, it was shown that p-cyme has stronger antinociceptive activity compared to other agents. Despite the strong antinociceptive effect, p-cymene showed no activity against NO but showed mild activity against superoxide in the SOD-like mechanism (Quintans-Júnior et al., 2013). P-cyme exerts a dose-dependent depressant impact on the CNS and has a suppressive effect on nociceptive responses evoked by chemical and thermal stimuli (Bonjardim et al., 2012).

The analgesic effect of p-cyme is short, and it has poor water solubility. In an experimental study, p-cymene/-cyclodextrin (β-CD) complex was used to increase the efficiency of p-cymene. As a result, it was reported that the antiinflammatory and analgesic activity of p-cymene was increased by this complex (Jde et al., 2013).

No sufficient information is available on the impacts of p-cymene on the peripheral nervous system. In an experimental study, p-cymene did not significantly alter the sciatic nerve compound action potential. However, it was demonstrated that thymol inhibited the neuronal excitability. Due to this reason thymol was accepted as a promising agent for therapeutic purposes (Barbosa et al., 2017).

17.22 Neuroprotective effects of carvacrol

As a monoterpenoid phenol, carvacrol or 2-methyl-5-isopropyl phenol has neuroprotective effects shown with several experimental studies (Hong et al., 2018). Fe2+ -induced neurotoxicity can be inhibited by carvacrol in neuroblastoma SH-SY5Y cells with a mechanism mediated by apoptosis attenuation by down-regulating the JNKNF/MAPK-κB signaling pathway (Cui et al., 2015).

Carvacrol has been reported to block TRP channels, subfamily M, member 7 (TRPM7) in hippocampal neurons, and the HEK cells overexpressing TRPM7 leading to neuroprotection in mice under focal ischemia (Yu et al., 2012; Parnas et al., 2009). As known, anoxic neuronal death is mediated by the TRPM7 as a key factor (Aarts et al., 2003).

Carvacrol has beneficial effects in protecting the brain from neonatal brain's hypoxic-ischemic injury by improving prosurvival signaling by restoring and reducing brain's infarct volume, and inhibiting proapoptotic signaling by decreasing caspase-3 activation and enhancing Bcl-2/Bax protein ratio. It also improved the behavioral outcomes. The TRPM7 channel function inhibition may lead to such neuroprotective effect (Chen et al., 2015).

The carvacrol's (25, 50 and 100 mg/kg) inhibitory effects were studied in LPS-induced memory impairment, and it was found that carvacrol provided significant improvement (Lee et al., 2015).

Carvacrol has been shown to reduce TNF-α and IL-1β in the prefrontal complex and hippocampus. Also, carvacrol has been shown to restore BDFN expression in hippocampus and attenuate toll-like receptor-4 levels. Carvacrol has been reported to has a dose-dependent beneficial effect on LPS-induced cognitive dysfunction, and optimal efficacy is achieved with 100 mg/kg.

Carvacrol has also been studied in models of epilepsy. Carvacrol inhibited early seizures in epilepsy and recurrent SE perforating path models in rats (Khalil et al., 2017). However, the subsequent chronic epilepsy development was not prevented by the carvacrol but preventing the cognitive function deterioration after post-SE. It was stated that carvacrol shows a reduction in memory impairment by inhibiting TRPM channels of carvacrol.

Carvacrol was shown to reduce global cerebral ischemia-induced hippocampal neuron death, reduce zinc translocation to the hippocampal pyramidal layer, reduce GCI-induced hippocampal oxidative damage and microglial activation, and inhibit the expression of TRPM7 channels after GCI (Hong et al., 2018). Carvacrol might have therapeutic efficacy by reducing intracellular zinc translocation in the post-GCI period.

The neuroprotective effects of carvacrol (73 mg/kg, intraperitoneally) and pomegranate in methotrexate-induced toxicity were studied in an experimental study. Both carvacrol and pomegranate reduced the proinflammatory response by reducing the total oxidant status, MDA, TNF-α, and IL-1β levels (Çelik et al., 2013). The effects of carvacrol and pomegranate were studied separately. The authors didn't perform any group looking for the presence of the synergistic effect of both agents.

The thymol and carvacrol's protective effects were evaluated in PC12 cells in terms of the Aβ25-35-induced cytotoxicity (Azizi et al., 2020). It was stated that these compounds' effect on Aβ25-35 was through reduction of oxidative damage and increased PKC activity as a protein related to memory. However, these compounds were studied separately, and the presence of any synergistic effect was not reported.

17.23 Conclusion

Most of the studies related with BC and BC seed oils are experimentally performed ones. In these studies, the neuroprotective efficacy of BC and seed oil has been demonstrated in general, but these studies have not reached the desired scientific levels. For example, recommended dose ranges for TQ, thymol, and carvacrol are not yet clear. Despite traditional long-term usages, studies related to these compounds are cell-derived effects mostly experimental and must be strengthened with human studies. Considering many points, it is obvious that the studies should continue for a long time.

References

Aarts, M., Iihara, K., Wei, W.L., Xiong, Z.G., Arundine, M., Cerwinski, W., MacDonald, J.F., Tymianski, M., 2003. A key role for TRPM7 channels in anoxic neuronal death. Cell 115 (7), 863–877. https://doi.org/10.1016/S0092-8674(03)01017-1.

Abd El Fattah, L.I., Zickri, M.B., Aal, L.A., Heikal, O., Osama, E., 2016. The effect of thymoquinone, α7 receptor agonist and α7 receptor allosteric modulator on the cerebral cortex in experimentally induced Alzheimer's disease in relation to MSCs activation. International Journal of Stem Cells 9 (2), 230–238. https://doi.org/10.15283/ijsc16021.

Ahmad, N., Ahmad, R., Al Qatifi, S., Alessa, M., Al Hajji, H., Sarafroz, M., 2020. A bioanalytical UHPLC based method used for the quantification of Thymoquinone-loaded-PLGA-nanoparticles in the treatment of epilepsy. BMC Chemistry 14 (1). https://doi.org/10.1186/s13065-020-0664-x.

Akgül, B., Aycan, İ.Ö., Hidişoğlu, E., Afşar, E., Yıldırım, S., Tanrıöver, G., Coşkunfirat, N., Sanlı, S., Aslan, M., 2021. Alleviation of prilocaine-induced epileptiform activity and cardiotoxicity by thymoquinone. DARU, Journal of Pharmaceutical Sciences 29 (1), 85–99. https://doi.org/10.1007/s40199-020-00385-2.

Akhondian, J., Kianifar, H., Raoofziaee, M., Moayedpour, A., Toosi, M.B., Khajedaluee, M., 2011. The effect of thymoquinone on intractable pediatric seizures (pilot study). Epilepsy Research 93 (1), 39–43. https://doi.org/10.1016/j.eplepsyres.2010.10.010.

Alhebshi, A.H., Gotoh, M., Suzuki, I., 2013. Thymoquinone protects cultured rat primary neurons against amyloid β-induced neurotoxicity. Biochemical and Biophysical Research Communications 433 (4), 362–367. https://doi.org/10.1016/j.bbrc.2012.11.139.

Alhebshi, A.H., Odawara, A., Gotoh, M., Suzuki, I., 2014. Thymoquinone protects cultured hippocampal and human induced pluripotent stem cells-derived neurons against α-synuclein-induced synapse damage. Neuroscience Letters 570, 126–131. https://doi.org/10.1016/j.neulet.2013.09.049.

Alhibshi, A.H., Odawara, A., Suzuki, I., 2019. Neuroprotective efficacy of thymoquinone against amyloid beta-induced neurotoxicity in human induced pluripotent stem cell-derived cholinergic neurons. Biochemistry and Biophysics Reports 17, 122–126. https://doi.org/10.1016/j.bbrep.2018.12.005.

Aliabadi, A., İzadi, M., Rezvani, E., smaeli-dejah, M., 2016. Effects of thymol on serum biochemical and antioxidant indices in kindled rats. International Journal of Medical Laboratory 3 (1), 43–49.

Al-Majed, A.A., Al-Omar, F.A., Nagi, M.N., 2006. Neuroprotective effects of thymoquinone against transient forebrain ischemia in the rat hippocampus. European Journal of Pharmacology 543 (1–3), 40–47. https://doi.org/10.1016/j.ejphar.2006.05.046.

Amiresmaeili, A., Roohollahi, S., Mostafavi, A., Askari, N., 2018. Effects of oregano essential oil on brain TLR4 and TLR2 gene expression and depressive-like behavior in a rat model. Research in Pharmaceutical Sciences 13 (2), 130–141. https://doi.org/10.4103/1735-5362.223795.

Atta, M.B., 2003. Some characteristics of nigella (Nigella sativa L.) seed cultivated in Egypt and its lipid profile. Food Chemistry 83 (1), 63–68. https://doi.org/10.1016/S0308-8146(03)00038-4.

Azizi, Z., Ebrahimi, S., Saadatfar, E., Kamalinejad, M., Majlessi, N., 2012. Cognitive-enhancing activity of thymol and carvacrol in two rat models of dementia. Behavioural Pharmacology 23 (3), 241–249. https://doi.org/10.1097/FBP.0b013e3283534301.

Azizi, Z., Salimi, M., Amanzadeh, A., Majelssi, N., Naghdi, N., 2020. Carvacrol and thymol attenuate cytotoxicity induced by amyloid β25-35 via activating protein kinase C and inhibiting oxidative stress in PC12 cells. Iranian Biomedical Journal 24 (4), 243–250. https://doi.org/10.29252/ibj.24.4.243.

Barbosa, R., Cruz-Mendes, Y., Silva-Alves, K.S., Ferreira-Da-Silva, F.W., Ribeiro, N.M., Morais, L.P., Leal-Cardoso, J.H., 2017. Effects of lippia sidoides essential oil, thymol, p-cymene, myrcene and caryophyllene on rat sciatic nerve excitability. Brazilian Journal of Medical and Biological Research 50 (12). https://doi.org/10.1590/1414-431X20176351.

Beyazcicek, E., Ankarali, S., Beyazcicek, O., Ankarali, H., Demir, S., Ozmerdivenli, R., 2016. Effects of thymoquinone, the major constituent of nigella sativa seeds, on penicillin-induced epileptiform activity in rats. Neurosciences 21 (2), 131–137. https://doi.org/10.17712/nsj.2016.2.20150781.

Bhandari, S.S., Kabra, M.P., 2014. To evaluate anti-anxiety activity of thymol. Journal of Acute Disease 136–140. https://doi.org/10.1016/S2221-6189(14)60030-5.

Bonjardim, L.R., Cunha, E.S., Guimarães, A.G., Santana, M.F., Oliveira, M.G.B., Serafini, M.R., Araújo, A.A.S., Antoniolli, Â.R., Cavalcanti, S.C.H., Santos, M.R.V., Quintans-Júnior, L.J., 2012. Evaluation of the anti-inflammatory and antinociceptive properties of p-cymene in mice. Zeitschrift Für Naturforschung C 67 (1–2), 15–21. https://doi.org/10.1515/znc-2012-1-203.

Braak, H., Braak, E., 2000. Pathoanatomy of Parkinson's disease. Journal of Neurology II3–II10. https://doi.org/10.1007/pl00007758.

Braak, H., Ghebremedhin, E., Rüb, U., Bratzke, H., Del Tredici, K., 2004. Stages in the development of Parkinson's disease-related pathology. Cell and Tissue Research 318 (1), 121–134. https://doi.org/10.1007/s00441-004-0956-9.

Butt, M.S., Sultan, M.T., 2010. Nigella sativa: reduces the risk of various maladies. Critical Reviews in Food Science and Nutrition 50 (7), 654–665. https://doi.org/10.1080/10408390902768797.

Cacabelos, R., 2017. Parkinson's disease: from pathogenesis to pharmacogenomics. International Journal of Molecular Sciences 18 (3). https://doi.org/10.3390/ijms18030551. PMID: 28273839; PMCID: PMC5372567.

Çelik, F., Gocmez, C., Bozkurt, M., Kaplan, I., Kamasak, K., Akil, E., Dogan, E., Guzel, A., Uzar, E., 2013. Neuroprotective effects of carvacrol and pomegranate against methotrexate-induced toxicity in rats. European Review for Medical and Pharmacological Sciences 17 (22), 2988–2993.

Çelik, F., Göçmez, C., Karaman, H., Kamaşak, K., Kaplan, I., Akil, E., Tufek, A., Guzel, A., Uzar, E., 2014. Therapeutic effects of thymoquinone in a model of neuropathic pain. Current Therapeutic Research - Clinical and Experimental 76, 11–16. https://doi.org/10.1016/j.curtheres.2013.11.001.

Chan, P.H., 2001. Reactive oxygen radicals in signaling and damage in the ischemic brain. Journal of Cerebral Blood Flow and Metabolism 21 (1), 2–14. https://doi.org/10.1097/00004647-200101000-00002.

Cheikh-Rouhou, S., Besbes, S., Hentati, B., Blecker, C., Deroanne, C., Attia, H., 2007. Nigella sativa L.: chemical composition and physicochemical characteristics of lipid fraction. Food Chemistry 101 (2), 673–681. https://doi.org/10.1016/j.foodchem.2006.02.022.

Chen, L., Zhao, L., Zhang, C., Lan, Z., 2014. Protective effect of p-cymene on lipopolysaccharide-induced acute lung injury in mice. Inflammation 37 (2), 358–364. https://doi.org/10.1007/s10753-013-9747-3.

Chen, W., Xu, B., Xiao, A., Liu, L., Fang, X., Liu, R., Turlova, E., Barszczyk, A., Zhong, X., Sun, C.L.F., Britto, L.R.G., Feng, Z.P., Sun, H.S., 2015. TRPM7 inhibitor carvacrol protects brain from neonatal hypoxic-ischemic injury. Molecular Brain 8 (1). https://doi.org/10.1186/s13041-015-0102-5.

Chen, Y., Wang, B., Zhao, H., 2018. Thymoquinone reduces spinal cord injury by inhibiting inflammatory response, oxidative stress and apoptosis via PPAR-γ and PI3K/Akt pathways. Experimental and Therapeutic Medicine 15 (6), 4987–4994. https://doi.org/10.3892/etm.2018.6072.

Cho, Y.W., Kim, K.T., 2019. The latest classification of epilepsy and clinical significance of electroencephalography. Journal of Neurointensive Care 1–3. https://doi.org/10.32587/jnic.2019.00094.

Cobourne-Duval, M.K., Taka, E., Mendonca, P., Soliman, K.F.A., 2018. Thymoquinone increases the expression of neuroprotective proteins while decreasing the expression of pro-inflammatory cytokines and the gene expression NFκB pathway signaling targets in LPS/IFNγ -activated BV-2 microglia cells. Journal of Neuroimmunology 320, 87–97. https://doi.org/10.1016/j.jneuroim.2018.04.018.

Cui, Z.W., Xie, Z.X., Wang, B.F., Zhong, Z.H., Chen, X.Y., Sun, Y.H., Sun, Q.F., Yang, G.Y., Bian, L.G., 2015. Carvacrol protects neuroblastoma SH-SY5Y cells against Fe2+-induced apoptosis by suppressing activation of MAPK/JNK-NF-κB signaling pathway. Acta Pharmacologica Sinica 36 (12), 1426–1436. https://doi.org/10.1038/aps.2015.90.

Cuijpers, P., Quero, S., Dowrick, C., Arroll, B., 2019. Psychological treatment of depression in primary care: recent developments. Curr Psychiatry Rep 21 (12), 129. https://doi.org/10.1007/s11920-019-1117-x.

Dariani, S., Baluchnejadmojarad, T., Roghani, M., 2013. Thymoquinone attenuates astrogliosis, neurodegeneration, mossy fiber sprouting, and oxidative stress in a model of temporal lobe epilepsy. Journal of Molecular Neuroscience 51 (3), 679–686. https://doi.org/10.1007/s12031-013-0043-3.

Deng, X.Y., Li, H.Y., Chen, J.J., Li, R.P., Qu, R., Fu, Q., Ma, S.P., 2015. Thymol produces an antidepressant-like effect in a chronic unpredictable mild stress model of depression in mice. Behavioural Brain Research 291, 12–19. https://doi.org/10.1016/j.bbr.2015.04.052.

DiSabato, D.J., Quan, N., Godbout, J.P., 2016. Neuroinflammation: the devil is in the details. Journal of Neurochemistry 139, 136–153. https://doi.org/10.1111/jnc.13607.

Dong, J., Zhang, X., Wang, S., Xu, C., Gao, M., Liu, S., Li, X., Cheng, N., Han, Y., Wang, X., Han, Y., 2021. Thymoquinone prevents dopaminergic neurodegeneration by attenuating oxidative stress via the Nrf2/ARE pathway. Frontiers in Pharmacology 11. https://doi.org/10.3389/fphar.2020.615598.

Dowlati, Y., Herrmann, N., Swardfager, W., Liu, H., Sham, L., Reim, E.K., Lanctôt, K.L., 2010. A meta-analysis of cytokines in major depression. Biological Psychiatry 67 (5), 446–457. https://doi.org/10.1016/j.biopsych.2009.09.033.

Ebrahimi, S.S., Oryan, S., Izadpanah, E., Hassanzadeh, K., 2017. Thymoquinone exerts neuroprotective effect in animal model of Parkinson's disease. Toxicology Letters 276, 108–114. https://doi.org/10.1016/j.toxlet.2017.05.018.

El Gazzar, M.A., 2007. Thymoquinone suppresses in vitro production of IL-5 and IL-13 by mast cells in response to lipopolysaccharide stimulation. Inflammation Research 56 (8), 345–351. https://doi.org/10.1007/s00011-007-7051-0.

Erkut, A., Cure, M.C., Kalkan, Y., Balik, M.S., Guvercin, Y., Yaprak, E., Yuce, S., Sehitoglu, I., Cure, E., 2016. Protective effects of thymoquinone and alpha-tocopherol on the sciatic nerve and femoral muscle due to lower limb ischemia-reperfusion injury. European Review for Medical and Pharmacological Sciences 20 (6), 1192–1202. http://www.europeanreview.org/.

Ermumcu, M., Şanlıer, N., 2017. Black cumin (Nigella sativa) and its active component of thymoquinone: effects on health. Journal of Food and Health Science 3 (4), 170–183.

Fahmy, H.M., Noor, N.A., Mohammed, F.F., Elsayed, A.A., Radwan, N.M., 2014. Nigella sativa as an anti-inflammatory and promising remyelinating agent in the cortex and hippocampus of experimental autoimmune encephalomyelitis-induced rats. The Journal of Basic & Applied Zoology 182–195. https://doi.org/10.1016/j.jobaz.2014.08.005.

Fan, C., Tian, F., Zhao, X., Sun, Y., Yang, X., Han, H., Pu, X., 2020. The effect of thymoquinone on the characteristics of the brain extracellular space in transient middle cerebral artery occlusion rats. Biological and Pharmaceutical Bulletin 43 (9), 1306–1314. https://doi.org/10.1248/bpb.b19-00991.

Farkhondeh, T., Samarghandian, S., Shahri, A., Samini, F., 2018. The neuroprotective effects of thymoquinone: a review. Dose-Response 16, 1559325818761455–1559 325818761455.

Firdaus, F., Zafeer, M.F., Ahmad, M., Afzal, M., 2018. Anxiolytic and anti-inflammatory role of thymoquinone in arsenic-induced hippocampal toxicity in Wistar rats. Heliyon 4 (6), e00650. https://doi.org/10.1016/j.heliyon.2018.e00650.

Fisher, R.S., Van Emde Boas, W., Blume, W., Elger, C., Genton, P., Lee, P., Engel, J., 2005. Epileptic seizures and epilepsy: definitions proposed by the International League Against Epilepsy (ILAE) and the International Bureau for Epilepsy (IBE). Epilepsia 46 (4), 470–472. https://doi.org/10.1111/j.0013-9580.2005.66104.x.

García, D.A., Bujons, J., Vale, C., Suñol, C., 2006. Allosteric positive interaction of thymol with the GABAA receptor in primary cultures of mouse cortical neurons. Neuropharmacology 50 (1), 25–35. https://doi.org/10.1016/j.neuropharm.2005.07.009.

Gilgun-Sherki, Y., Melamed, E., Offen, D., 2004. The role of oxidative stress in the pathogenesis of multiple sclerosis: the need for effective antioxidant therapy. Journal of Neurology 251 (3), 261–268. https://doi.org/10.1007/s00415-004-0348-9.

Gökce, E.C., Kahveci, R., Gökce, A., Cemil, B., Aksoy, N., Sargon, M.F., Kısa, Ü., Erdoğan, B., Güvenç, Y., Alagöz, F., Kahveci, O., 2016. Neuroprotective effects of thymoquinone against spinal cord ischemia-reperfusion injury by attenuation of inflammation, oxidative stress, and apoptosis. Journal of Neurosurgery: Spine 24 (6), 949–959. https://doi.org/10.3171/2015.10.spine15612.

González-Trujano, M.E., Hernández-Sánchez, L.Y., Muñoz Ocotero, V., Dorazco-González, A., Guevara Fefer, P., Aguirre-Hernández, E., 2017. Pharmacological evaluation of the anxiolytic-like effects of Lippia graveolens and bioactive compounds. Pharmaceutical Biology 55 (1), 1569–1576. https://doi.org/10.1080/13880209.2017.1310908.

Gülşen, I., Ak, H., Çölçimen, N., Alp, H.H., Akyol, M.E., Demir, I., Atalay, T., Balahroǐlu, R., Raǐbetli, M., 2016. Neuroprotective effects of thymoquinone on the Hippocampus in a rat model of traumatic brain injury. World Neurosurgery 86, 243–249. https://doi.org/10.1016/j.wneu.2015.09.052.

Gülşen, İ., Ak, H., Kara, M., Gökalp, A., Akyol, V., Koçak, Ö.F., Rağbetli, M.Ç., 2016. Timokinonun akut periferik sinir hasarı üzerine akut etkisi: deneysel bir çalışma. Ulusal Travma ve Acil Cerrahi Dergisi 22 (6), 526–530. https://doi.org/10.5505/tjtes.2016.40204.

Haeseler, G., Maue, D., Grosskreutz, J., Bufler, J., Nentwig, B., Piepenbrock, S., Dengler, R., Leuwer, M., 2002. Voltage-dependent block of neuronal and skeletal muscle sodium channels by thymol and menthol. European Journal of Anaesthesiology 19 (8), 571–579. https://doi.org/10.1097/00003643-200208000-00005.

Hamdan, A.M., Al-Gayyar, M.M., Shams, M.E.E., Alshaman, U.S., Prabahar, K., Bagalagel, A., Diri, R., Noor, A.O., Almasri, D., 2019. Thymoquinone therapy remediates elevated brain tissue inflammatory mediators induced by chronic administration of food preservatives. Scientific Reports 9 (1). https://doi.org/10.1038/s41598-019-43568-x.

Hong, D., Choi, B., Kho, A., Lee, S., Jeong, J., Kang, B., Kang, D., Park, K.-H., Suh, S., 2018. Carvacrol attenuates hippocampal neuronal death after global cerebral ischemia via inhibition of transient receptor potential melastatin 7. Cells 231. https://doi.org/10.3390/cells7120231.

Hosseini, M., Zakeri, S., Khoshdast, S., Yousefian, F.T., Rastegar, M., Vafaee, F., Kahdouee, S., Ghorbani, F., Rakhshandeh, H., Kazemi, S., 2012. The effects of Nigella sativa hydro-alcoholic extract and thymoquinone on lipopolysaccharide - induced depression like behavior in rats. Journal of Pharmacy and Bioallied Sciences 4 (3), 219–225. https://doi.org/10.4103/0975-7406.99052.

Hosseinzadeh, H., Parvardeh, S., 2004. Anticonvulsant effects of thymoquinone, the major constituent of Nigella sativa seeds, in mice. Phytomedicine 11 (1), 56–64. https://doi.org/10.1078/0944-7113-00376.

Hosseinzadeh, H., Parvardeh, S., Asl, M.N., Sadeghnia, H.R., Ziaee, T., 2007. Effect of thymoquinone and Nigella sativa seeds oil on lipid peroxidation level during global cerebral ischemia-reperfusion injury in rat hippocampus. Phytomedicine 14 (9), 621–627. https://doi.org/10.1016/j.phymed.2006.12.005.

Howren, M.B., Lamkin, D.M., Suls, J., 2009. Associations of depression with c-reactive protein, IL-1, and IL-6: a meta-analysis. Psychosomatic Medicine 71 (2), 171–186. https://doi.org/10.1097/PSY.0b013e3181907c1b.

Ismail, N., Ismail, M., Mazlan, M., Latiff, L.A., Imam, M.U., Iqbal, S., Azmi, N.H., Ghafar, S.A.A., Chan, K.W., 2013. Thymoquinone prevents β-amyloid neurotoxicity in primary cultured cerebellar granule neurons. Cellular and Molecular Neurobiology 33 (8), 1159–1169. https://doi.org/10.1007/s10571-013-9982-z.

Javed, H., Azimullah, S., Meeran, M.N., Ansari, S.A., Ojha, S., 2019. Neuroprotective effects of thymol, a dietary monoterpene against dopaminergic neurodegeneration in rotenone-induced rat model of Parkinson's disease. International Journal of Molecular Sciences 20 (7). https://doi.org/10.3390/ijms20071538.

Jde, M., Santos, Bonjardim, L., Almeida, J., Gelain, D., Araújo, A., 2013. Quintans-Júnior LJ. Improvement of p-cymene antinociceptive and anti-inflammatory effects by inclusion in β-cyclodextrin. Phytomedicine 20 (5). https://doi.org/10.1016/j.phymed.2012.12.009. Epub.

Jukic, M., Politeo, O., Maksimovic, M., Milos, M., Milos, M., 2007. In vitro acetylcholinesterase inhibitory properties of thymol, carvacrol and their derivatives thymoquinone and thymohydroquinone. Phytotherapy Research 21 (3), 259–261. https://doi.org/10.1002/ptr.2063.

Kabir, Y., Shirakawa, H., Komai, M., 2019. Nutritional composition of the indigenous cultivar of black cumin seeds from Bangladesh. Progress in Nutrition 21, 428–434. https://doi.org/10.23751/pn.v21i1-S.6556.

Kandeil, M.A., Gomaa, S.B., Mahmoud, M.O., 2020. The effect of some natural antioxidants against cisplatin-induced neurotoxicity in rats: behavioral testing. Heliyon 6 (8), e04708. https://doi.org/10.1016/j.heliyon.2020.e04708.

Kaur, P., Sharma, S., 2018. Recent advances in pathophysiology of traumatic brain injury. Current Neuropharmacology 16 (8), 1224–1238. https://doi.org/10.2174/1570159X15666170613083606.

Khalil, A., Kovac, S., Morris, G., Walker, M.C., 2017. Carvacrol after status epilepticus (SE) prevents recurrent SE, early seizures, cell death, and cognitive decline. Epilepsia 58 (2), 263–273. https://doi.org/10.1111/epi.13645.

Kilinc, E., Dagistan, Y., Kukner, A., Yilmaz, B., Agus, S., Soyler, G., Tore, F., 2018. Salmon calcitonin ameliorates migraine pain through modulation of CGRP release and dural mast cell degranulation in rats. Clinical and Experimental Pharmacology and Physiology 45 (6), 536–546. https://doi.org/10.1111/1440-1681.12915.

Koyuncu Irmak, D., Kilinc, E., Tore, F., 2019. Shared fate of meningeal mast cells and sensory neurons in migraine. Frontiers in Cellular Neuroscience 13. https://doi.org/10.3389/fncel.2019.00136.

Lee, B.H., Nam, T.G., Park, W.J., Kang, H., Heo, H.J., Chung, D.K., Kim, G.H., Kim, D.O., 2015. Antioxidative and neuroprotective effects of volatile components in essential oils from Chrysanthemum indicum Linné flowers. Food Science and Biotechnology 24 (2), 717–723. https://doi.org/10.1007/s10068-015-0093-0.

Lee Mosley, R., Benner, E.J., Kadiu, I., Thomas, M., Boska, M.D., Hasan, K., Laurie, C., Gendelman, H.E., 2006. Neuroinflammation, oxidative stress, and the pathogenesis of Parkinson's disease. Clinical Neuroscience Research 6 (5), 261–281. https://doi.org/10.1016/j.cnr.2006.09.006.

Levy, D., Labastida-Ramirez, A., MaassenVanDenBrink, A., 2019. Current understanding of meningeal and cerebral vascular function underlying migraine headache. Cephalalgia 39 (13), 1606–1622. https://doi.org/10.1177/0333102418771350.

Li, J.Z., Liu, C.Z., Sato, T., 2016. Novel antitumor invasive actions of p-cymene by decreasing MMP-9/TIMP-1 expression ratio in human fibrosarcoma HT-1080 cells. Biological and Pharmaceutical Bulletin 39 (8), 1247–1253. https://doi.org/10.1248/bpb.b15-00827.

Miller, A.H., 2009. Mechanisms of cytokine-induced behavioral changes: psychoneuroimmunology at the translational interface. Brain, Behavior, and Immunity 23 (2), 149–158. https://doi.org/10.1016/j.bbi.2008.08.006.

Mishra, R.K., Baker, M.T., 2014. Seizure prevention by the naturally occurring phenols, carvacrol and thymol in a partial seizure-psychomotor model. Bioorganic and Medicinal Chemistry Letters 24 (23), 5446–5449. https://doi.org/10.1016/j.bmcl.2014.10.028.

Mohamed, A., Shoker, A., Bendjelloul, F., Mare, A., Alzrigh, M., Benghuzzi, H., Desin, T., 2003. Improvement of Experimental Allergic Encephalomyelitis (EAE) by thymoquinone; an oxidative stress inhibitor. Biomedical Sciences Instrumentation 39, 440–445.

Monczor, M., 2005. Diagnosis and treatment of Alzheimer's disease. Current Medicinal Chemistry - Central Nervous System Agents 5 (1), 5–13. https://doi.org/10.2174/1568015053202723.

Munoz, L., Ammit, A.J., 2010. Targeting p38 MAPK pathway for the treatment of Alzheimer's disease. Neuropharmacology 58 (3), 561–568. https://doi.org/10.1016/j.neuropharm.2009.11.010.

Nagoor Meeran, M.F., Javed, H., Al Taee, H., Azimullah, S., Ojha, S.K., 2017. Pharmacological properties and molecular mechanisms of thymol: prospects for its therapeutic potential and pharmaceutical development. Frontiers in Pharmacology 8. https://doi.org/10.3389/fphar.2017.00380.

Noor, N.A., Fahmy, H.M., Mohammed, F.F., Elsayed, A.A., Radwan, N.M., 2015. Nigella sativa amliorates inflammation and demyelination in the experimental autoimmune encephalomyelitis-induced Wistar rats. International Journal of Clinical and Experimental Pathology 8 (6), 6269–6286. http://www.ijcep.com/files/ijcep0008732.pdf.

Oskouei, Z., Akaberi, M., Hosseinzadeh, H., 2018. A glance at black cumin (Nigella sativa) and its active constituent, thymoquinone, in ischemia: a review. Iranian Journal of Basic Medical Sciences 21 (12), 1200–1209. https://doi.org/10.22038/ijbms.2018.31703.7630.

Parente, M.S.R., Custódio, F.R., Cardoso, N.A., Lima, M.J.A., Melo, T.S., Linhares, M.I., Siqueira, R.M.P., Nascimento, A.Á.D., Catunda Júnior, F.E.A., Melo, C.T.V., 2018. Antidepressant-like effect of *Lippia sidoides* CHAM (verbenaceae) essential oil and its major compound thymol in mice. Scientia Pharmaceutica 27 (8). https://doi.org/10.3390/scipharm86030027.

Parker, D., Marino, V., Ong, J., 2014. Pharmacological actions of thymol and an analogue at GABAB autoreceptors. Clinical and Experimental Pharmacology and Physiology 41 (9), 623–627. https://doi.org/10.1111/1440-1681.12278.

Parnas, M., Peters, M., Dadon, D., Lev, S., Vertkin, I., Slutsky, I., Minke, B., 2009. Carvacrol is a novel inhibitor of Drosophila TRPL and mammalian TRPM7 channels. Cell Calcium 45 (3), 300–309. https://doi.org/10.1016/j.ceca.2008.11.009.

Poorgholam, Y., Hajebrahimi, Z., 2018. Thymoquinone recovers learning function in a rat model of Alzheimer's disease. Avicenna J Phytomed 8 (3), 188–197.

Procaccini, C., De Rosa, V., Pucino, V., Formisano, L., Matarese, G., 2015. Animal models of multiple sclerosis. European Journal of Pharmacology 759, 182–191. https://doi.org/10.1016/j.ejphar.2015.03.042.

Quintans-Júnior, L., Moreira, J.C.F., Pasquali, M.A.B., Rabie, S.M.S., Pires, A.S., Schröder, R., Rabelo, T.K., Santos, J.P.A., Lima, P.S.S., Cavalcanti, S.C.H.,

Araújo, A.A.S., Quintans, J.S.S., Gelain, D.P., 2013. Antinociceptive activity and redox profile of the monoterpenes (+) -Camphene, p -cymene, and geranyl acetate in experimental models. ISRN Toxicology 1–11. https://doi.org/10.1155/2013/459530.

Radad, K.S., Al-Shraim, M.M., Moustafa, M.F., Rausch, W.D., 2015. Neuroprotective role of thymoquinone against 1-methyl-4-phenylpyridinium-induced dopaminergic cell death in primary mesencephalic cell culture. Neurosciences 20 (1), 10–16. http://www.neurosciencesjournal.org/PDFFiles/Jan15/05NSJJan2015NeuroprotectivePg10-16.pdf.

Razavi, B.M., Hosseinzadeh, H., 2014. A review of the effects of Nigella sativa L. and its constituent, thymoquinone, in metabolic syndrome. Journal of Endocrinological Investigation 37 (11), 1031–1040. https://doi.org/10.1007/s40618-014-0150-1.

Sammi, S.R., Trivedi, S., Rath, S.K., Nagar, A., Tandon, S., Kalra, A., Pandey, R., 2017. 1-Methyl-4-propan-2-ylbenzene from thymus vulgaris attenuates cholinergic dysfunction. Molecular Neurobiology 5468–5481. https://doi.org/10.1007/s12035-016-0083-0.

Sancheti, J., Shaikh, M.F., Chaudhari, R., Somani, G., Patil, S., Jain, P., Sathaye, S., 2014. Characterization of anticonvulsant and antiepileptogenic potential of thymol in various experimental models. Naunyn-Schmiedeberg's Archives of Pharmacology 387 (1), 59–66. https://doi.org/10.1007/s00210-013-0917-5.

Sedaghat, R., Roghani, M., Khalili, M., 2014. Neuroprotective effect of thymoquinone, the Nigella sativa bioactive compound, in 6-hydroxydopamine-induced hemi-parkinsonian rat model. Iranian Journal of Pharmaceutical Research 13 (1), 227–234. http://ijpr.sbmu.ac.ir/?_action=showPDF&article=1385&_ob=5a0dd6b85e8ff8c8e772036c4a07f07a&fileName=full_text.pdf.

Seifi-Nahavandi, B., Yaghmaei, P., Ahmadian, S., Ghobeh, M., Ebrahim-Habibi, A., 2020. Cymene consumption and physical activity effect in Alzheimer's disease model: an in vivo and in vitro study. Journal of Diabetes and Metabolic Disorders 19 (2), 1381–1389. https://doi.org/10.1007/s40200-020-00658-2.

Shao, Y., Feng, Y., Xie, Y., Luo, Q., Chen, L., Li, B., Chen, Y., 2016. Protective effects of thymoquinone against convulsant activity induced by lithium-pilocarpine in a model of status epilepticus. Neurochemical Research 41 (12), 3399–3406. https://doi.org/10.1007/s11064-016-2074-y.

Shao, Y.Y., Li, B., Huang, Y.M., Luo, Q., Xie, Y.M., Chen, Y.H., 2017. Thymoquinone attenuates brain injury via an antioxidative pathway in a status epilepticus rat model. Translational Neuroscience 8 (1), 9–14. https://doi.org/10.1515/tnsci-2017-0003.

Shin, E.J., Jeong, J.H., Chung, Y.H., Kim, W.K., Ko, K.H., Bach, J.H., Hong, J.S., Yoneda, Y., Kim, H.C., 2011. Role of oxidative stress in epileptic seizures. Neurochemistry International 59 (2), 122–137. https://doi.org/10.1016/j.neuint.2011.03.025.

Silachev, D.N., Plotnikov, E.Y., Zorova, L.D., Pevzner, I.B., Sumbatyan, N.V., Korshunova, G.A., Gulyaev, M.V., Pirogov, Y.A., Skulachev, V.P., Zorov, D.B., 2015. Neuroprotective effects of mitochondria-targeted plastoquinone and thymoquinone in a rat model of brain ischemia/reperfusion injury. Molecules 20 (8), 14487–14503. https://doi.org/10.3390/molecules200814487.

Sveinbjornsdottir, S., 2016. The clinical symptoms of Parkinson's disease. Journal of Neurochemistry 139. https://doi.org/10.1111/jnc.13691. Epub.

Tabeshpour, J., Mehri, S., Abnous, K., Hosseinzadeh, H., 2019. Neuroprotective effects of thymoquinone in acrylamide-induced peripheral nervous system toxicity through MAP-Kinase and apoptosis pathways in rat. Neurochemical Research 44 (5), 1101–1112. https://doi.org/10.1007/s11064-019-02741-4.

Taka, E., Mazzio, E.A., Goodman, C.B., Redmon, N., Flores-Rozas, H., Reams, R., Darling-Reed, S., Soliman, K.F.A., 2015. Anti-inflammatory effects of thymoquinone in activated BV-2 microglial cells. Journal of Neuroimmunology 286, 5–12. https://doi.org/10.1016/j.jneuroim.2015.06.011.

Tian, F., Liu, R., Fan, C., Sun, Y., Huang, X., Nie, Z., Zhao, X., Pu, X., 2020. Effects of thymoquinone on small-molecule metabolites in a rat model of cerebral ischemia reperfusion injury assessed using MALDI-MSI. Metabolites 10 (1). https://doi.org/10.3390/metabo10010027.

Tuppo, E.E., Arias, H.R., 2005. The role of inflammation in Alzheimer's disease. International Journal of Biochemistry and Cell Biology 37 (2), 289–305. https://doi.org/10.1016/j.biocel.2004.07.009.

Türköz, D., Türköz, A., Önger, M.E., Kuruoğlu, E., 2021. Stereological and histopathological evaluation of the effect of thymoquinone on peridural fibrosis following laminectomy in rats. Turkish Journal of Medical Sciences 51 (1), 375–381. https://doi.org/10.3906/sag-2006-147.

Ullah, I., Badshah, H., Naseer, M.I., Lee, H.Y., Kim, M.O., 2015. Thymoquinone and vitamin C attenuates pentylenetetrazole-induced seizures via activation of GABAB1 receptor in adult rats cortex and Hippocampus. NeuroMolecular Medicine 17 (1), 35–46. https://doi.org/10.1007/s12017-014-8337-3.

Üstün, N., Aras, M., Ozgur, T., Bayraktar, H.S., Sefil, F., Ozden, R., Yagiz, A.E., 2014. Timokinon deneysel spinal kord yaralanmalı sıçanlarda spinal kord hasarını azaltır. Ulusal Travma ve Acil Cerrahi Dergisi 20 (5), 328–332. https://doi.org/10.5505/tjtes.2014.05021.

Valerio, A., Boroni, F., Benarese, M., Sarnico, I., Ghisi, V., Bresciani, L.G., Ferrario, M., Borsani, G., Spano, P., Pizzi, M., 2006. NF-κB pathway: a target for preventing β-amyloid (Aβ)-induced neuronal damage and Aβ42 production. European Journal of Neuroscience 23 (7), 1711–1720. https://doi.org/10.1111/j.1460-9568.2006.04722.x.

Velagapudi, R., El-Bakoush, A., Lepiarz, I., Ogunrinade, F., Olajide, O.A., 2017. AMPK and SIRT1 activation contribute to inhibition of neuroinflammation by thymoquinone in BV2 microglia. Molecular and Cellular Biochemistry 435 (1–2), 149–162. https://doi.org/10.1007/s11010-017-3064-3.

Velagapudi, R., Kumar, A., Bhatia, H.S., El-Bakoush, A., Lepiarz, I., Fiebich, B.L., Olajide, O.A., 2017. Inhibition of neuroinflammation by thymoquinone requires activation of Nrf2/ARE signalling. International Immunopharmacology 48, 17–29. https://doi.org/10.1016/j.intimp.2017.04.018.

Wang, X., Chen, Y., Wang, Q., Sun, L., Li, G., Zhang, C., Huang, J., Chen, L., Zhai, H., 2017. Support for natural small-molecule phenols as anxiolytics. Molecules 22 (12). https://doi.org/10.3390/molecules22122138.

Xiao, X.Y., Zhu, Y.X., Bu, J.Y., Li, G.W., Zhou, J.H., Zhou, S.P., 2016. Evaluation of neuroprotective effect of thymoquinone nanoformulation in the rodent cerebral ischemia-reperfusion model. BioMed Research International. https://doi.org/10.1155/2016/2571060.

Xu, Z.H., Wang, C., Fujita, T., Jiang, C.Y., Kumamoto, E., 2015. Action of thymol on spontaneous excitatory transmission in adult rat spinal substantia gelatinosa neurons. Neuroscience Letters 606, 94–99. https://doi.org/10.1016/j.neulet.2015.08.042.

Yalçin, S.S., Ak, H., Aydin, V., Serter, M., Göçmen, A.Y., 2013. Akut periferik sinir yaralanmasi{dotless}ni{dotless}n tedavisinde nikergolinin akut dönem etkilerinin değerlendirilmesi. Journal of Neurological Sciences 30 (4), 651–659. http://jns.dergisi.org/pdf/pdf_JNS_706.pdf.

Youdim, K.A., Deans, S.G., 1999. Beneficial effects of thyme oil on age-related changes in the phospholipid C20 and C22 polyunsaturated fatty acid composition of various rat tissues. Biochimica et Biophysica Acta - Molecular and Cell Biology of Lipids 1438 (1), 140–146. https://doi.org/10.1016/S1388-1981(99)00045-1.

Youdim, K.A., Deans, S.G., 2000. Effect of thyme oil and thymol dietary supplementation on the antioxidant status and fatty acid composition of the ageing rat brain. British Journal of Nutrition 83 (1), 87–93. https://doi.org/10.1017/s000711450000012x.

Yu, H., Zhang, Z., Chen, J., Hua, P.A., Qian, F., et al., 2012. Carvacrol, a food additive, provides neuroprotection on focal cerebral ischemia/reperfusion injury in mice. PLoS One 7, e33584.

Yvon, Y., Guy Raoelison, E., Razafindrazaka, R., Randriantsoa, A., Romdhane, M., Chabir, N., Guedri Mkaddem, M., Bouajila, J., 2012. Relation between chemical composition or antioxidant activity and antihypertensive activity for six essential oils. Journal of Food Science 77 (8), H184–H191. https://doi.org/10.1111/j.1750-3841.2012.02812.x.

Zhang, F., Jiang, L., 2015. Neuroinflammation in Alzheimer's disease. Neuropsychiatric Disease and Treatment 30 (11), 243–256. https://doi.org/10.2147/NDT.S75546.

Zhong, W., Chi, G., Jiang, L., Soromou, L.W., Chen, N., Huo, M., Guo, W., Deng, X., Feng, H., 2013. P-cymene modulates in vitro and in vivo cytokine production by inhibiting MAPK and NF-κB activation. Inflammation 36 (3), 529–537. https://doi.org/10.1007/s10753-012-9574-y.

CHAPTER EIGHTEEN

Black cumin (*Nigella sativa*) to treat bronchial asthma, ischemia, and cardiovascular diseases

Ahmed Abdul-Sabour Ahmed Bader
Department of Clinical Pharmacology, Faculty of Medicine, Menoufiya University, Shibin el Kom, Egypt

18.1 Introduction

18.1.1 *Nigella sativa* or black cumin to treat bronchial asthma

Chronic inflammatory reactions, inflammatory mediators, and oxidative stress play a central role in the pathogenesis of many of lung disorders such as asthma, tracheitis, and chronic obstructive pulmonary diseases (Aggarwal et al., 2019; Sethi et al., 2019). Many preclinical and human studies reported that *Nigella sativa* (NS) and in addition to its constituents are of significant value in the treatments of such lung diseases, which could be attributed to the curative and protective properties of NS as bronchodilation, antihistaminic, antiinflammatory, and effective immunomodulatory actions on many of lung diseases (Koshak, Koshak et al., 2017; Gholamnezhad et al., 2019; Noorbakhsh et al., 2018).

It was found that NS extract has bronchodilator, preventive, and curative therapeutic effects on chronic asthma patients that was confirmed by results of many clinical trials that showed a reduction of asthma symptoms, with a significant decrease in serum biomarkers of asthma, and together with increased peak expiratory flow by 5%—75% and decreased fractional exhaled NO and serum IgE and effective antiinflammation on mast cells by the inhibition of histamine release and has no toxic effect on mast cell by treatment with NS preparations (Koshak, Koshak et al., 2017; Gholamnezhad et al., 2019; Ikhsan et al., 2018; Salem et al., 2017; Koshak, Wei et al., 2017). These findings provide evidence for the potential benefits of NS oral supplementation in the clinical management of bronchial asthma.

Biochemistry, Nutrition, and Therapeutics of Black Cumin Seed
ISBN: 978-0-323-90788-0
https://doi.org/10.1016/B978-0-323-90788-0.00018-4

Other clinical study showed that NS oil and active constituent (α-hederin) have significant improvement of responsiveness of trachea with a significant antiinflammatory activity that was explained to a decrease in the release of histamine and leukotrienes while an increase in the PGE2 from the mast cells and perfused lungs in experimental animal model of allergic asthma (Ikhsan et al., 2018; Saadat et al., 2015; Keyhanmanesh et al., 2013; Boskabady et al., 2011). These antiasthmatic effects are in accordance to other different clinical studies that reported that NS showed an improvement of clinical symptoms and pulmonary function as well as various asthma biomarkers (Salem et al., 2017; Boskabady et al., 2010). So, future clinical studies on a wide scale with allocating a large number of patients needed to elucidate the effects of black cumin for the management of many of lung diseases.

18.2 NS to treat ischemia and cardiovascular diseases

Experimental use of oral NS at dose of 400 mg/kg in the experimental study of L-NAME-induced hypertensive state in rats reported that there were an improvement in arterial blood pressure (BP), systolic blood pressure (SBP), diastolic blood pressure (DBP), and serum lactate dehydrogenase (LDH), increase in serum nitric oxide (NO), and suggested that antiischemic effect could be related to the significant decrease in blood lipids and the increase in serum NO (Tasar et al., 2012; Murugesan et al., 2012).

Experimental study of renovascular ischemic hypertension in rats using NS oil (i.p) 0.2 mL/kg showed decrease in SBP, tissue malondialdehyde (MDA), luminol, and lucigenin chemiluminescence, increase in tissue Na+ and K+-ATPase and in plasma NO, decrease in plasma creatine kinase (CK), low-density lipoprotein (LDL), and symmetric dimethylarginine, and suggested that NS oil could have a therapeutic antihypertensive effect against renovascular hypertension in ischemic hypertensive rat (Sayed et al., 2009).

The intragastrical use of NS (150 mg/kg body weight [BW]) for a period of 15 days in experimental study of myocardial infarction in rats produced by isoproterenol injection showed marked decrease in serum aspartate and alanine transaminases (AST and ALT), LDH, CK, and tissue lipid profile of triglycerides (TGs), cholesterol, free fatty acids in NS-treated rat groups, and suggested that NS offered protective effects against rat experimental myocardial infarction (Jaarin et al., 2015).

In the experimental study of hypercholesterolemia in rabbits, the use of NS per oral in a dose of 5 g/kg/day showed significant blood lipid-lowering effects when compared to placebo and also has a favorable significant increase in serum high-density lipoprotein (HDL) and concluded that the dietary supplement of NS for long-term use is beneficial for primary prevention of hypercholesterolemia and atherosclerosis and may have a therapeutic potential actions for ischemic patients with coronary artery disease (Paul et al., 2010).

Moreover, the intraperitoneal use of NS oil (0.2 mL/kg), before ischemia and before reperfusion in experimental study of hepatic ischemia/reperfusion injury in rats showed a significant decrease in serum liver enzymes AST, ALT, and LDH in treated rats in comparison with non-treated rats, and recommended that NS oil treatment can protect the rat liver against ischemia reperfusion injury (Yildiz et al., 2008). Moreover, in the experimental ischemic study of reperfusion injury of skeletal muscle in rats, the NS aqueous (1, 1.5, and 2 g/kg) and ethanolic extracts (1.6, 2.4, and 3.2 g/kg) showed an increase of MDA levels in the rat skeletal muscle and therefore inhibiting lipid peroxidation following ischemia-reperfusion injury due to antioxidant properties and free radical scavenging of NS (Hosseinzadeh et al., 2007). Also, NS aqueous (1 g/kg) and ethanolic (1.6 g/kg) extracts in an experimental study of four-vessel occlusion model in rats resulted in significant reduction in neuronal cell injuries of rat hippocampus and suggested NS seed extracts could have curative effects against cerebral ischemia (Hosseinzadeh et al., 2006).

Interestingly, the injection of NS at different doses (0.5, 1, and1.5 mg/kg BW) in adrenaline-induced dyslipidemia and left ventricular dysfunctions in rats and oral NS seed oil capsules with the dose (900 mg/oral per day) for eight weeks in patients with coronary artery disease showed significant decrease in serum levels of LDL, TGs, very low-density lipoprotein, and total cholesterol (TC) with significantly increase in free radical scavenging activity and with attenuation of ischemic cardiomyocyte damage, and suggested that NS having antidyslipidemic, antioxidative, and cardioprotective effects on heart with cardiac ischemia and left ventricular hypertrophy (Ali et al., 2013; El-Deeb et al., 2010; Tasawar et al., 2011).

In a clinical study of patients with mild hypertension and patients with ischemic heart disease, the use of NS seed extract 100 mg/kg and 200 mg/kg, orally per day for three months, has significant reduction in SBP and DBP (dose-dependent), reduction in total and LDL cholesterol, and beneficial reductive effects on blood glucose levels especially in diabetic patients, with a significant increase in serum HDL and a decrease in serum TG after regular intake of NS confirming that NS seed extract could be exerting antihypertensive and hypolipidemic, antiischemic effects, which may be beneficial for ischemic hypertensive diseases by decreasing serum TC and LDL cholesterol for ischemic heart disease after regular intake for several months (Dehkordi and Kamkhah, 2008; Zaki et al., 2005; Bader, 2015). Additionally, black cumin and its active component, thymoquinone (TQ), exhibited a reduction in oxidative stress via calcium channel blockade and increasing urine output activity which might have been linked to reduction in BP (Leong et al., 2013).

18.3 Conclusion

So, based on majority of these reports exploring the antiischemic properties of NS that protecting antioxidants, antiinflammatory, bronchodilator, and antihistaminic effects on the lungs, and also, protecting antioxidants, antiinflammatory, and antiischemic effects on heart and other vital organs against ischemia that decreasing incidence of bronchial asthma and myocardial infraction and ischemic reperfusion injuries on different body organs in experimental animal models, which supported by results of clinical trials that providing a clear evidence that both the NS oil and its active ingredients, in particular TQ, possessing reproducible antioxidant, antiasthmatic, and antiischemic effects, suggesting that the regular use of NS seed is recommended for prevention of bronchial asthma, many of ischemic cardiovascular diseases as a promising basis of natural antiasthmatic, antihypertensive, and antianginal drugs based on majority of results and reports on animal models and clinical studies in such diseases.

References

Aggarwal, T., Wadhwa, R., Thapliyal, N., Sharma, K., Rani, V., Maurya, P.K., 2019. Oxidative, inflammatory, genetic, and epigenetic biomarkers associated with chronic obstructive pulmonary disorder. Journal of Cellular Physiology 234, 2067–2082.

Ali, Y., Islam, M.S., Alam, A.H.M., Rahman, M.A.A., Mamun, A., Hossain, M.K., Hossain, K.M.M., Parvin, M.S., Rashid, M., 2013. Inhibitory effects of *Nigella sativa* seed extract on adrenaline-induced dyslipidemia and left ventricular hypertrophy in rats. Journal of Scientific Research 5 (2), 325–334.

Bader, A.A., 2015. Anti-ischemic properties of Nigella sativa against cardiac and noncardiovascular ischemia. International Journal of Pharmacology and Toxicology 5 (1), 53–61.

Boskabady, M.H., Mohsenpoor, N., Takaloo, L., 2010. Antiasthmatic effect of Nigella sativa in airways of asthmatic patients. Phytomedicine 17 (10), 707–713.

Boskabady, M.H., Keyhanmanesh, R., Khamneh, S., Ebrahimi, M.A., 2011. The effect of Nigella sativa extract on tracheal responsiveness and lung inflammation in ovalbumin-sensitized Guinea pigs. Clinics 66 (5), 879–887.

Dehkordi, F.R., Kamkhah, A.F., 2008. Antihypertensive effect of *Nigella sativa* seed extract in patients with mild hypertension. Fundamental & Clinical Pharmacology 22 (4), 447–452.

El-Deeb, H., Elrehany, M.M., Elbawb Elsaid, M.E., Younis, H.A., 2010. Biochemical effect of *Nigella sativa* on patients with coronary artery disease. El-Minia Medical Bulletin 21 (1), 25–36.

Gholamnezhad, Z., Shakeri, F., Saadat, S., Ghorani, V., Boskabady, M.H., 2019. Clinical and experimental effects of Nigella sativa and its constituents on respiratory and allergic disorders. Avicenna Journal of Phytomedicine 9, 195–212.

Hosseinzadeh, H., Jaafari, M.R., KhoeiAR, Rahmani, M., 2006. Anti-ischemic effect of Nigella sativa L. seed in male rats. Iranian Journal of Pharmaceutical Research 1, 53–58.

Hosseinzadeh, H., Moghim, F.F., Mansou, S.M., 2007. The Effect of *Nigella sativa* seed extracts on ischemic reperfusion in rat skeletal muscle. Pharmacologyonline 2, 326–335.

Ikhsan, M., Hiedayati, N., Maeyama, K., Nurwidya, F., 2018. Nigella sativa as an anti-inflammatory agent in asthma. BMC Research Notes 11 (1), 1–5. Article no. 744.

Jaarin, K., Foong, W.D., Yeoh, M.H., et al., 2015. Mechanisms of the antihypertensive effects of Nigella sativa oil in L-NAME induced hypertensive rats. Clinics 70 (11), 751–757.

Keyhanmanesh, R., Bagban, H., Nazemieh, H., Bavil, F.M., Alipour, M.R., 2013. The main relaxant constituents of Nigella Sativa methanolic fraction on Guinea pig tracheal chains. Iranian Journal of Allergy, Asthma and Immunology 12 (2), 136–143.

Koshak, A., Koshak, E., Heinrich, M., 2017. Medicinal benefits of Nigella sativa in bronchial asthma: a literature review. Saudi Pharmaceutical Journal 25, 1130–1136.

Koshak, A., Wei, L., Koshak, E., Wali, S., Alamoudi, O., Demerdash, A., Qutub, M., Pushparaj, P.N., Heinrich, M., 2017. Nigella sativa supplementation improves asthma control and biomarkers: a randomized, double-blind, placebo-controlled trial. Phytotherapy Research 31, 403–409.

Leong, X.F., Rais Mustafa, M., Jaarin, K., 2013. Nigella sativa and its protective role in oxidative stress and hypertension. Evidence-Based Complementary and Alternative Medicine 2013. Article ID 120732, 9 pages.

Murugesan, M., Ragunath, M., Prabu, T., Nadanasabapathi, S., Sakthivel, M., Manju, V., 2012. Protective role of black cumin (*Nigella sativa*) on isoproterenol induced myocardial infarction in rats. International Journal of Pharmacology and Clinical Sciences 1, 45–53.

Noorbakhsh, M.-F., Shaterzadeh-Yazdi, H., Hayati, F., Samarghandian, S., Farkhondeh, T., 2018. Protective effects of thymoquinon on pulmonary disorders in experimental studies. Tanaffos 17, 211.

Paul, S., Paul, K.K., Palodhi, S., Dutta, S., 2010. Effect of atorvastatin and black seed (*Nigella sativa*) in experimentally induced hypercholesterolemia in rabbits. Pharmacologyonline 2, 842–849.

Saadat, S., Mohammadi, M., Fallahi, M., keyhanmanesh, R., Aslani, M.R., 2015. The protective effect of α-hederin, the active constituent of Nigella sativa, on tracheal responsiveness and lung inflammation in ovalbumin-sensitized Guinea pigs. The Journal of Physiological Sciences 65 (3), 285–292.

Salem, A.M., Bamosa, A.O., Qutub, H.O., Gupta, R.K., Badar, A., Elnour, A., Afzal, M.N., et al., 2017. Effect of Nigella sativa supplementation on lung function and inflammatory mediators in partly controlled asthma: a randomized controlled trial. Annals of Saudi Medicine 37 (1), 64–71. https://doi.org/10.5144/0256-4947.2017.64.

Sayed, H.M., El-Latif, H.A.A., Eid, N.I., Elsayed, A.Z., El-Kader, E.M.A., 2009. Potential antihypertensive and antioxidative effects of *Nigella sativa* seeds or biomass and Syzygium aromaticum extracts on L-NAME-induced hypertensive rats. Egyptian Journal of Pharmaceutical Sciences 50, 127–146.

Sethi, G.S., Dharwal, V., Naura, A.S., 2019. Immunological basis of Oxidative stress-induced lung inflammation in asthma and COPD. In: Oxidative Stress in Lung Diseases. Springer, Berlin/Heidelberg, Germany, pp. 195–223.

Tasar, N., Sehirli, O., Yigner, O., 2012. Protective effects of *Nigella sativa* against hypertension-induced oxidative stress and cardiovascular dysfunction in rats. Marmara Pharmaceutical Journal 16 (2), 141–149.

Tasawar, Z., Siraj, Z., Ahmad, N., Lashari, M.H., 2011. The effects of *Nigella sativa* (Kalonji) on lipid profile in patients with stable coronary artery disease in multan, Pakistan. Pakistan Journal of Nutrition 10 (2), 162–167.

Yildiz, F., Coban, S., Terzi, A., Ates, M., Aksoy, N., Cakir, H., Ocak, A.R., Bitiren, M., 2008. *Nigella sativa* relieves the deleterious effects of ischemia reperfusion injury on liver. World Journal of Gastroenterology 14 (33), 5204–5209.

Zaki, A., Kandil, S., Abaz, M., Ragab, M., El-Attar, H., Zaki, M., Wanas, H., 2005. Study of the effect of *Nigella sativa* seeds on serum level of glucose and Some lipoprotein Fractions in patients with ischemic heart disease. Journal of the Medical Research Institute 26 (1), 15–19.

CHAPTER NINETEEN

Potential medicinal properties of black cumin (*Nigella sativa*) in traditional medicine

Sarya F. Amin
Indigenous Knowledge and Heritage Center, Ghibaish College of Science and Technology, Ghibaish, Sudan

Abbreviations

CMM Chinese Materia Medica
CAM Complementary and Alternative Medicine
N. sativa *Nigella sativa*
TCM Traditional Chinese Medicine
TIM Traditional Indian Medicine
TPM Traditional Persian System
WMS Whole System of Medicine
WHO World Health Organization

19.1 Introduction

Unsegment of different ways, materials, and procedures beyond the common medical care is known as complementary and alternative medicine. Likewise, it may be regarded as the Medicare scheme which utilizes group of practices, knowledge, and skills earned out of the informal medical disciplines (Gurmu et al., 2017). The World Health Organization (WHO) had defined traditional medicine as the overall combination of the knowledge, skills, and practices emanated from theories, beliefs, and experiences primeval various cultures, which are used in the maintenance, prevention, diagnosis, improvement, or treatment of physical and mental health, regardless of being rational or not (World Health Organization, 2019). Another definition by Ibrahim, et al. (2017) *considered traditional medicine as a systematic healthcare framework with antique origin, cultural ligature, skilled menders, and a theoretical paradigm* (Ibrahim et al., 2017). There are noticeable terminologies express traditional medicine interchangeably such as: complementary medicine, alternative medicine, or conventional medicine among countries (World Health Organization, 2019). Ros et al. (2018) reported that high-

Biochemistry, Nutrition, and Therapeutics of Black Cumin Seed
ISBN: 978-0-323-90788-0
https://doi.org/10.1016/B978-0-323-90788-0.00014-7

income countries habitually call it "complementary" or "alternative" medicine while simply known as traditions in low- and middle-income countries (Ros et al., 2018).

19.1.1 The journey of traditional medicine

From first principles, traditional medicine initiated on the basis of natural representation of diseases in conjunction with patient's history. In those days, physicians endeavor to differentiate diseases by perceptible manifestations, for instance: temperament and meridian, thus, treated herbaceously (Jafari et al., 2020).

Culturally talking, almost all nations had their traditional healing methods. Away from the exact timing of Traditional Chinese Medicine (TCM) establishment; it is known to be originated through ongoing work and scrutiny, targeting the use of herbs on medicinal functions, that is, Chinese Materia Medica (CMM). According to Lemonnier et al. (2017), CMM deemed to be the most comprehensive encyclopedia integrating the medicinal uses of herbs in TCM. It is gleaned by Li Shi Zhen who enclosed approximately over 1800 inputs with greater than 11,000 blueprints. Other techniques alongside CMM in the TCM ideology and rituals were: acupuncture, moxibustion, tuna massage, and qigong exercise (Lemonnier et al., 2017). Kampo medicine is the name of Japanese Traditional Herbal Medicine. Kampo verbatim meaning is "method from the Han period of ancient China" alluding to its stemming from primeval China. As yet, Kampo Japanese Traditional Herbal Medicine is being distinctive by means of its protracted advancement bygone time (Watanabe et al., 2011). The transmission of the TCM to Japanese islands took place in the 15th century via the Korean peninsula. Sooner afterward, the first medical regulations were enacted (Lemonnier et al., 2017). The Chinese herbal medicine as branch of TCM was dominated among cultural areas of Western countries while TCM was more widespread than Kampo in the United States (Watanabe et al., 2011). Korean traditional medicine also linked to TCM but slightly differs. It arose from the two prime philosophical Chinese folklores Taoism and Confucianism. Taoism used to intensify the interaction between the external factors (i.e., nature, universe, and environment) and well-being. Contrariantly, Confucianism deals with the individuals' innermost solidities and debilities (Lemonnier et al., 2017). By that means, both Korean traditional medicine and TCM subsumed usage of numerous treatment methods. Nevertheless, distinctiveness of the Korean traditional medicine was through the Confucianism approach, and the fact that the human body is entangled physically and mentally (Yoo et al., 2012). A solemn subclass of the Traditional Indian Medicine is Ayurveda. Ayurveda labeled as a Whole System of Medicine (WMS), which was applied for thousands of years in the South Asian countries; hence, it was seen as the ancient WMS worldwide. This term is literally translated from Sanskrit to "knowledge of life," or it could be interpreted as "systematic knowledge of the lifespan." Interestingly, the WHO had totally approved Ayurveda as a medical science on par with TCM (Kessler et al., 2013). Nowadays, it is used as the primary health-care system in the Indian countryside for around 70%–80% (Lemonnier et al., 2017). Adu-Gyamfi et al. (2019) articulated that traditional medicine served as the supremely reachable wellness structure for certain Africans (Adu-Gyamfi et al. (2019). In the opinion of Mozaffarpur et al. (2020), another of the earliest recognized medicinal systems is Traditional Persian Medicine, which had nearly 14,000 authentic references (Mozaffarpur et al., 2020). Of others, the traditional medicine in the Arabic regions and Islamic culture comprised copious amounts of practices, chiefly with respect to the herbal traditional medicine. Notably, herbs' use dated back to the Sumerians, the oldest pharmacy was in Baghdad where doctors used to prescribe herbs in miscellaneous forms, and a group of brilliant physicians were celebrated (e.g., Al-Razi, Ibin Sina, and Albetar). Islamic tradition is prominent in getting soul and body simultaneously, usage of earmarked foods like date and honey, and practices like "Alhijama" (i.e., cupping) (Ibrahim et al., 2017). The new science-based medicine was introduced by Hippocrates (Lemonnier et al., 2017). Following the recommended integration of traditional medicine into the health-care system by the WHO, an intensifying need for a broad stratagem concerning the study of the medicinal plants, plus the establishment of research and training centers was declared by the World Health Assembly (Gupta and Ahmed, 2018).

19.2 The miracle herb *Nigella sativa*

Under the umbrella of traditional medicine, medicinal herbs are highly popular and favorable ways of primary treatments either in the developing countries or globally, in addition to their cultural and historical benefits. What causes people to choose them: the soothing, synergistic, and nutritive effects. Some of these herbs are: *Allium sativum, Crocus sativus, Zataria multiflora, Thymus vulgaris*, and *Nigella sativa* (black seed) (Qureshi et al., 2017). Throughout all the medicinal herbs; *N. sativa* had glorious situate due to the fact that it is the top evidence-based plant all over the map (Al-Naggar et al., 2017). The name of *N. sativa* was taken over from the Latin word nigellus, which means black, apparently elucidating its black seeds (Anbarasu et al., 2018). *N. sativa* has numerous nomenclatures such as black seeds in East Africa, Middle East, South Asia, and Mediterranean countries (Habib and Choudhry, 2021), black cumin in English, Habat-ul-Sauda or Habbat-ul-Barakah (seed of blessing) in Arabic (Ahmad and Alkreathy, 2018), jintan hitam in Indonesia (Lokeswara et al., 2019), black caraway and Kalonji in South Asia (Dajani et al., 2016), and Kalongi in India (Qayyum et al., 2020). What is more, in line with its vast religious history, *N. sativa* is regarded as a miraculous plant (Ermumcu and Şanlıer, 2017). It is mentioned as the curative black cumin in the Holy Bible (Bensiameur-Touati et al., 2017), whereas in the Islamic religion, the Prophet Muhammad (peace be upon him) said: "Hold on to the use of the habbat alsauda—*N. sativa*—for indeed it has a remedy for every disease except death (Ikhsan et al., 2018). Additionally, it is so-called Melanthion of Hippocrates and Dioscorides and as Glitch of Pliny (Maideen, 2020). *N. sativa* is a dicotyledon (Mammad et al., 2017), anural efflorescence plant (Javadi et al., 2018), and a member of Ranunculaceae family. It is characterized by a height of 8—12 inches, pinnated and segmented leaves, solitary, self-pollinated, white and blue flowers, and the plenty triangular seeds encapsuled within the fruits (Pop et al., 2020). It is sown in the Middle East, Eastern Europe, Western Asia (Ahmad et al., 2020), North Africa (Bensiameur-Touati et al., 2017), Iran (Pop et al., 2020), Southern Europe, India, Pakistan, Syria, Saudi Arabia, and Turkey (Ermumcu and Şanlıer, 2017). *N. sativa* is loaded with components including: both fixed and essential oils (Eid et al., 2017), saponin, alkaloids, various proteins (Arazmjoo et al., 2021), antioxidants—especially thymoquinone (Maghsoudi and Haj-allahyari, 2018), vitamins (A, B, C), minerals (calcium, potassium, zinc, iron), and saturated and unsaturated fatty acids (Elshama, 2018). It is customarily used as spice (Koshak et al., 2018), flavoring additive (Elshama, 2018), food preservative (Mammad et al., 2017), carminative, condiment, aromatic, and can be eaten with honey, sweet foods, bread, and cheese (Bensiameur-Touati et al., 2017), with addition to spicing yogurt, marinades, sauces, and salads (Tiji et al., 2021).

19.3 Medicinal properties of *Nigella sativa* in traditional medicine

Owing to the distinguished position of *N. sativa* at all levels, its biological and therapeutic features have been massively addressed by researchers. It has been revealed to be used as diuretic, appetite stimulant, milk production augmentation for lactating mothers, antihypertensive, antidiabetic, immunomodulatory, analgesic, antimicrobial, anthelmintic, antiinflammatory, spasmolytic, bronchodilator, gastroprotective, hepatoprotective, emmenagogue, renal, and antioxidant. Furthermore, the seeds considered a remedy for bronchitis, asthma, diarrhea, skin disorders (Begum and Mannan, 2020), hypertension, eczema, inflammation, rheumatism, jaundice, gastrointestinal problems (Saleh et al., 2018), diabetes (Pelegrin et al., 2019), warts (Maideen, 2020), cough, nasal congestion, headache, dental pain, menstrual disorders, impotence (Samadipour et al., 2020) stomachache, abdominal pain, colic, and general fatigue (Koshak et al., 2018). Kaur et al. (2018) accounted that *N. sativa* could contribute to lowering blood glucose level and body weight through its ability to control the intestinal uptake of the glucose and minimization of hepatic gluconeogenesis (Kaur et al., 2018). Also, its supplementary use in the form of aqueous extraction alongside with copper nanoparticles synthesized serves as neuroprotective in the sake for curing nervous disorders (Yan et al., 2020).

Aforesaid, *N. sativa* renowned in Unani, Ayurveda, Siddha, Chinese, and other traditional medicine systems (Al-Naggar et al., 2017). The use of *N. sativa,* its seeds, and oil was disseminated of long decades ago. Imam Ibn Qayyim Al-Jawziyya prescribed *N. sativa* for lessening gasping and hard breathing (Qayyum et al., 2020). Hippocrates used the seeds and oil to strengthen liver, treat poisonous bites, abscesses, skin rashes, head infections, and common cold. Pedanius Dioscorides used the oil alone for headache, toothache, nasal congestion and to destroy enterozoa, while Ibni Sina used it for metabolism stimulation and to allaying asthenia and lethargy. Interestingly, by the time of the pharaohs, *N. sativa's* oil was intrinsic for the royal health and beauty of Queen Cleopatra, and its remnants were found in Tutankhamun's tomb. *N. sativa* has been used as traditional treatment with all its forms (i.e., powder, oil, and extract) (Ermumcu and Şanlıer, 2017). *N. sativa* reported to be used for obesity, hypertension, and diabetes in Unani traditional medicine and Islamic medicine (Leisegang et al., 2021) to treat respiratory, gastrointestinal, rheumatic, and inflammatory disorders, and cancer in traditional Jordanian folklore, various respiratory disorders in Morocco, Pakistan, and Southern Europe (Tiji et al., 2021), rheumatoid arthritis, diabetes, asthma, and dyslipidemia in the Middle East and Southeast Asia (Namazi et al., 2018). In Sri Lanka, *N. sativa* played a crucial role in dermatological treatments as antiacne topical formula to heal acne vulgaris (Nawarathne et al., 2019), besides the anticancer and antiproliferative effects which stated in Unani, Ayurveda, and Chinese medicine (Mollazadeh et al.,

2017). Conformably, it is used in traditional Arabic herbal medicine for asthma, gastrointestinal dysfunctions, conjunctivitis, eczema, intrinsic hemorrhage, hypertension, and pyrexia (Bakal et al., 2017). Another method of using *N. sativa's* seeds is boiling. Boiled seeds were taken postmeals to alleviate symptoms accompanied malaria including chills, headache, nausea, and lack of appetite by indigenous Arabs, Asians, and Bantus of the Kenyan Coastal Province (Udu et al., 2021). In Northern Africa, they treated skin diseases (eczema), upper respiratory tract diseases (e.g., bronchitis, cough, and influenza), fever, headache, and rheumatic diseases by using *N. sativa.* Among Arabian and African countries, oil has been used for treating allergies (Eid et al., 2017). Further, in the Gulf region of Arabia, it has been utilized to relieve toothache, obviate radiation damage, postsurgical adhesions, and lethargy (Qayyum et al., 2020).

19.4 Conclusion

N. sativa and other medicinal plants were favored to be used by most of the communities throughout the ages for its availability, affordability, and considerably safety. Therewith, a profound validation with regards to dosages and possible interactions to food, nutrients, and other drugs should be addressed immensely in comorbid patients and high-vulnerable individuals. Moreover, its credible properties as immune booster and antiviral might have robust advantage in facing the concurrent pandemic.

References

Adu-Gyamfi, S., Oware, R., Awuah, D.B., 2019. Interest groups, issue definition and the politics of traditional medicine in Ghana. Nordic Journal of African Studies 28 (4), 19-19.

Ahmad, A., Alkreathy, H.M., 2018. Comparative biochemical and histopathological studies on the efficacy of metformin and Nigella sativa oil against thioacetamide-induced acute hepatorenal damage in rats. Biomedical Research (0970-938X) 29 (15), 3106–3116.

Ahmad, R., Ahmad, N., Mohd, A.M.I.R., AlJHISI, F., Alamer, M.H., Al-Shaban, H.R., Alsultan, B.M., Alsadah, Z.A., Aldawood, N.A., Chathoth, S., Aslam, K.H.A.N., 2020. Variation in Nigella sativa quality and its standardization via instrumental analysis: a study based on geographical origin. Notulae Botanicae Horti Agrobotanici Cluj-Napoca 48 (3), 1141–1154.

Al-Naggar, R.A., Osman, M.T., Mohamed, I.N., Aripin, K.N.B.N., Abdulghani, M.A., 2017. Effect of Nigella sativa supplementation on human lipids: systematic. Journal of Applied Pharmaceutical Science 7 (04), 213–219.

Anbarasu, S., Sundar, R., Manigundan, K., Rajasekar, T., Shamya, M., Joseph, J., 2018. Evaluating the anti-mycobacterial activity of Nigella sativa seed extracts. Der Pharmacia Lettre 10 (7), 1–9.

Arazmjoo, S., Es-haghi, A., Mahmoodzadeh, H., 2021. Evaluation of anti-cancer and antioxidant properties of nanoemulsions synthesized by Nigella Sativa L. tincture. Nanomedicine Journal 8 (1), 57–64.

Bakal, S.N., Bereswill, S., Heimesaat, M.M., 2017. Finding novel antibiotic substances from medicinal plants—antimicrobial properties of Nigella sativa directed against multidrug resistant bacteria. European Journal of Microbiology and Immunology 7 (1), 92–98.
Begum, S., Mannan, A., 2020. A review on Nigella sativa: a marvel herb. Journal of Drug Delivery and Therapeutics 10 (2), 213–219.
Bensiameur-Touati, K., Kacimi, G., Haffaf, E.M., Berdja, S., Aouichat-Bouguerra, S., 2017. In vivo subacute toxicity and antidiabetic effect of aqueous extract of Nigella sativa. Evidence-based Complementary and Alternative Medicine 2017, 8427034.
Dajani, E.Z., Shahwan, T.G., Dajani, N.E., 2016. Overview of the preclinical pharmacological properties of Nigella sativa (black seeds): a complementary drug with historical and clinical significance. Journal of Physiology & Pharmacology 67 (6), 801–817.
Eid, A.M., Elmarzugi, N.A., Abu Ayyash, L.M., Sawafta, M.N., Daana, H.I., 2017. A review on the cosmeceutical and external applications of Nigella sativa. Journal of Tropical Medicine 2017, 7092514.
Elshama, S.S., 2018. The preventive and curative role of Nigella sativa in poisoning cases. Journal of Clinical and Experimental Toxicology 2 (2), 18–24. https://doi.org/10.4066/2630-4570.015.
Ermumcu, M.Ş.K., Şanlıer, N., 2017. Black cumin (Nigella sativa) and its active component of thymoquinone: effects on health. Food and Health 3 (4), 170–183.
Gupta, B.M., Ahmed, K.M., 2018. Research on Nigella sativa: a scientometric assessment of global publications' output during 1989–2018. International Journal of Pharmaceutical Investigation 8 (4), 173–181.
Gurmu, Endale, A., Teni, F.S., Tadesse, W.T., 2017. Pattern of traditional medicine utilization among HIV/AIDS patients on antiretroviral therapy at a university hospital in northwestern Ethiopia: a cross-sectional study. Evidence-based Complementary and Alternative Medicine 2017, 1724581.
Habib, N., Choudhry, S., 2021. HPLC quantification of thymoquinone extracted from Nigella sativa L.(Ranunculaceae) seeds and antibacterial activity of its extracts against Bacillus species. Evidence-based Complementary and Alternative Medicine 2021, 6645680.
Ibrahim, I.R., Hassali, M.A., Saleem, F., FAl, H.T., 2017. The history of traditional medicine in the Arab region. Computer Science 82, 138–142.
Ikhsan, M., Hiedayati, N., Maeyama, K., Nurwidya, F., 2018. Nigella sativa as an anti-inflammatory agent in asthma. BMC Research Notes 11 (1), 1–5.
Jafari, M., Wang, Y., Amiryousefi, A., Tang, J., 2020. Unsupervised learning and multipartite network models: a promising approach for understanding traditional medicine. Frontiers in Pharmacology 11, 1319.
Javadi, S.M.R., Hashemi, M., Mohammadi, Y., MamMohammadi, A., Sharifi, A., Makarchian, H.R., 2018. Synergistic effect of honey and Nigella sativa on wound healing in rats. Acta Cirurgica Brasileira 33, 518–523.
Kaur, G., Invally, M., Khan, M.K., Jadhav, P., 2018. A nutraceutical combination of Cinnamomum cassia & Nigella sativa for Type 1 diabetes mellitus. Journal of Ayurveda and Integrative Medicine 9 (1), 27–37.
Kessler, C., Wischnewsky, M., Michalsen, A., Eisenmann, C., Melzer, J., 2013. Ayurveda: between religion, spirituality, and medicine. Evidence-based Complementary and Alternative Medicine 2013, 952432.
Koshak, A.E., Yousif, N.M., Fiebich, B.L., Koshak, E.A., Heinrich, M., 2018. Comparative immunomodulatory activity of Nigella sativa L. preparations on proinflammatory mediators: a focus on asthma. Frontiers in Pharmacology 9, 1075.
Leisegang, K., Almaghrawi, W., Henkel, R., 2021. The effect of Nigella sativa oil and metformin on male seminal parameters and testosterone in Wistar rats exposed to an obesogenic diet. Biomedicine & Pharmacotherapy 133, 111085.

Lemonnier, N., Zhou, G.B., Prasher, B., Mukerji, M., Chen, Z., Brahmachari, S.K., Noble, D., Auffray, C., Sagner, M., 2017. Traditional knowledge-based medicine: a review of history, principles, and relevance in the present context of P4 systems medicine. Progress in Preventive Medicine 2 (7), e0011.

Lokeswara, A.W., Afaratu, K., Prihastama, R.A., Farida, S., 2019. Antihypertensive effects of Nigella sativa: weighing the evidence. International Journal of Applied Pharmaceutics 11, 135−139.

Maghsoudi, H., Haj-allahyari, S., 2018. Anti-inflammatory effect of alcoholic extract of Nigella sativa L on bovine fibroblast-like synoviocyte and THP-1. International Journal of Contemporary Research and Review 9 (02), 20181−20191.

Maideen, N.M.P., 2020. Prophetic medicine-Nigella sativa (black cumin seeds)—potential herb for COVID-19? Journal of Pharmacopuncture 23 (2), 62.

Mammad, Z., Mammad, K., Aqeil, T., Kribii, A., Ounine, K., 2017. Antibacterial and antioxidant activity of Nigella sativa. International Journal of Innovation and Scientific Research 31 (1), 167−172.

Mollazadeh, H., Afshari, A.R., Hosseinzadeh, H., 2017. Review on the potential therapeutic roles of Nigella sativa in the treatment of patients with cancer: involvement of apoptosis:- black cumin and cancer. Journal of Pharmacopuncture 20 (3), 158.

Mozaffarpur, S.A., Naseri, M., Kamalinejad, M., Shams, M.A., Memariani, Z., Moeini, R., Gorji, N., Moradi Farahani, A., Shirafkan, H., 2020. Nine steps to discover new medicines from traditional sources: the example of Persian medicine. Journal of Alternative & Complementary Medicine 26 (5), 365−368.

Namazi, N., Larijani, B., Ayati, M.H., Abdollahi, M., 2018. The effects of Nigella sativa L. on obesity: a systematic review and meta-analysis. Journal of Ethnopharmacology 219, 173−181.

Nawarathne, N.W., Wijesekera, K., Wijayaratne, W.M.D.G.B., Napagoda, M., 2019. Development of novel topical cosmeceutical formulations from Nigella sativa L. with antimicrobial activity against acne-causing microorganisms. The Scientific World Journal 2019, 5985207.

Pelegrin, S., Galtier, F., Chalançon, A., Gagnol, J.P., Barbanel, A.M., Pélissier, Y., Larroque, M., Lepape, S., Faucanié, M., Gabillaud, I., Petit, P., 2019. Effects of Nigella sativa seeds (black cumin) on insulin secretion and lipid profile: a pilot study in healthy volunteers. British Journal of Clinical Pharmacology 85 (7), 1607−1611.

Pop, R.M., Trifa, A.P., Popolo, A., Chedea, V.S., Militaru, C., Bocsan, I.C., Buzoianu, A.D., 2020. Nigella sativa: valuable perspective in the management of chronic diseases. Iranian Journal of Basic Medical Sciences 23 (6), 699.

Qayyum, M., Ashiq, K., Tanveer, S., Bajwa, M.A., Shah-Rukh, A., Jahangir, N., 2020. Review on phytochemical evaluation and extraction of Nigella sativa (Kalongi) with pharmacological and traditional applications. International Journal of Biosciences 16 (3), 231−241.

Qureshi, A.S., Rehan, S., Enbergs, H., 2017. Nigella sativa seed extract affects granulocyte phagocytosis and lymphocytes proliferation in goats. Pakistan Veterinary Journal 37, 411−414.

Ros, B., Lê, G., McPake, B., Fustukian, S., 2018. The commercialization of traditional medicine in modern Cambodia. Health Policy and Planning 33 (1), 9−16.

Saleh, F.A., El-Darra, N., Raafat, K., El Ghazzawi, I., 2018. Phytochemical analysis of Nigella sativa L. utilizing GC-MS exploring its antimicrobial effects against multidrug-resistant bacteria. Pharmacognosy Journal 10 (1), 99−105.

Samadipour, E., Rakhshani, M.H., Kooshki, A., Amin, B., 2020. Local usage of Nigella sativa oil as an innovative method to attenuate primary dysmenorrhea: a randomized double-blind clinical trial. Oman Medical Journal 35 (5), e167.

Tiji, S., Benayad, O., Berrabah, M., El Mounsi, I., Mimouni, M., 2021. Phytochemical profile and antioxidant activity of Nigella sativa L growing in Morocco. The Scientific World Journal 2021, 6623609.

Udu, R., Oyweri, J., Gathirwa, J., 2021. Antimalarial activity of Nigella sativa L. seed extracts and selection of resistance in plasmodium berghei ANKA in a mouse model. Journal of Pathogens 2021, 6165950.

Watanabe, K., Matsuura, K., Gao, P., Hottenbacher, L., Tokunaga, H., Nishimura, K., Imazu, Y., Reissenweber, H., Witt, C.M., 2011. Traditional Japanese Kampo medicine: clinical research between modernity and traditional medicine—the state of research and methodological suggestions for the future. Evidence-based Complementary and Alternative Medicine 2011, 513842.

World Health Organization, 2019. WHO Global Report on Traditional and Complementary Medicine 2019. World Health Organization.

Yan, W., Liu, Y., Mansooridara, S., Kalantari, A.S., Sadeghian, N., Taslimi, P., Zangeneh, A., Zangeneh, M.M., 2020. Chemical characterization and neuroprotective properties of copper nanoparticles green-synthesized by Nigella sativa L. seed aqueous extract against methadone-induced cell death in adrenal phaeochromocytoma (PC12) cell line. Journal of Experimental Nanoscience 15 (1), 280–296.

Yoo, J., Lee, E., Kim, C., Lee, J., Lixing, L., 2012. Sasang constitutional medicine and traditional Chinese medicine: a comparative overview. Evidence-based Complementary and Alternative Medicine 2012, 980807.

Index

Note: Page numbers followed by '*f*' indicate figures those followed by '*t*'.

A

Acetylcholinesterase (AchE) inhibitors, 215
Active compounds (ACs), 46–47
Adenosine monophosphate-activated protein kinase (AMPK), 202
5′-Adenosine monophosphate (AMP)-activated protein kinase (AMPK), 154
Agricultural yield, 22–23
Agronomy
  adaptation, 3
  essential oils, 3
  origin, 3
  pharmacological functions, 2–3
  production, 4
  properties, 1–2
  varieties, 4
Alkali extraction with isoelectric precipitation (AE-IP), 52
Allergic rhinitis, 129–130
Alpha-hederin (α-hederin), 137, 164
Alzheimer's disease (AD), 209–210, 215
Animal feed
  fish feed, 121, 122f
  nutritional composition, 119–120, 120f
  phytochemical profiling, 120–121
  phytogenic feed additives, 119
  poultry breeds feed, 122–124, 122f
  ruminants feed, 122f, 124
Anthocyanins, 183
Antiatherogenic effects, 149–150
Antibacterial activities, 115, 190–192
Anticancer activity, 137–138
  alpha-hederin, 164
  carvacrol, 165
  longifolene, 165
  phytochemical compounds, 162, 163f
  prevalence and mortality, 161–162
  synthetic compounds, 162
  thymol, 164–165
  thymoquinone, 162–164
Antidiabetic properties, 134–136
Antifungal properties, 2, 115, 133–134
Antihypercholesterolemic activity, 144–146
Antihypercholesterolemic agent, 153–156
Antihyperglycemic agent, 153–156
Antihyperglycemic effects, 150–152
Antihyperlipidemic properties, 115
Antiinflammatory activity, 91, 114
Antiinfluenza virus activity, 196
Antimalarial activities, 197–198
Antimicrobial efficacy, 130–133
Antimicrobial properties, 2
Antioxidants, 173t–175t
  processing methods, 181–182
  properties, 65, 91, 172–181
  ROS, 171–172
Antiparasitic properties, 115
Antiviral properties, 115
Anxiety, 215–216
Aroma-active compounds, 80, 81t
Arylesterase, 153
2,20-azino-bis-3-ethylbenzothiazoline-6-sulfonic acid (ABTS), 176–178

B

Bacillus subtilis, 190
Bioactive compounds
  climate change impact, 13–14, 14f
  ecology, 14–15
  essential oils, 65–66
  evaluation of, 48
  fertilizers, 15–16
  growing conditions, 14–15
  human ailments, 11
  irrigation, 15
  minerals, proteins, and sugars content, 12–13, 13t
  primary fatty acids, 48
  saturated fatty acids (SFAs), 48
  soil type, 15–16

Bioactive compounds (*Continued*)
thymoquinone (TQ), 12, 12f
water management, 15
Biochemical components, 28f–30f, 29
Biomolecules, 63–65
Black cumin seed protein hydrolysates (BCSPH), 55–56
Black seed oil (BSO)
antibacterial activities, 190–192
antimalarial activities, 197–198
antimicroorganism activities, 192–196
chemical compounds, 190
extraction of, 190
neuroprotective effects. *See* Neuroprotective effects
Botanical description, 72
Bronchial asthma, 231–232

C

Calcitonin gene-related peptide (CGRP), 205
Candida albicans, 134
Carbohydrates, 28f, 31–32
Carvacrol (COL), 165, 220–221
Cerebellar granule neurons (CGNs), 209–210
Cerebral ischemia, 211–212
Chemical composition, 113
Chinese Materia Medica (CMM), 238–239
Cisplatin (CP), 162–164, 206
Climate change, 13–14, 14f
Cold press (CP) extraction method, 176
Column chromatography (CC), 106–107, 106f
Conventional food processing techniques
decoction processing technology, 97–98
hydro/steam distillation, 93f, 95–97
infusion, 97–98
maceration, 97–98
mechanical oil extraction and processing technology, 98–100, 99f–100f
percolation, 97–98
solvent extraction technology
advantages, 94
on composition, 92–93, 93f
disadvantages, 94
Soxhlet extraction technology, 93f, 94–95
Coronary artery disease (CAD), 148
Coronary heart disease, 128
Coronavirus 2019 (COVID-19), 170–171
COVID-19, 12, 12f
C-reactive protein (CRP) levels, 203
Crystallization, 106f, 108
Cytotoxic test, 191

D

Decoction processing technology, 97–98
Dementia, 216
Depression, 205–206, 216
Diabetes-related proinflammatory cytokine, 134–135
Diabetic neuropathy, 135–136
Dihydrochalcones, 183
Dihydroflavonols, 183
3-(4,5-Dimethylthiazol-2-yl)-2,5-diphenyltetrazolium bromide assay, 194
2, 2-Diphenyl-1-picrylhydrazyl (DPPH), 176–178, 182

E

Ecology, 14–15
Enzyme-assisted extraction technology (EAE), 104–105
Epilepsy, 206–208, 217
Escherichia coli, 191
Essential oils (EOs), 180, 194
benefits, 81–82
extraction and composition, 74–78, 76t–80t
functional properties, 62–66
Experimental allergic/autoimmune encephalomyelitis (EAE), 212
Extracellular matrix (ECM) protein, 130

F

Fasting blood glucose (FBG), 148
Fast protein liquid chromatography (FPLC)-based fractionation techniques, 52
Fatty acid (FA), 45, 49, 49t–50t

Ferric reducing antioxidant power (FRAP) assays, 176, 178
Fertilizers, 15–16
Fish feed, 121, 122f
Folin–Ciocalteu method, 177
Folk medicine, 7–10, 113
Food processing techniques
  conventional, 92–100, 93f, 99f–100f
  cumin seeds varieties, 89–90, 89f
  medicinal properties, 89–90, 90f
  natural food coloring additives, 90–91
  novel/advanced, 101–106, 102f
  pharmacological properties, 89–90, 90f
  purification techniques, 106–108, 106f
Food revitalization, 65
Food value addition, 65
Fourier-transform infrared spectroscopy (FTIR), 193
Functional properties
  botanical overview, 62
  food application, 64–66
  medicinal application, 62–64
  press waste material, 66–67
Fusarium oxysporum, 134

G

Gallic acid equivalents (GAEs), 177
Gas chromatography-mass spectrometry (GC-MS), 132–134
Glucose homeostasis, 153
gp91phox and p40phox protein, 202
Gram-positive bacteria, 131
Growth, 22–23
Growth performance (GP)
  effect of diets, 45
  feeding application, 45
  nutrient digestibility, and immunity, 45
  shape and color, 45, 46f

H

Hashimoto thyroiditis (HT), 127–128
Hepatic fibrosis, 130
Hepatic stellate cells (HSCs), 130
Herbal drugs, 62–63
High-density lipoprotein (HDL), 128, 144–145
High-density lipoprotein cholesterol (HDL-C), 144–145
High-performance liquid chromatography with a photodiode array detector (HPLC–PDA), 183–184
High-speed countercurrent chromatography (HSCCC), 106f, 108
Hind limb extension (HLE), 217
Hippocrates, 241–242
Human pluripotent stem cell (hiPSC), 209
Hydrodistillation (HD), 74–76, 76t–77t
Hydro/steam distillation, 93f, 95–97
6-Hydroxydopamine, 208
Hypercholesterolemia, 128–129, 144–146, 149–150, 232–233
Hyperglycemia, 151, 153–155
Hyperlipidemia, 147
Hypocholesterolemic effects, 147–149
Hypolipidemic effects, 147

I

Immunoglobulins, 134–135
Infusion, 97–98
Inhabitation, 65–66
Insulin resistance syndrome, 129
Irrigation, 15
Ischemic brain injury, 211–212
Ischemic cardiovascular diseases, 232–234

L

Lipid profile, 128–129
Lipids, 28f, 33–35, 49
Lipopolysaccharide (LPS)-induced neuroinflammation, 202–203
Liquid chromatography paired with electrospray ionization-quadrupole time-of-flight and mass spectrometry (LC-ESI-QTOF/MS), 182–183
Liver fibrosis, 130
Longifolene (LON), 165
Low-density lipoprotein (LDL), 128, 144–145
Low-density lipoprotein cholesterol (LDL-C), 144–145

M

Maceration, 97–98
Macrophomina phaseolina, 134
Mechanical oil extraction and processing technology, 98–100, 99f–100f
Medicinal herbs, 239–240
Medicinal properties
  in cultures, 8–10
  food processing techniques, 89–90, 90f
  history, 7–8
Meningeal mast cells (MMCs), 205
Methanol extracts, 180
1-Methyl-4-(propane-2-yl) benzene (p-cymene), 219–220
Microwave-assisted extraction (MAE), 103–104
Microwave radiation, 181
Migraine, 205
Minerals, 28f, 33, 33t–34t, 47
Morphology, 20–22, 21f, 73–74, 73f
Multiple sclerosis (MS), 212–213
Murine cytomegalovirus (MCMV) infections, 196

N

Neuroinflammation, 201–204
Neuroprotective effects
  carvacrol, 220–221
  essential components, 201
  p-cymene, 219–220
  thymol
    Alzheimer's disease, 215
    anxiety, 215–216
    definition, 215
    dementia, 216
    depression, 216
    epilepsy, 217
    nervous system, 218–219
    Parkinson's disease, 217–218
  thymoquinone
    Alzheimer's disease, 209–210
    cerebral ischemia, 211–212
    depression, 205–206
    epilepsy, 206–208
    migraine, 205
    multiple sclerosis, 212–213
    neuroinflammation, 201–204
    Parkinson's disease, 208–209
    peripheral nerve injury, 214
    spinal cord injury, 213–214
    traumatic brain injury (TBI), 212
Nigellothionin, 133–134
Novel/advanced food processing methods
  enzyme-assisted extraction technology (EAE), 104–105
  microwave-assisted extraction (MAE), 103–104
  supercritical fluid extraction (SFE), 105–106
  ultrasound-assisted extraction (UAE), 101–103, 102f
N. sativa seeds polysaccharides (NSSPs), 135
Nutrients, 64–66
  composition, 119–120, 120f
  value, 28f, 30–31, 31f

O

Oil globules, 46, 47f
Oral Glucose Tolerance Test, 150
Oxidative stability, 180–181
Oxidative stress, 146, 171–172, 203–204

P

Paraoxonase 1 (PON1) enzymes, 153
Parkinson's disease (PD), 208–209, 217–218
Penicillin-induced epilepsy model, 207
Percolation, 97–98
Peripheral blood mononuclear cells (PBMCs), 194
Peripheral nerve injury, 214
Phenolic compounds and constituents, 172–184
Phenolic content, 181–182
Phenolic-rich fractions (PRFs), 180
Physiology, 22–23
Phytochemical profiling, 120–121
Plant-derived natural antioxidants, 169–170
Plasmodium sp., 197
Polycystic ovarian syndrome, 153
Polyphenols, 172
Polysaccharides, 28f, 63

classification, 35
emulsifying ability index (EAI), 37
emulsifying stability index (ESI), 37
extraction, 37–39, 38f
foaming capacity (FC), 37
foam stability (FS), 37
oil-holding capacity (OHC), 36–37
structural attributes, 36
water-holding capacity (WHC), 36–37
Polyvinyl alcohol (PVA)-Nigella nanofibrous mat (PNNM), 136
Positive allosteric modulator (PAM), 210
Potent antiprotein tyrosine phosphatase (PTP1B) activity, 52
Poultry breeds feed, 122–124, 122f
Preparative supercritical fluid chromatography (SFC), 106f, 107–108
Press waste material, 66–67
Production methods, 4, 19–20, 72
Proinflammatory cytokines, 202–203
Proteins, 28f, 32, 32t
alkali extraction with isoelectric precipitation (AE-IP), 52
amino acids profile, 53–55, 54t
concertation, 53–55, 54t
dwarf mutants, 50–51
film technology, 52
hydrolysate, 55–56
isoelectric point (pI) and pH, 51
limitations, 51
research efforts and specialization, 51
shape and color, 53, 53f
Pulsed electric fields (PEFs), 45
Purification processing techniques, 106–108, 106f

Q

Q-Exactive/Orbitrap Proteomic MS/LC-MS, 204

R

Radical scavenging activity test, 176–177
Reproductive system disorder, 129
RT2 polymerase chain reaction (PCR) array, 203–204
Ruminants feed, 122f, 124

S

Saturated fatty acids (SFAs), 47–49
Semipreparative high-performance liquid chromatography (HPLC), 106f, 107
Silver nanoparticles (AgNPs), 135–136, 138
Sirtuin 1 (SIRT1), 202
SKOV-3 cell, 137
Sodium nitrite (SN)-induced brain injury, 203
Solvent extraction (SE) technology
advantages, 94
on composition, 92–93, 93f
disadvantages, 94
essential oils, 74–76, 76t–77t
Soxhlet extraction technology, 93f, 94–95
Spinal cord injury, 213–214
Status epilepticus (SE), 206–207
Steam distillation (SD), 74–76, 76t–77t
Streptozotocin-induced diabetes, 134–135
Streptozotocin (STZ)-induced rats, 151–152
Supercritical fluid extraction (SFE), 74–76, 76t–77t, 105–106, 176, 179
Synthetic antioxidants, 170
Systolic blood pressure, 148

T

Taxonomy, 19, 19t, 73
Therapeutic potential
history, 127
insulin resistance syndrome, 129
lipid profile and hypercholesterolemia, 128–129
nonclinical study
anticancer properties, 137–138
antidiabetic properties, 134–136
antifungal properties, 133–134
antimicrobial efficacy, 130–133
liver fibrosis, 130
wound healing properties, 136–137
reproductive system disorder, 129
rhinitis, 129–130
thyroid status, 127–128

Thin-layer chromatography, 193
Thymol (THY), 164–165
 Alzheimer's disease, 215
 anxiety, 215–216
 definition, 215
 dementia, 216
 depression, 216
 epilepsy, 217
 nervous system, 218–219
 Parkinson's disease, 217–218
Thymoquinone (TQ), 130, 137, 146, 155–156, 162–164, 192–193, 201–204
 Alzheimer's disease, 209–210
 antihyperlipidemic properties, 115
 antiinflammatory activity, 114
 antimicrobial properties, 115
 cerebral ischemia, 211–212
 chemical composition, 113
 COVID-19, 12, 12f
 curative characteristics, 114
 depression, 205–206
 epilepsy, 206–208
 migraine, 205
 multiple sclerosis, 212–213
 neuroinflammation, 201–204
 Parkinson's disease, 208–209
 peripheral nerve injury, 214
 properties, 44, 44f, 63–64
 spinal cord injury, 213–214
 traumatic brain injury (TBI), 212
Thyroid status, 127–128
Tibb-e-Nabawi, 45
Total cholesterol (TC), 128
Total flavonoid content (TFC), 177, 179, 182
Total phenolic content (TPC), 177–180, 182
Traditional Chinese Medicine (TCM), 238–239
Traditional medicine, 238–242
Traumatic brain injury (TBI), 212
Triacylglycerides (TGs), 128
Trichophyton rubrum, 133
Type 1 diabetes, 154–155
Type-2 diabetes mellitus (DM), 128–129, 152

U

Ultrasound-assisted extraction (UAE), 101–103, 102f
Ultraviolet-visible spectroscopy, 193
Unsaturated fatty acids (USFAs), 47–48, 64–65

V

Very low-density lipoprotein (VDL), 128
Vitamins, 28f, 33, 33t, 47

W

Water management, 15
Whole System of Medicine (WMS), 238–239
Wistar rats, 146
World Health Organization (WHO), 237–238
Wound healing properties, 136–137

FSC
www.fsc.org
MIX
Papier | Fördert gute Waldnutzung
FSC® C083411

Kolibri 360 GmbH
Ferdinand-Jühlke-Straße 7
99095 Erfurt, Deutschland
produktsicherheit@kolibri360.de